TIDES

A Primer for Deck Officers and Officer of the Watch Exams

Philip M. Smith

Routledge
Taylor & Francis Group

LONDON AND NEW YORK

First published 2019
by Routledge
2 Park Square, Milton Park, Abingdon, Oxon OX14 4RN

and by Routledge
52 Vanderbilt Avenue, New York, NY 10017

Routledge is an imprint of the Taylor & Francis Group, an informa business

British Library Cataloguing-in-Publication Data
A catalogue record for this book is available from the British Library

Library of Congress Cataloging-in-Publication Data
Names: Smith, Phil (Teacher of navigation), author.
Title: Tides : a primer for deck officers and officer of the watch
    exams / Philip M. Smith. Other titles: Tides, a primer for deck
    officers and OOW exams | Primer for deck officers and officer
    of the watch exams
Description: First edition. | Routledge : Abingdon, Oxon ; New York,
    NY, [2018] | Includes bibliographical references and index.
Identifiers: LCCN 2018033805 | ISBN 9781138366305
    (hardback : alk. paper) | ISBN 9781138674752 (pbk. : alk.
    paper) | ISBN 9781315561110 (ebook)
Subjects: LCSH: Navigation—Study and teaching. | Tides.
Classification: LCC VK559 .S567 2018 | DDC 623.89—dc23
LC record available at https://lccn.loc.gov/2018033805

ISBN: 978-1-138-36630-5 (hbk)
ISBN: 978-1-138-67475-2 (pbk)
ISBN: 978-1-315-56111-0 (ebk)

Typeset in Bembo
by Swales & Willis Ltd, Exeter, Devon, UK

This book is the second in a series of primers for the Officer of the Watch and is dedicated to the memory of Brian Norton and to Rosemary Norton, a lifelong friend who continues to provide encouragement, and support to my whole family in whatever endeavours they have embarked upon.

Huge appreciation goes to my wife Helen, who once again has been commandeered for proof-reading of this script as she spurs me on and to my children who have had to suffer the continual presence of 'An Author' about the house.

My thanks also go to Egle Zigaite at Routledge who has had to endure the wait for the final copy, but without whose help and the success of the initial publication of the *Terrestrial Navigation* primer, I would not have got to this point – Cheers! I still owe you one.

CONTENTS

FIGURES

ABOUT THE AUTHOR

Philip M. Smith is a Senior Lecturer at Warsash Maritime Academy (WMA). After over 22 years at sea navigating, for the most part in a traditional way, he now teaches Officer of the Watch and Cadets Terrestrial Navigation.

Phil joined the Pacific Steam Navigation Company of Liverpool as a Deck Cadet in the early 1970s, and after qualification, remained with them until 1987. During his career, he saw the end of the general cargo steam-ship era and witnessed the advent of the container age and the arrival of sophisticated electronic navigation systems on ever larger tonnage and more specialised vessels.

His career spanned service on General Cargo Vessels, Bulk-Carriers, Passenger Ro-Ro Ferries, Reefer and Container Vessels trading worldwide. After serving in the North Sea on Platform Supply Vessels, he joined Cable & Wireless Marine and ended his sea-going career with Global Marine in 2003 after working on several major international shore-end connections, lays and cable repairs throughout the world. During this period, he added to his experience by becoming proficient in the use of Dynamic Positioning systems and using increasingly more accurate electronic navigational equipment and hydrographic survey techniques. Phil has never bored of the need for learning and has a Master's Degree in International Management of Shipping, Ports and the Environment as well as a second MSc. in Marine Surveying.

He has been a Senior Lecturer at Warsash Maritime Academy (WMA) for the past 15 years, during which time he has taught across the spectrum of courses, specialising in Terrestrial Navigation, Tides, Meteorology and Preparation for Oral Examinations for the Officers of The Watch (OOW) and Chief Mates. His experience in offshore operations found him well placed to deliver units on Safety and Offshore Technologies on the BSc. programme in Nautical Studies at Southampton Solent University.

During his time at Warsash, he has produced a portfolio of practice examination papers for his students to prepare for their final OOW examinations. A natural extension of this was to write a series of Primers for revision which, as a learning tool, would provide study guides either as preparation for, or use between college phases at sea, to improve their chance of success in their final examinations.

PREFACE

The requirement for Deck Officers to acquire the skills to pass MCA written examinations requiring tidal calculations has changed little over the past 40 years. There are increasing numbers of students passing the Higher Certificates, Diplomas and more recently Foundation Degree courses offered by the Maritime Training establishments but unlike these academic examinations, the Maritime & Coastguard Agency (MCA) still require a 65% achievement in a final examination.

The hurdle of achieving a sufficiently high pass mark has proved to be a stumbling block to many on the way to gaining the MCA certification for the Officer of the Watch Certificate of Competency.

This Tides Primer is the second in a series of guides and is intended to assist those studying to gain Deck Officer Qualification as Officer of the Watch. It is written with the aim of providing a study and revision guide for students, Cadets or Seniors, with helpful hints where not to lose valuable marks as a result of making errors in principle which incur a heavy loss of marks in examinations. Principle errors are those considered to be so fundamental to the calculation, that incorrect application or omission renders the answer wildly inaccurate or at worse, unsafe. The majority of examination errors are the result of carelessness, examination nerves, or simply a failure to understand or question the value that the calculator is giving.

This Primer will contain basic revision of the tidal theory together with questions and worked solutions, explaining of the mathematical techniques and highlighting the most common errors made by students in exams. It will cover all the areas of tidal calculations in the syllabi required for Higher National Certificate (HNC), Higher National Diploma (HND), Foundation Degree (FdSc) and SCOTVEC qualification. It will cover both European and Pacific tidal calculations for Standard and Secondary Ports, with worked examples of the calculations and test papers at the required standard expected in final examinations. Whilst all questions have been written to reflect realism, the final solutions **ARE NOT to be used for Navigation** as the calculations rely on out of date tidal information and fictitious hydrographic references to charted features which are used only to demonstrate the method of calculating these examples.

INTRODUCTION

All mariners must master the technique of calculating the heights of tides (HOTs) as this is fundamental to ensuring that during all times of a voyage, there is sufficient depth of water (DOW) to provide a safe clearance under the hull. Understanding tides and the important role they play in ensuring safe and successful voyage planning cannot be under-estimated. Over the years, this has been demonstrated graphically with some spectacular incidents of groundings. Many of these incidents have resulted in serious damage to the ships involved as well as to the environment and in the worst cases have resulted in loss of life.

It is for this reason that such a high emphasis has been placed on a student's ability to correctly calculate heights of tide whilst following a course of study for the Officer of the Watch and subsequent examinations, whether through a Foundation Degree, HNC or HND programme of study. Nowadays many Global Positioning Systems (GPS) and Electronic Chart Display and Information Systems (ECDIS) equipment have the tidal information accessible at the touch of a button but this has not always been the case. The traditional 'handimatic' method of calculation is still required in order to satisfy the Maritime & Coastguard Agency (MCA). The advantage of this is that once you appreciate the mechanics of the calculation taking place within the machine, then you can develop the ability to check it in the case where critical decisions depend on accurate information. The electronic programme in use simply performs the mechanical function for which it has been designed and does not allow for any safety factors, or other external influences, affecting the predictions for HOT and hence safety clearances required on the day. These could be as a result of squat, list, change in water density or aspects of the weather, all of which have a bearing on that all important Under-Keel Clearance (UKC).

Already you are starting to meet terms which will become common in all calculations and in this first chapter, it will be worth spending the time at this stage explaining the meanings of the common terminology so that there is no confusion as you progress through the Primer.

All the Tide Tables will have information giving predicted values for heights of tides and values for standard reference levels. It is imperative before getting too much further that these levels and the relationships they have with each other are explained clearly. Different values will be found in different areas of Tide Tables and it is vitally important that not only that the terminology is fully understood but equally where the correct values for this data is to be found within the tables. For this Tides Primer, the Admiralty Tide Tables (ATT) will be used for tidal calculations although different authorities will use much the same nomenclature and layout in their own publications.

Whilst some students find this subject difficult, I prefer to see it not as difficult just new and different. Once the terminology to be used has been clarified, the subject is basically reduced to two simple questions:

A) What is the *Height of Tide* at a given time?
B) What *time* will the tide be a given height?

A closer study of the tidal curves when they are covered, later in the next chapters, will reveal that these two answers are in fact, the **only** two values that can be extracted from the tidal curves. If you enter with a time required, the tidal curves will give the height for that time. If you enter with a height of tide, the curve will give the time when this height occurs.

It is however, the process involved in getting the correct values to enter the curve with in the first place, where all the difficulties start to appear. In examinations, failure to use the tidal information correctly repeatedly causes mistakes which can be very costly.

For academic purposes, the mark required to achieve Pass at HNC/HND and Foundation Degree level is 40% but, as with other safety subjects within the Nautical Science syllabus, the MCA requires a higher safety standard and in their view at least 65% is considered to be an acceptable mark in order to achieve a Pass.

ACKNOWLEDGEMENTS

Extracts of the Admiralty Time Tables NP 201a, NP201b and NP 206 as well as examples using Tidal Stream Atlas NP 233 are reproduced with the kind permission of the United Kingdom Hydrographic Office, for use with calculations.

DISCLAIMER

The opinions of the author expressed in this book are his alone and are not to be taken as those of any organisation with which he has a connection. Although this is a tides primer it does not comprehensively deal with all aspects of tidal calculations, merely those areas which will allow students to satisfy the examination syllabi for OOW and Chief Mate certification. Great care has been taken with calculations to avoid errors but the book is designed to give practice using worked examples for comparison to those required as a navigator. It states clearly in the Preface that under no circumstances should any of the worked calculations be used for navigation at sea as they simply offer realism to the exercises. All of the ports used have been used simply to offer a sense of realism to the calculations but actual Charted Depths, Charted Heights and features used in the calculations have not been checked for their actual existence. They are only to be used to demonstrate the method for calculations and use of any values given in this publication are for training purposes only and would render their use as completely unsafe if used in navigation.

1

TERMINOLOGY USED IN TIDAL CALCULATIONS

1.1 Planetary influences on tides

In order that simple but costly mistakes made by extracting incorrect figures from the Tide Tables can be avoided, it is important that all the relevant terminology to be encountered is clarified. Many people have a broad idea of what the tides are caused by but as a mariner, a slightly deeper understanding is required other than that they are caused by the Moon.

They are indeed caused by the Moon but it is a combination of the effects of the gravitational forces exerted upon the liquid surface of the earth by both the Sun *and* the Moon which influence the tidal pattern of movement. Between the Sun, the Earth and the Moon there is an attractive force of gravity holding the Earth and all the other planets in orbit around the Sun. Tides result as a consequence of the combined gravitational forces of the Moon, and to a lesser degree, the Sun, acting upon the liquid surface of the Earth. As the Moon is closer to the Earth it exerts much more of an influence on the tides than the Sun which is 93 million miles away. It is possible using Newton's Laws of Gravity and Motion to quantify how much less the Sun's gravity affects the Earth's tides.

Newton's Law of Motion states that:

$$\text{Acceleration}\,(a) = \frac{\text{Force}\,(F)}{\text{Mass}\,(M)}$$

Newton's Law of Gravity states that:

A body of mass M exerts a gravitational attraction G on a unit of mass at a distance (r)

$$Fg = \frac{G \times M}{r^2}$$

(Where G is universal gravitational Konstant)

Comparing gravitational attraction of the Sun on the Earth with that of the Moon on the Earth:

Mass of the Sun = 27 million times that of the mass of the Moon

Distance from Sun to the Earth = 390 times the distance from the Moon to the Earth

Thus: $\dfrac{\text{Fg (Sun)}}{\text{Fg (Moon)}} = \dfrac{27 \times 10^6}{390^2}$

$= 178$ times that of the Moon

From this, it appears that the gravitational effect of the Sun on the Earth is much greater than that of the Moon. It is a commonly assumed fact which most people know however, that it is the Moon which has more effective influence on the tides than the Sun. This is perfectly correct and the reason for this is because it is just the *proportion* of the gravitational force (GF) *NOT* balanced by centripetal acceleration (Ac) in the Earth's orbital motion, that produces the tides. This unbalanced portion is proportional to the inverse cube of the distances rather than the inverse *square* of the distance from the Earth.

Because gravity is still proportional to the mass, we can see that forces causing the tides are approximately:

$$Fg = \frac{G \times M}{r^2}$$

$$= \frac{178}{390}$$

This shows that the gravitational effect of the Sun on tides is equal to 0.46 of the effects caused by the Moon.

1.2 Tidal terminology explained

1.2.1 Chart legend

The basic unit of measurement for tidal calculations is metres and the information extracted from British Admiralty charts will also be in metres. It is important however, that the navigator always consults the Legend of every chart which confirms the units of measurements used for navigational and tidal data on that particular edition of the chart.

As already highlighted, it is extremely important that the terminology used throughout this guide is well understood right from the outset. It is equally important that the relationship between all the various terms and levels used in the tidal calculations are clearly distinguished in order that solutions to tide questions can be solved with a high degree of accuracy.

There are in fact only two questions to be answered in tidal examinations:

A) What is the *Height of Tide* at a given time? e.g. 21:30 hours tonight.
B) What *time* will the tide be a given height? e.g. 5.5 metres.

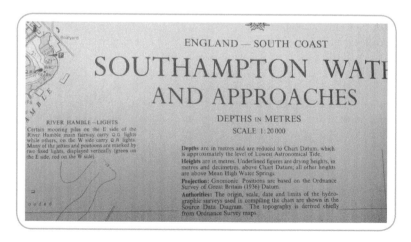

FIGURE 1.1   Admiralty chart legend notes on maritime charts

In order to answer these two questions, we need simply to look up the correct pair of tides relevant to the time of day required, apply some relatively simple maths and with the aid of the tidal curves provided in the Tide Tables, both of these questions can be solved.

1.2.2 Tide

A tide may be described as the regular and predictable movement of water governed by the relative positions of the astronomical bodies, primarily, the Sun and the Moon. It is regarded as the vertical movement of water which rises and then falls.

1.2.3 Flood tide

The term refers to the time during which the tide comes in, normally over about a six hour period after Low Water (LW) until it reaches its maximum height above Chart Datum at High Water (HW).

1.2.4 Ebb tide

The term refers to the time during which the tide recedes, normally over about a six hour period after HW until it reaches its lowest height above Chart Datum at LW.

1.2.5 Slack Water

This term refers to the short period immediately after HW or the period after LW when there is no vertical change in depth of water and the tide effectively stands still. It is this time when it is ideal to swing vessels around in turning basins within ports, as there will be no loss of Under-Keel Clearance (UKC) during the swing as there is no vertical tidal movement of water level. This is sometimes referred to as the 'Stand' for similar reasons.

1.2.6 Current

The horizontal movement of water brought about by a number of different conditions but primarily as a result of wind's influence on the direction, or topography of the surrounding features of the land. Currents may vary in magnitude or even direction depending on the season, differences in sea water temperature or salinity.

1.2.7 Height of Tide (HOT)

The predicted Height of Tide is found in tidal publications for High and Low Water and the HOT found at these times, indicates the height in metres measured above the reference point Chart Datum, to the water level.

Height of Tide is required to be calculated at any point in time in order to determine the more important value which is the Under-Keel Clearance between the keel and the seabed.

1.2.8 The causes of tides

The planet Earth revolves around the Sun in an elliptical orbit and it maintains its position along this path as a result of the gravitational pull exerted on it by the Sun. The Moon revolves around the Earth and once again a constant orbit is maintained as a result of the gravitational force between the Earth and the Moon. As the Moon is closer to the Earth, the gravitational force between them effects the liquid surface of the Earth, the oceans, more than that of the Sun. The Moon revolves around the Earth approximately once every 29 days. This varies slightly due to the changing proximity of the Sun as the Earth continues along its orbital path. The Earth takes 365 days to complete one orbit of the Sun and regularly, all three bodies are aligned resulting in a combination of gravitational forces all acting in the same direction. It is this maximum combined effect of gravitational forces which leads to extra high tides known as Spring Tides and produces greater ranges of tides over the daily tidal cycle.

The point at which the cycle begins, when the Earth, Moon and Sun are in the same vertical plane, is called the New Moon and is the first of the Phases which the Moon will pass through as it completes its orbital cycle around the Earth of one lunar month. As the Sun is projecting light on the surface of the Moon facing away from the Earth, at New Moon the planet is invisible to an observer on Earth.

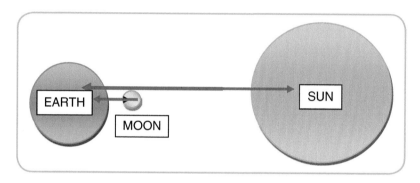

FIGURE 1.2   Phases of the Moon – New Moon – Spring Tides

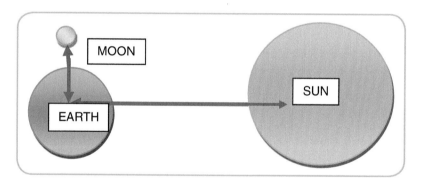

FIGURE 1.3   Phases of the Moon – Half Moon – First Quarter – Neap Tides

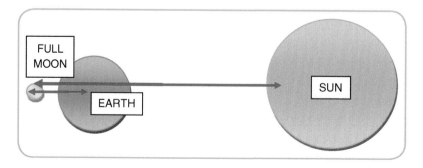

FIGURE 1.4   Phases of the Moon – Full Moon – Spring Tides

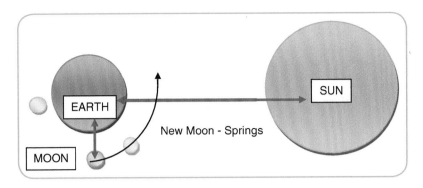

FIGURE 1.5 Phases of the Moon –Third Quarter – Neap Tides

On day two of the cycle and each day onwards, the Moon, the Earth and Sun get further out of alignment and the combined gravitational effect between them gradually diminishes. After about seven days the Earth, Moon and Sun will be about 90° out of alignment with each other and the combined gravitational forces are now at their lowest. This phase is referred to as the First Quarter and as a result, the lowest High Water and the highest Low Water levels are achieved. This is known as the Neap Tides which occur when the Moon is said to be in quadrature with the Sun (Fig 1.3). During this period, the Moon will start being visible and grow from a slim crescent towards Half Moon when it reaches the First Quarter Phase.

After reaching the First Quarter in the cycle the Moon continues its orbit around the Earth and as it does so, the directional forces of gravity start to realign themselves again. As the Moon approaches Full Moon, at a position in opposition to the Sun, once again the maximum combined gravitational forces between the Earth, Sun and Moon produce another set of Spring Tides. At this point the shape visible has changed again through that of a Gibbous Moon as it passes through the First Quarter towards a Full Moon, to a position when the full face of the Moon is exposed by the Sun's rays shining directly onto it (Fig 1.4).

The cycle continues, resulting in decreasing strength in the combined gravitational forces as the Moon gets further out of alignment with the Sun. At the Third Quarter, it once again reaches 90° to the direction of the Sun i.e. in quadrature, producing another set of Neap Tides. The shape of the Moon enters into the shadow of the Earth and reduces back towards Half Moon at the Third Quarter.

In the Final Quarter, with the planets realigning once again, the combined gravitational forces between the planets increase, until the Moon completes its lunar orbit and is back in line with the Sun. At this point the planetary pull on the oceans is once again at the strongest returning to the period of Spring Tides.

1.2.9 Spring Tides

The period during Spring Tides produce the highest levels of High Water and the lowest levels of Low Water resulting in the maximum range of tides over the period. The maximum effects occur at Full Moon and at New Moon. Here, the combined forces of gravity caused by the Sun and Moon acting on the sea surface of the Earth are at a maximum. This will be when the Moon is either in conjunction or opposition with the Earth and the Sun.

1.2.10 Neap Tides

Neap Tides are the lowest levels of High Water and the highest levels of Low Water resulting in the smallest range of heights of Tides. Neap Tides occur over the period during the First or the Third

Quarter phases of the lunar month. They are the resultant of the combined forces of gravity caused by the Moon and the Sun acting at right angles to each other, producing the least combined gravitational effect on the sea surface of the Earth. At the First and Third Quarter Phases, the Moon is described as being in quadrature with the Sun.

1.2.11 Chart Datum

As with all scientific data the values are actually misleading unless related to a reference point. Tidal information in Tide Tables are generally referenced to Chart Datum, the level which the tide will not normally fall below. This often but not always coincides with the height of Lowest Astronomical Tide (LAT). Another way of considering Chart Datum is as the level where the Height of Tide is Zero.

In the UK and where the United Kingdom Hydrographic Office is the Authority, Chart Datum approximates to the same level for LAT (Nautical Institute, 2008). This ordnance level is based on the average value of Mean Sea Level (MSL) at Newlyn, Cornwall, which aligns itself closely to the levels used in the land levelling system used in England, Scotland, Wales and some of the closer UK offshore islands. Northern Ireland which uses the different Ordnance Datum at Belfast is not aligned to the same level of MSL (Newlyn). Other datum's such as Mean Low Low Water (MLLW) or Mean Low Water Springs (MLWS) may be used and it will be important for the mariner to once again refer to the Legend of the chart in use (See Fig 1.1). The Legend will clarify which Datum is relevant to the charted depths on the chart and the port to which tidal data on the chart applies.

1.2.12 The tidal cycles

Depending on the area of the world, there are generally three defined categories of tidal cycles. The most common to Europe is called a Semi-Diurnal or 'twice daily' cycle of tides resulting in a sequence of High Water to Low Water then back to High Water, once roughly every 12 hours. There will be two High Waters and two Low Waters over the period of a lunar day.

The Earth completes one full rotation about its axis roughly once every 24 hours. We have already considered the effects of gravity on raising the water levels, depending on the relative positions of the planets, and this generally causes the water levels to 'bulge' towards the direction of the gravitational pull. This in turn, as there is only a finite amount of water covering the planet, will reduce the levels elsewhere on the surface of the Earth and the maximum effect will be seen at Spring Tides. As the Moon progresses on its orbital path around the Earth, the strength of the Sun's gravitational force will diminish so that the heights of tides directly related to the positional 'Bulge' will vary on a daily basis. This force will be at its weakest during the period of quadrature. This is the reason why at these times the High Waters will be the least and the Low Waters will be at their highest, because the sea surface is not subjected to such strong combined forces coincident with the First Quarter Phase of the Moon which results in the Neap Tides.

1.2.13 Succession of tides

In Figure 1.6 we can see a situation where at 00:00 hours the planets are all aligned and the gravitational effect forms a bulge towards the Moon. This will produce a High Water at the observer's position. As the Earth rotates, six hours later, the observer is now 90 degrees from the 00:00 hours position in the most affected part of the 'bulge'. At this position there is a marked reduction in the water level and this will produce the first Low Water.

Continued rotation of the Earth once again places the observer back in the area of highest influence of the gravitational effect, and produces a second High Water. As the Earth continues spinning then once again the observer is in a position where the levels are lower and this gives the second Low Water. This involves the rotation of 270 degrees which takes about 18 hours before completing the cycle over 24 hours during which there will have been two High Waters and two Low Waters occurring – the Semi-Diurnal Cycle.

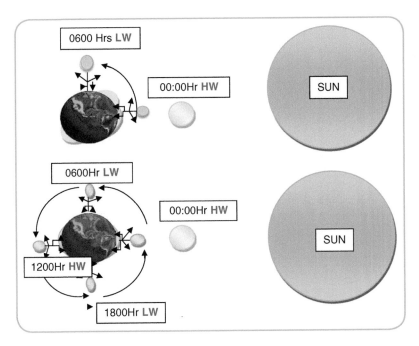

FIGURE 1.6   Semi-Diurnal Cycle of tides over a 24 hour day i.e. one rotation of the Earth

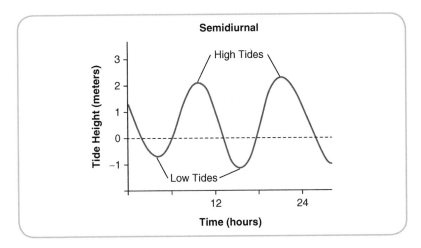

FIGURE 1.7   Semi-Diurnal tidal cycle

1.2.14 The Semi-Diurnal Cycle

The Semi-Diurnal Cycle gives rise to two High Waters and two Low Waters tides of approximately equal levels over a period of a lunar day.

1.2.15 Diurnal Tidal Cycle

During the course of this Primer we will address the two main methods of calculation for tides, so as to enable the calculation of both European Tides and Pacific Tides. The process will be very similar

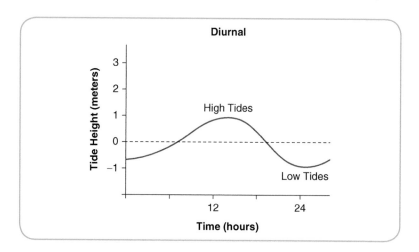

FIGURE 1.8 Diurnal Tidal Cycle

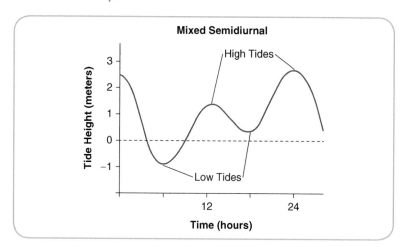

FIGURE 1.9 Mixed Semi-Diurnal Tidal Cycle

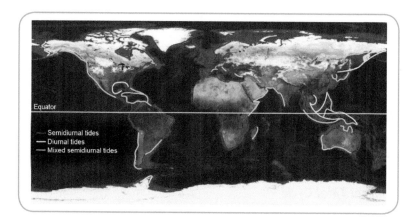

FIGURE 1.10 Distribution of tidal cycles on earth (NOAA, 2018)

but varies slightly due to the way that the movement of water in the coastal regions of both oceans differ. This difference gives rise to a cycle which is much less regular in the Pacific Ocean, and in some cases results in only one High and one Low Water in the period of a lunar day which is known as a Diurnal Cycle.

1.2.16 Mixed Semi-Diurnal Cycles

The third cycle is a combination of both cycles being a mixed cycle which can result in two Highs and two Low Waters but of differing heights over the lunar day.

The reason for these variations lies with the differences in the oceans from where effectively the tidal flow comes from, and returns back to, as it floods and ebbs. In Europe the water supplied to the coastal strip and rivers comes from the Atlantic Ocean and flows back into the ocean as it ebbs. The Atlantic Ocean covers a vast area but there is nothing much in it except the Azores and Canary Islands. This enables a fairly uninterrupted flow of water as the earth spins. It takes roughly six hours to flood and six hours to ebb in a regular and predictable cycle.

The movement of the water supplied to ports and the coastal strip of countries bordering the Pacific Ocean on the other hand, experience very different tidal flows. The Pacific Ocean is very deep and much larger than the Atlantic. The main factor for this is that the Pacific Ocean contains thousands of islands both on its coastal strip, especially in Southeast Asia as well as the myriad of South Pacific island chains in the middle. These islands in many cases are quite small in surface area but are actually vast pinnacles of rock rising up from the sea floor. These pinnacles of rock act in the same way as baffles, which deflect and slow the movement of water past them. The Earth, of course, is still rotating at the same speed but the water, as a result of all the islands, is now moving at a much slower speed. It therefore takes longer than the regulated flow of six hours in and out again, as experienced in the Atlantic regions of Europe. The reduced movement of water, in some areas, results in less than two tides a day with much more, or less, than the six hours between successive High and Low Waters. The cycle of only one High Water and one Low Water occuring during the period of a lunar day is known as a Diurnal Cycle and as such, requires an alternative method of calculation for times and heights of tides, from many ports with the Semi-Diurnal pattern of tides.

1.2.17 Lowest Astronomical Tide (LAT)

Lowest Astronomical Tide is the lowest predicted level of tide based on a combination of both the average meteorological and astronomical conditions experienced over time, and can fall, exceptionally, below Chart Datum. This would be as a result of a combination of factors all coinciding to produce a particularly low tide, but not normally expected to achieve that level as the norm. Offshore winds combined with high barometric pressure during Spring Tides could be such a case leading to very low tides falling below the level of Chart Datum. LAT is just another reference level, in this case the very lowest recorded Height of Tide and as such, is again referenced to Chart Datum.

LAT may be at a level below chart datum, but in UK waters it generally coincides with Chart Datum effectively making the lowest ever tide height expected to be zero metres in height. There may be ports where LAT falls regularly below zero as a result of the topography of the seabed but the datum for heights of tides will still be referenced to Chart Datum and the Height of Tide predicted as a negative value signifying it is below the Chart Datum.

1.2.18 Highest Astronomical Tide (HAT)

Highest Astronomical Tide (HAT) is the highest ever level of tide based on a combination of both the average meteorological and astronomical conditions experienced over time and will be higher than the height of Mean High Water Springs (MHWS) which is only an average value. This could be as a result of a combination of factors all coinciding to produce a particularly high tidal prediction but not normally expected to achieve that level as the norm. Onshore winds combined with low barometric pressure during Spring Tides could be such a case leading to very high tides occurring as a result.

Highest Astronomical Tide should be the safest reference level to use for recording charted heights of either bridges or cables on charts. This would invariably allow greater than the required margin of safety, as the tide would only on extreme combination of circumstances ever reach that height. Currently as Admiralty charts are replaced by new editions of that chart, their legends will clarify which reference point is being used for charted height. HAT is just another reference level, or maximum Height of Tide, and as such is again referenced as height in metres measured above Chart Datum. Values for HAT levels for both Standard and Secondary Ports can be found in the Admiralty Tide Tables in *Table V: Part 1 for Standard Ports and in Part 2 for Secondary Ports*.

TOP TIP

In time, all the Admiralty charts will be replaced and amended to show that HAT is the standard reference level taken for bridges and cables or any other obstacle for which a vertical clearance would be needed to pass under.

Other terrestrial objects which ships do not pass under do not pose the same threat and objects such as lighthouses, beacons and masts will remain referenced above the height of Mean High Water Springs. It will be important to always check if the charted height values on a particular chart have been corrected and relevant to HAT, or still remain in relation to the height of Mean High Water Springs.

FIGURE 1.11 Charted Height datum information given on Admiralty charts for calculating height of a terrestrial object or vertical obstruction above sea level at any given time

1.2.19 Mean High Water Springs (MHWS)

The mean height used is the average height of the highest two consecutive High Water levels recorded over a 24 hour period at *Spring Tides*. MHWS is the reference point for terrestrial features on charts such as headland promontories, masts and lighthouses. In the case of lighthouses, the measurement for charted heights is taken from the level of MHWS up to the focal point of the light, otherwise to the top of the geographical feature.

1.2.20 Mean Low Water Springs (MLWS)

The mean height used is the average height of the two lowest consecutive low water levels over a 24 hour period recorded at *Spring Tides*.

1.2.21 Mean High Water Neaps (MHWN)

The mean height used is the average height of the highest two consecutive high water levels recorded over a 24 hour period during *Neap Tides*.

1.2.22 Mean Low Water Neaps (MLWN)

The mean height used is the average height of the highest two consecutive high water levels recorded over a 24 hour period during *Neap Tides*.

The values for the heights of MHWS, MLWS, and MLWN and MHWN are found tabulated in Part II of the Admiralty Tide Tables and we will revisit these when we consider tidal information in a separate chapter, referring to the use of Tide Tables.

1.2.23 Charted Height (Ch.Ht)

The charted height of an object is the height in metres that the top of it measures above the reference level. The reference level above which it is measured will be identified in the legend for a particular chart but will usually be either the height of MHWS or HAT.

If the chart is a new publication, the legend will state: *all vertical clearance heights (charted heights of bridges, cables) are measured above HAT. All other charted heights are above MHWS.* (See Fig 1.11).

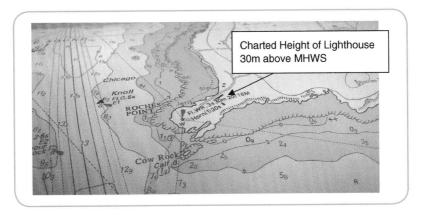

Charted Height of Lighthouse
30m above MHWS

FIGURE 1.12 Charted Heights as depicted on an Admiralty chart will be referenced to MHWS

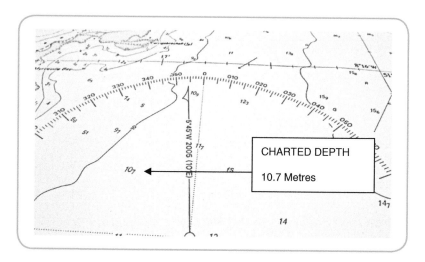

CHARTED DEPTH

10.7 Metres

FIGURE 1.13 Charted depth as depicted on Admiralty charts usually referenced below Chart Datum

1.2.24 Charted Depth (Ch. Depth)

The term Charted Depth refers to the depths measured down from the reference point Chart Datum, to the sea-bed. These depths are represented on the Chart in sea areas as metres and decimetres e.g. 10_7 (See Fig 1.13).

1.2.25 Depth of Water (DOW)

Depth of Water (DOW) is a combination of the HOT and Charted Depth. HOT is calculated from the Tide Tables and Charted Depths are displayed on the chart.

1.2.26 Under-Keel Clearance (UKC)

Determination of the Under-Keel Clearance should be of the utmost concern to the mariner and will vary with the vessels' draft as the ship progresses through the voyage. It is the distance measured from below the keel to the seabed.

1.2.27 Draft

Draft is the depth of the vessel below the waterline measured in metres from the keel upwards. In order to determine UKC then:

e.g/ DOW = HOT + Charted Depth = 3.0m + 5m = 8m

UKC = DOW − Draft = 8m − 5m = 3.0m UKC

Under-Keel Clearance (UKC) = Depth of water − Draft

e.g. DOW = HOT + Charted Depth = 3.0m + 5m = 8m

UKC = DOW − Draft = 8m − 5m = 3.0m UKC

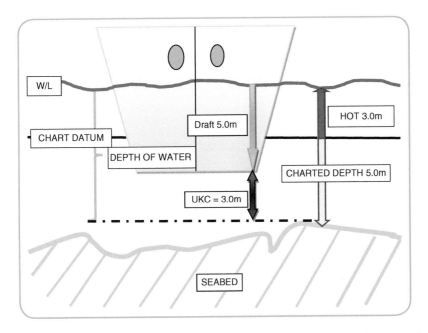

FIGURE 1.14 Diagram relating Charted Depth, DOW and Draft and HOT to derive UKC

1.2.28 Drying Heights

Drying Heights are marked in green on the chart to draw attention to the fact that at low water the seabed becomes exposed and is therefore a hazard to navigation. These areas may be mud, sand or rocks and the amount which they are exposed is indicated as a height measurement ABOVE Chart Datum in metres and decimetres e.g. 0_6 (0.6m). Drying Heights are distinguished from Charted Depths by the symbol having a line below it, referring to the fact that the measurement is ABOVE Chart Datum.

e.g. DOW = HOT + Charted Depth = 8.0m − 1.0m = 7.0m

UKC = DOW − Draft = 7.0m − 6.0m = 1.0m UKC

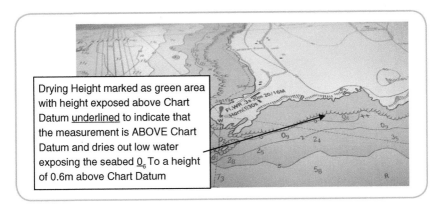

Drying Height marked as green area with height exposed above Chart Datum underlined to indicate that the measurement is ABOVE Chart Datum and dries out low water exposing the seabed $\underline{0}_6$ To a height of 0.6m above Chart Datum

FIGURE 1.15 Drying Heights as depicted on an Admiralty chart will be referenced above Chart Datum

1.2.29 Air-draft

In just the same way as we should be concerned about the amount of water below the keel, the same concern must be appreciated for safety clearances above the ship as it passes below obstructions such as cables

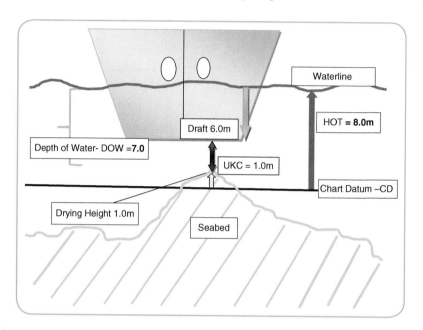

FIGURE 1.16 Drying Heights and their effect on the UKC

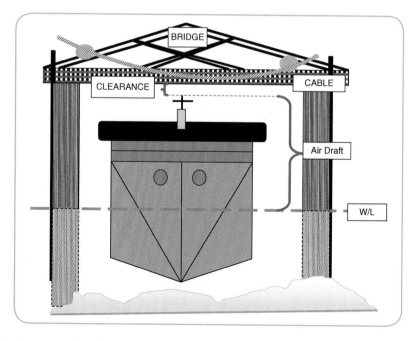

FIGURE 1.17 Air-draft and clearances beneath bridges and cables

and bridges. Air-draft is the distance in metres measured from the waterline to highest structural part of the ship, i.e. masts, funnels, top masts, aerials or other items which may come into contact with bridges or cables and cause damage to either the structure or the vessel. In the case of cables, the Charted Height measurement is taken to refer to the lowest part of the catenary of the cable: the lowest dip of the 'Sag'.

1.2.30 Clearance

Clearance is the term used to define the measurement in metres measured from the top of the highest extremity of the ships' structure to the underside of the obstruction. In the case of a bridge this includes from the underside of any fitted painting gantries permanently fixed and capable of traversing the whole of the underside of the bridge. In the case of cables, the measurement is taken to refer to the lowest part of the catenary of the cable.

1.2.31 Other miscellaneous terminology – times of day

We have so far covered most of the technical terms which you will encounter daily and are required to understand as a routine part of the job for an Officer of the Watch. When it comes to the examinations there are several more important terms which you need to be clear upon when you encounter them in the questions. They may seem obvious but generally, at least two candidates in a cohort will fail by not fully appreciating the difference between a *morning, afternoon* and an *evening tide*. If you fail to take care to check the time of day you are required to either pass over or under an obstruction, you will extract the wrong pair of tides from the Tide Tables. This will result in your answer being wrong from the first line of your calculation onwards.

TOP TIP

As far as the examination is concerned, if the question requires the earliest time in the morning and you have an answer neatly underlined which says that the earliest time in the morning will be 18:30 hours. . . you can see what is coming next – it will make a nonsense of your efforts.

It needs to be emphasised that in tidal calculations we use a 24 hour clock. The *morning* starts after midnight 00:00 hours, until 12:00 hours. The *afternoon* is from 12:00 hours until about 18:00 hours when it is getting dark and the *evening* will be from about 18:00 hours until midnight – 24:00 hours.

TOP TIP

Equally important when reading a question and solving tide problems, I have noticed that there is sometimes confusion between the meaning of the words *earliest and latest.*

1.2.32 Earliest and latest times

These two words provide another source of costly examination mistakes as they will have different meanings depending on what it is you are attempting to find out.

If a question asks you to find the *earliest* time that there will be sufficient water to safely sail, then this suggests it refers to the earliest time of the day which will be any time after midnight i.e. in the morning.

Likewise if the question asks to find the *latest* time you can sail it again seems fairly obvious that the latest time in the day will be most likely be in the early evening or at night. The difficulty with *earliest* and *latest* is that, depending on what you have to do, the tide will be either rising or falling.

If you are trying to find the *latest* time to get underneath a bridge or cable, then you should consider the question carefully and ensure you extract a pair of tides from the tables which is **a flood tide**. This should seem obvious as during any time after, the tide is rising further and it will no longer be possible to pass under the bridge after this *latest* time calculated.

When trying to calculate the *latest* time to get over a shallow patch or obstruction however, the latest time must occur on **an ebbing tide**, so choosing a flood tide will immediately be an error in principle.

The same principle applies for choosing the correct pair of tides when a problem requires the *earliest* time to pass underneath a bridge or cable to be calculated. Clearly you need to look for the appropriate pair of **ebbing tides** in this case. The *earliest* time to cross a shallow sandbank however, must be somewhere during **a rising tide**, and you should be careful to choose an appropriate pair of tide times from the tides tables.

These words are further complicated by the fact that you may be required to find the **latest** time on a **morning flood** to pass under a bridge. This time, the word **latest** has been defined as being in the **morning** where without the addition of this one word, as we have already discussed, the *latest* would have been in the early evening or night. Take care to ensure that you not only choose the correct time of day, but as the question has also specified, it must be the last **flood** tide in the **morning**.

2

THE ADMIRALTY TIDE TABLES

2.1 Tide Table coverage

The contents of the Admiralty Tide Tables (ATTs) are published each year providing worldwide coverage of ports and harbours in nine volumes.

Volume 1A...............United Kingdom – English Channel to R. Humber (including Scilly Isles, Channel Islands and European Channel ports).

Volume 1B...............United Kingdom and Ireland Humber (excluding Scilly Isles, English Channel to R. Humber, Channel Islands and European Channel ports).

Volume 2.................. North Atlantic Ocean and Arctic Regions.

Volume 3.................. Indian Ocean.

Volume 4.................. South Pacific Ocean.

Volume 5.................. South China Sea and Indonesia.

Volume 6.................... North Pacific Ocean.

Volume 7.................. South West Atlantic and South America.

Volume 8.................South East Atlantic, West Africa and Mediterranean.

The Tables are split into four parts as follows:

Part I. This is the main body of the tables giving all the daily tidal predictions for times and heights of High and Low Water at *STANDARD PORTS* in that area of coverage. All the Standard Ports are listed on the inside front cover of the volume. The tables give the predictions for the entire year starting on the 1st of January through until the end of December.

Part II. This section gives the tables of corrections to be applied to the listed tidal predictions at the appropriate Standard Port in order to calculate the times and heights of predictions at the *SECONDARY PORTS*.

Part III. This part contains the *HARMONIC CONSTANTS* to be applied when the use of the simplified harmonic method of calculating tidal predictions is required.

Part IIIa. This part contains the tables of *HARMONIC CONSTANTS* for the *tidal streams* to be applied to use of the simplified harmonic method for calculating tidal predictions.

2.2 How the Tide Tables work

The Tide Tables provide detailed examples of how to calculate the times and height of tide at intermediate times between consecutive Low and High Waters including instructions on use of the graphs and associated tables of corrections. The values listed for times and heights are predictions based on historical data which is all iterated by a computer programme but using average values for particular locations. Over years, the accumulation of hundreds of thousands of items of recorded data refines the averages which are then used to produce the final predictions.

It is important to realise that the published information in the Tide Tables is only as accurate as the data and averages used to produce them. As the term implies, the listed values are still only *predicted* values and in reality, on the day, if any of the relevant parameters are above or below the averages used, there will be a slight discrepancy in *actual* time or height of a tide with that which was *predicted* using pure averages.

2.3 Meteorological effects of tidal predictions

These differences in values and the effect that they can have can be understood more easily if one considers the meteorological data used to compile the predictions. There are several meteorological influences which will bring about differences between the predicted and the actual values for heights of tides including the following, which are also fully explained in the front of ATT volumes.

1.   Barometric air pressure.
2.   Wind direction and force.
3.   Rainfall.
4.   Seiches.
5.   Storm surges.
6.   Negative surges.

2.3.1 Barometric air pressure

When making the predictions for tides, the value for average barometric pressure is used for any given location. For the UK, the average air pressure at sea level is generally regarded as being about 1012mb (millibars). When looking at a synoptic weather chart there will be areas marked as HIGH and LOW which refer to the pressure being either above or below this average value. This has an effect on the

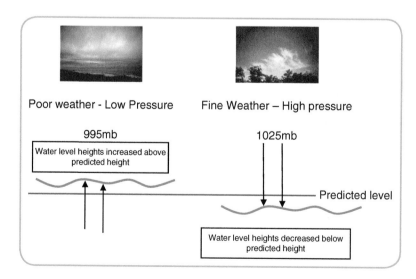

FIGURE 2.1   Effects of atmospheric pressure on the actual height of tide

sea level because if the pressure is higher e.g. 1025mb, than those used in the predictions, the actual heights of tides will be **lower** than those predicted.

If the pressure is lower than the average value used in the predictions e.g. 995mb, then this will allow actual heights to be higher than the heights predicted in the tide tables.

2.3.2 Wind speed and direction

The actual tide levels on a particular day will differ from the predictions if the wind speed or direction is different to the averages which have been used in the calculation. For any particular area, the average wind speeds and direction have been used in the calculation for that particular month e.g. August – south-westerly force 3. If the actual wind however is easterly force 8 then the actual heights will be different from the predictions; instead of the sea being blown onshore, the water will be carried in the opposite direction and actual tide levels will be lower than predicted as water is carried offshore. As a rule, onshore winds above average speed increase the actual height of tides, and offshore winds above average speed will decrease the actual heights of tides below those of the predictions around coastal areas and ports.

2.3.3 Rainfall

Depending on the area of the world, values for average rainfall will vary enormously. Due to exceptional weather systems such as monsoons or hurricanes, extra water is shed into areas exceeding the average expected amounts resulting in flooding. Even though heavy rainfall has not been experienced on the coastal area, inland flooding will eventually work its way down through the rivers and out into estuaries and the coastal regions. As a result of more water in the system, the actual levels experienced will rise above those predicted. If there is a drought however and less water in the system than expected, actual height levels experienced will be less than predicted.

Factors such as these should be considered when deciding whether or not there are suitable safety factors of either sufficient water below the keel, or height clearance above the superstructure, with all the various meteorological effects influencing the actual height of the water levels.

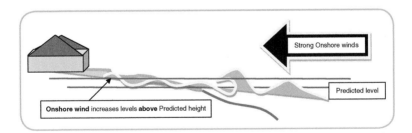

FIGURE 2.2 Effects of onshore wind on tidal levels

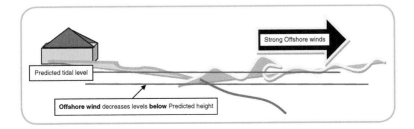

FIGURE 2.3 Effects of offshore wind on tidal levels

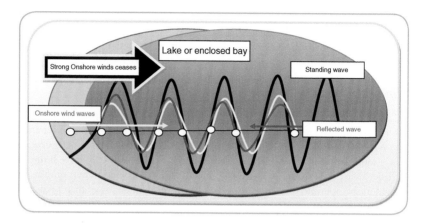

FIGURE 2.4 Seiche effects on a lake or an enclosed bay or harbour after abrupt change in wind or atmospheric conditions

2.3.4 Seiches

These are described in ATTs as: 'Abrupt changes in meteorological conditions such as the passage of an intense depression or a line squall may cause oscillations in sea level.' (ATT, Volume 1A, 2018)

Seiches are set up by strong winds which produce waves oscillating back and forth typically in an enclosed coastal bay or inlet. When strong winds blow, water is blown to one side of a bay piling up higher water levels on the leeward shore with lower levels as a result on the windward side. When it stops blowing or changes direction the water rebounds in the opposite direction. It continues to rebound back and forth and can become synchronised. The two waves travelling towards each other combine to form a standing wave with a much longer wavelength and higher amplitude, with the largest oscillations at each end of the body of water. Seiches can cause a change in height from a few centimetres to sometimes more than a metre. These are especially common in lakes but the same effect may be seen in enclosed ports and harbours where their shape and size make them very susceptible to small seiches especially in winter (Hydrographer of the Navy, 2017).

2.3.5 Storm surges

Storm surges are most often associated with extremes of weather such as hurricanes, cyclones and typhoons. The low pressure allows for the rise in sea level over the concentrated area at the centre of the system

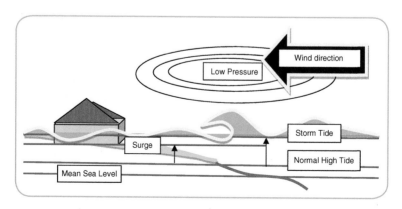

FIGURE 2.5 Effects of storm surges on tidal height predictions

which is driven along as the low pressure moves. As the topography of the seabed shallows approaching the coast, the water piles up higher creating a storm surge. This will be compounded at times of High Water and more especially when combined with Spring Tides where extensive flooding can result.

2.3.6 Negative storm surges

Although storm surges generally are associated with excessive increases in coastal water levels, if the direction of movement of the low pressure system is away from the coast, this can lead to unusually low water levels along a coastline. Reduced depths of water will be potentially hazardous to large vessels with low Under-Keel Clearances operating in these coastal waters. These are fairly common in the southern North Sea and low lying shallow areas such as the Thames Estuary where a warning service has been established to promulgate information to mariners regarding the onset of these events.

2.4 Seismic activity

In areas of the world's oceans which are subjected to seismic activity such as submarine earthquakes and volcanoes, waves with a long wavelength will be radiated from the centre of the activity as a result. These waves can travel at tremendous speeds possibly as much as 500 miles per hour over vast distances. At sea in deep water, these will be almost undetectable and have no effect on ships which are transiting the area. When these waves meet the continental shelf however, and the sea bed shallows rapidly, the waves slow down and grow in energy. The tops of the waves travel faster than the bottom causing them to build in height and arrive on the coastal strip as a dangerous and highly destructive ocean wave called a wave train. This excess water arriving leads to widespread flooding of the low lying coastal areas. These waves are more commonly known as **tsunamis,** from the Japanese for 'harbour wave', and are most common around the tectonic plates in the Pacific Ocean regions known as the 'Ring of Fire'. A series of Ocean Data Acquisition System Buoys have been stationed at sea around areas of the tectonic activity to measure increases in sea level and these relay information to forewarn of any impending tsunamis, so that coastal populations can be evacuated to higher ground in good time.

3

THE USE OF TIDAL CURVES

3.1 Standard and Secondary Ports

For ease of learning these study notes will be split into two groups of calculations – those for Standard Ports and then once the process is better understood, progressing to discuss calculations for the Secondary Ports.

Standard Ports are those ports which are classed as major ports, or hubs of marine activity providing either deep water, suitable for commercial cargo ships and fishing activity, or considered important for their proximity to industry and transportation links. As a result, the Admiralty have constructed the Tide Tables in such a way that these ports have their own calendar of predictions for times and Heights of Tide, for each day and month of the year. Alongside these predictions is a set of Tidal Curves depicting the characteristic of the tide used for the calculation of intermediary times and heights of tide. If the time of High Water is known, use of the curves then allow the user to calculate the Height of Tide at any time during either side of the published times. This enables mariners to ensure that there is a satisfactory Depth of Water (DOW) for safe passage, with sufficient Under-Keel Clearance (UKC) at all times.

To use the tables easily, all the Standard Ports are listed on the inside cover of the tables so that the user can go straight to the pages appropriate for that port. The predictions for times of High and Low Water will be printed for each day and month together with the corresponding predicted heights of the tide for those times. (Fig 3.1). In order to find times and heights in between the tabulated information, the most common method used is referred to as the 'Slopy Line' Method. This will be covered in detail later in this chapter.

3.2 The Tidal Curves

There are two scales on the 'X' axis of the tidal curves. One is for Time Intervals of up to six hours either side of High Water, with graduations for every 10 minutes. The second scale is at the top and mirrored again at the bottom of the graph to the left of the tidal curves which is for recording the heights of High and Low Water in metres. This scale is also used to obtain the heights of tides at any given times in between the two.

3.3 Effect of the topography of the seabed on the Tidal Curves

You will notice that the graph has two Tidal Curves; one consisting of a solid line and the other a pecked line. The solid line represents the tidal characteristic for the port during a Spring Tide and the other represents the characteristic followed during a Neap Tide. The characteristic shape of each curve

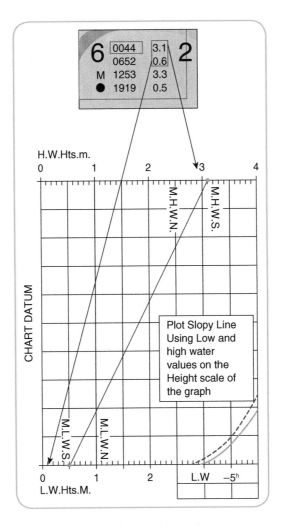

FIGURE 3.1 How the extracted tidal data relates to the graph

changes slightly with each port and is governed by the topography of the seabed. Different types of seabed profiles will give a different shape to the curve as the rate of increase or decrease of height of tide is different at each port. An example of this can be explained if you compare a port in a fjord, or where the rock formation provides a steep sided shoreline to a port in a wide estuary (Fig 3.4). In this case, as the tide comes in, the increase in Height of Tide per hour is fairly uniform. In an estuary however, which has a central river channel and then a wide silted and gradually rising seabed (as in Fig 3.4), the rate of increase of Height of Tide per hour is non-linear. This gives rise to a different shape tidal graph which is applicable to each port. Whatever the shape however, the graphs are referenced to High Water with a distinct peak in the centre of the graph representing High Water and the interval in hours before, during the flood tide and after, over the ebb tide along the axis at the bottom.

Referring back to Figure 3.1, you will notice that there is a black circle below the day in question, in this case M (Monday). This indicates that there is a New Moon and this means that this is a period when Spring Tides are occurring. Spring Tides occur approximately every two weeks and by working out the daily predicted range, it is easy to decide whether you are approaching Springs or Neaps from

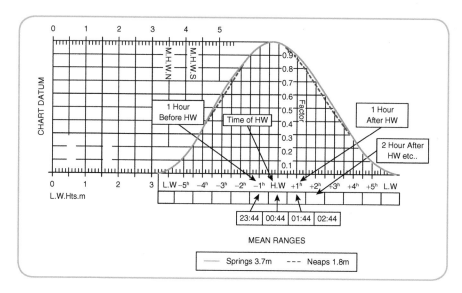

FIGURE 3.2 Tide Table curve – use of the interval scale

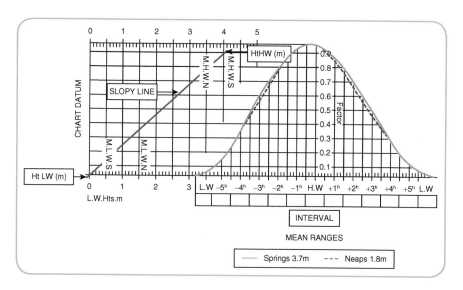

FIGURE 3.3 Creating the Slopy Line

the mean ranges shown on the tidal curves. Most of the Standard Port graphs are constructed in the same way with the interval scale along the bottom referenced to High Water. The tidal heights are plotted on the scale on the top and bottom of the page along the left hand side of the graph.

As previously mentioned there are only two questions in Tides:

A) What is the *Height of Tide* at a given time?
B) What *time* will the tide be a given height?

The graph is designed to solve either of these two questions and nothing else. It works on the principle that if you have one parameter (Height of Tide) and enter the curve with this, the tidal curve for the

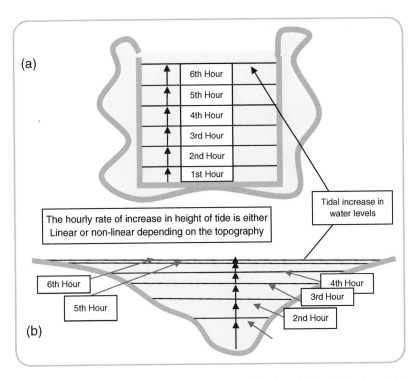

FIGURE 3.4 Seabed topography affecting the rate of change of heights of tide and the shape of the curve

port will give you the other, namely the TIME it will occur. It will come in the form of either an interval before, or after the time of High Water.

3.4 How to use the graph is simple

For Standard Ports, the following will help to follow the process:

1. Select the appropriate pair of tides from the Tide Tables taking care to select the correct time of day and either the required ebb or flood tide depending on what you ar trying to achieve. Incorrect selection of data is a principle error and will immediately cost you 50% of the marks in an exam as well as making the answer hopelessly wrong. Tides are considered a safety subject by the MCA and as such, a mistake made by using incorrect data could end in the vessel foundering, hitting the span of a bridge or fouling an overhead power cable.
2. Enter the Low Water height in metres along the height scale at the bottom left hand side of the graph and similarly along the top left hand side for the High Water height on the scale provided.
3. Join the two with a straight line taking care to construct the line accurately. This method is known by the highly technical term of the 'SLOPY LINE' method.
4. Enter the graph with the height of tide required along the left hand height scale dropping a vertical line intersecting with the 'Slopy Line' and then construct a horizontal line to cross the curve.
5. Where the line meets the curve, drop a vertical line down to the scale along the bottom of the graph and apply the interval read to either the time before, or after the appropriate High Water time chosen from the tide tables.
6. On most graphs you will notice that there are in fact two curves: one a solid line – the SPRING curve, and the other, a pecked line, which represents the NEAP curve.

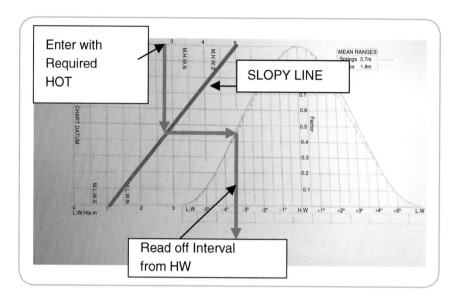

FIGURE 3.5 Using the tidal curves

7. This means that when you cross the curve when entering with a height you in fact cross both curves giving you a value at Springs and a different one during Neaps.

Generally you will be somewhere between the two and in order to drop the vertical line in the correct place a system called the Percentage Springs method is used.

3.5 Percentage Springs method of calculation

The Percentage Springs method requires the use of one formula and this is the only formula you need to remember in tidal calculations. After attempting several calculations, this formula will be firmly implanted in your memory as you will need to use it every time you work out the height of a tide problem. I think of the term 'Percentage Springs' as an abbreviation of the *percentage of the spring difference between the two curves*.

The formula used is as follows:

$$\%Sp = \left\{ \frac{PR - Np(R)}{Sp(R) - Np(R}\right\} \times 100$$

FIGURE 3.6 Percentage Springs formula

We have already referred to these ranges earlier in the definitions, but you need to know where to find them in the Tide Tables.

PR = Predicted Range

Np(R) = Neap Range

Sp(R) = Spring Range

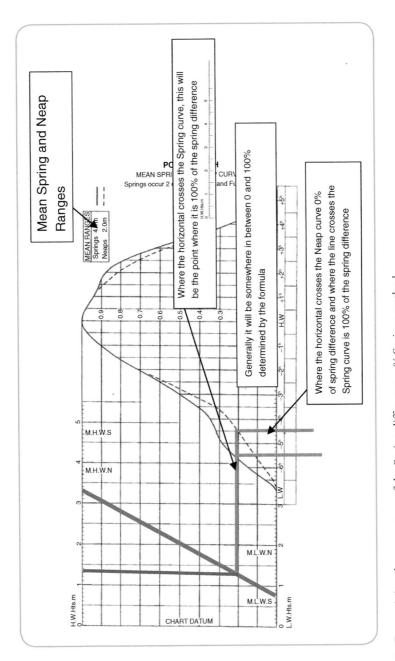

Mean Spring and Neap Ranges

MEAN RANGES
Springs 4.0m
Neaps 2.0m

PO... ...H
MEAN SPRI... ...CURV...
Springs occur 2 ...and Fu...

Where the horizontal crosses the Spring curve, this will be the point where it is 100% of the spring difference

Generally it will be somewhere in between 0 and 100% determined by the formula

Where the horizontal crosses the Neap curve 0% of spring difference and where the line crosses the Spring curve is 100% of the spring difference

M.H.W.S
M.H.W.N
M.L.W.N
M.L.W.S

H.W.Hts.m
CHART DATUM
L.W.Hts.m

H.W.Hts.m

FIGURE 3.7 Determining the percentage of the Spring difference – % Springs method

TOP TIP

If you look at any Standard European Port tidal curve, you will see a couple of boxes marked with the MEAN SPRING and MEAN NEAP RANGES. These will always be found in the same place – by the curves. It is important that these are not confused with the value of the MHWS for that Port. The Mean Ranges are the average *differences between High and Low Water heights* during Spring and Neap Tide periods. This is something completely different to the MHWS value which is an average of all the highest *Heights of Tides* which were recorded during Spring Tides.

The PREDICTED Range is simply the difference between the predicted heights of High and Low Water of the relevant pair of tides you have chosen from the daily page in the tide tables. The resultant fraction is multiplied by 100 to bring it to a percentage of the difference between the two curves. This relates directly to the position between the two curves needed in order to drop a vertical line and read of the height from the scale.

The percentage will normally lie for any particular day, somewhere between 0% and 100% and irrespective of whether crossing the curves vertically when trying to find the time interval, or horizontally to find the height of tide, it is *always* measured from the Neap curve TOWARDS the Spring curve which represents 100% of the Spring difference.

3.6 Understanding the Percentage Springs value

There will be some occasions when the Percentage Springs value will come out as over 100% and on others the Percentage Springs value will be a negative value. This is perfectly acceptable providing that you recognise what this in fact means. If the Percentage Springs is over 100% then you must be on Spring Tides and use the Spring curve. Conversely if the percentage springs value is negative, then it must be Neap Tides and you can only use the Neap curve. Anything either side of the two curves is simply fresh air, so do not try to estimate outside the curves. This area is only there to provide a guide for constructing your plot. The reason for these uncharacteristic values is because the Neap and Spring ranges that you are using are MEAN Spring and Neap Tide values and on *your* day, the Predicted Range (PR) is higher or less than the values printed on the appropriate tidal graph from which you took your values and then used in the formula.

In the following example, the Standard Port is Portsmouth where the mean Spring range marked on the tidal curve is 4.1m and the mean Neap range is 2.0 metres.

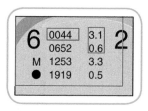

FIGURE 3.8 Extracting data from the daily page

On the 6th May using the formula to calculate the Percentage Springs value for the morning ebb tide:

PR (Predicted Range) = (3.1 − 0.6) = 2.5

Np (R) Neap Range = 2.0m

Sp (R) Spring Range = 4.1m

Using the formula Fig 3.6

$$\frac{2.5 - 2.0}{4.1 - 2.0} \times 100 \quad = \quad \frac{0.5}{2.1} \times 100 \quad \boxed{= 24\%}$$

If PR is greater than Spring Range

$$\frac{4.2 - 2.0}{4.1 - 2.0} \times 100 \quad = \quad \frac{2.2}{2.1} \times 100 \quad \boxed{= 104\%}$$

MUST ONLY USE **SPRING** CURVE

If PR is less than the Neap Range

$$\frac{1.8 - 2.0}{4.1 - 2.0} \times 100 \quad = \quad \frac{-0.2}{2.1} \times 100 \quad \boxed{= -9\%}$$

MUST ONLY USE **NEAP** CURVE

FIGURE 3.9 The use of the curves when % Springs value lies outside 0% or 100%

3.7 Using the information to create the plot on the tidal curves

In order to find the answer to any tidal problem, the following three parameters must be known before you can enter the tidal curves:

TOP TIP

1. HOT / Interval
2. SLOPY LINE
3. % SPRINGS

FIGURE 3.10 Three elements required before entering tidal curves

If any of them are incorrect or omitted, then whatever figures are extracted from the graph will result in the answer being wrong. In examinations this risks a principle error deduction of at least 50% of the available marks as the mistake could have a serious impact on the safety of the ship.

TOP TIP

Before starting a calculation it is useful to have an *aide memoire* to help make sure that you do not make basic and costly mistakes (Fig 3.10).

1. If it is the Height of Tide that is sought, the tidal curve must be entered with the time interval before or after High Water. Alternatively, if it is the time when a given Height of Tide is sought, the curves must be entered with the Height of Tide required.
2. In order to obtain the values required, the values for High and Low Water must be known in order to construct the 'Slopy Line'.
3. In order to find how far between the Neap and Spring curves to draw the line to read the values from the height or the interval scales, the Percentage Spring value for that day must also be calculated.

The last thing to consider before entering the tidal curve will be whether the tide is flooding or ebbing as this dictates which side of High Water you will initially enter the curve. Almost all of the curves are referenced to High Water so the left hand side will be the flood side and the right hand side will be the ebb.

TOP TIP

However, whilst this holds good for all curves, the exception is at Southampton.

If you look at the tidal curve for Southampton in the UK (ATT Volume 1) you will notice that the graph in this case is referenced to LOW Water so the left hand side of the curve is before LOW water, in other words the ebb side. The right hand side of the curve rises from Low Water to High Water thus representing the flood.

3.8 The effect of the Isle of Wight on tides at Southampton

Care must be taken when using the Southampton or Secondary Ports associated with Southampton, that the normal procedure for use of the curve is reversed.

The reason Southampton is arranged in this way is due to its proximity to the Solent, a narrow strait dividing the Isle of Wight from the mainland. As the tide floods and ebbs, the water in Northern Europe is effectively supplied from and returns to, the Atlantic Ocean. As the incoming flow meets the UK land mass, it splits into two main streams. One stream flows north and creates a circulation into the North Sea over the top of the British Isles. The other flows along the south coast of UK flowing eastwards along the English Channel.

As the flood reaches the Isle of Wight it is diverted through the West Solent channel up the Estuary to Southampton. A couple of hours after the ebb has commenced, the area then receives a boost

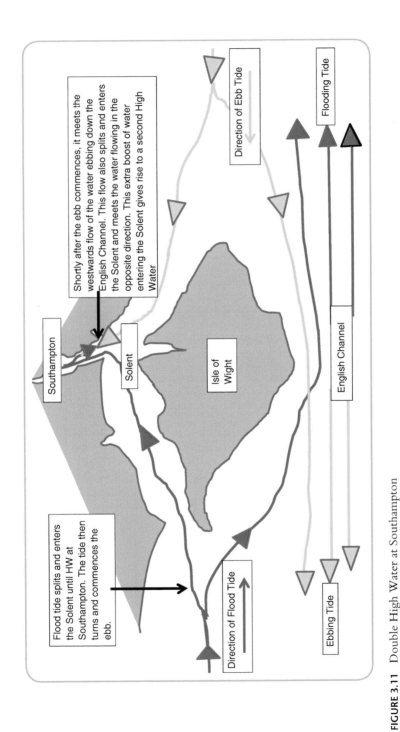

Shortly after the ebb commences, it meets the westwards flow of the water ebbing down the English Channel. This flow also splits and enters the Solent and meets the water flowing in the opposite direction. This extra boost of water entering the Solent gives rise to a second High Water

Southampton

Solent

Isle of Wight

Direction of Ebb Tide

Flooding Tide

English Channel

Flood tide splits and enters the Solent until HW at Southampton. The tide then turns and commences the ebb.

Direction of Flood Tide

Ebbing Tide

FIGURE 3.11 Double High Water at Southampton

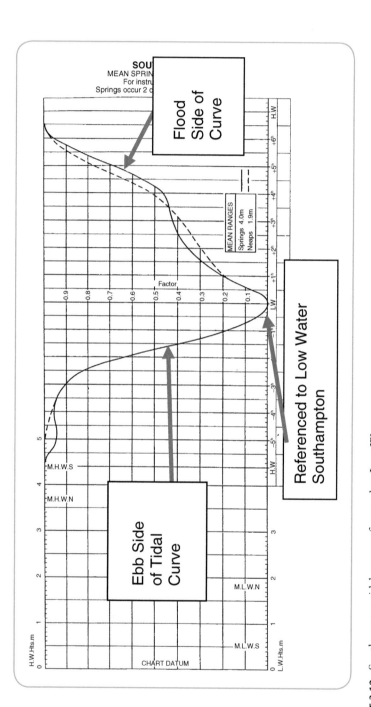

FIGURE 3.12 Southampton tidal curve referenced to Low Water

of extra water which is supplied by the flow ebbing westwards along the English Channel. This ebb flow, on meeting the Isle of Wight is again diverted northwards via the East Solent Channel and this extra boost of water gives to an increase in height of tide at Southampton and effectively provides the port with a second high water. This phenomenon has contributed to the success of the port which can provide suitable depths of water over a longer period of time than other Channel ports.

As there are two peaks in the tidal curve, to avoid ambiguity over which peak to use for the correct time interval, the curve is referenced to Low Water as the tidal curve has only one distinct Low Water which is well defined on the tidal curve used.

4

PREPARATORY STANDARD
PORT EXERCISES

To find the Height of Tide at a given time.

If you are trying to find the Height of Tide then, as the graph will only tell you two things, you will need to enter the curve with the appropriate time interval to find this Height of Tide.

4.1 Milford Haven – example 1

Find the Height of Tide at Milford Haven at 06:00 GMT on the morning of the 6th of September. Bearing in mind the things that you have to determine before you get to use the graph, the first thing will be the INTERVAL.

Find the appropriate day of the month and look for the pair of tides either side of 06:00. Note that the Zone Time is given on the top left hand side of the Standard Port calendar page of predictions.

All the predicted times of tides are always given in local time at that Standard Port. Local Time, Zone Time and Standard Time are all expressions of the same thing which will be the time it would say on your watch if you were there at that port i.e. local time.

Choose the pair of tides either side of 06:00 GMT (GMT, UTC and UT all mean the same thing – Greenwich Mean Time).

The appropriate predicted tides are:

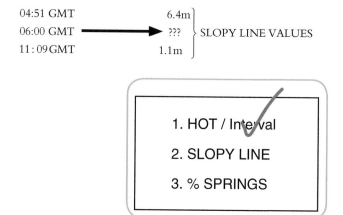

```
04:51 GMT                6.4m ⎫
06:00 GMT  ──────────▶   ???   ⎬ SLOPY LINE VALUES
11:09 GMT                1.1m ⎭
```

```
1. HOT / Interval ✓
2. SLOPY LINE
3. % SPRINGS
```

FIGURE 4.1   What you need to find before entering curves for any solution

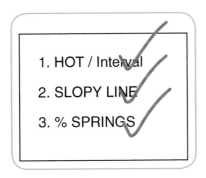

FIGURE 4.2   What you need to find before entering Milford Haven tidal curves – example 1

From this you can ascertain the Time Interval to use. Note that the graph for Milford Haven is referenced to the Time of High Water. The interval between 04:51 to 06:00 is a time difference of one hour and nine minutes after High Water. This is expressed on the scale as a time interval of **PLUS** 1 Hour 9 Minutes. Time *after* High Water is defined as *positive* (+) and time before High Water is *minus* (−) on the interval scale.

In this case the right hand side of the curve is the ebb side.

The only other value you will need is the Percentage Spring value:

$$\%\text{Sp} = \frac{\text{PR} - \text{Np(R)}}{\text{Sp(R)} - \text{Np(R)}} \times 100$$

PR = Predicted Range is the difference between the predicted High and Low Waters.

i.e. 6.4 − 1.1 = 5.3m.

Spring Range [Sp(R)] from the Milford Haven tidal curve is 6.3m and the Neap Range [Np(R)] is 2.7m.

$$\%\text{Sp} = \frac{5.3 - 2.7}{6.4 - 2.7} \times 100 \quad = \frac{2.6}{3.7} \times 100 = \boxed{70\%}$$

The only thing to check now is which side of the curve to plot on. Is the tide flooding or ebbing? It clearly is on the ebb at 06:00 GMT therefore you will need to plot on the right hand side of the curve, the ebb side.

Firstly construct the 'Slopy Line' by plotting LW 1.1m and HW 6.4m using the height scale on the left hand side of the graph. You need the graph to tell you the HOT at 06:00, therefore you must enter the curve with the Time Interval of +1 hour 9 minutes.

After you have plotted the 'slopy line' and entered at the correct time interval, you will notice that in this question, that the vertical line only crosses one curve. There are in fact two curves, but the Spring curve in superimposed upon the Neap curve which has merged and they both follow the same characteristics of the tide at this point.

In this case, although the Percentage Springs were calculated as 70%, at this point of the curve, there is no percentage difference between the curves and you simply construct your horizontal to the 'slopy line' where the interval intersects with the curve. When the line constructed crosses two separate curves, then best judgement of 70% between the two would have been required. The answer is 5.8m.

4.2 Milford Haven – example 2

In order to find the *time* when the Height of Tide would be 5.8m then the same process is required. You will still need to find the same three parameters except that this time you know the height of tide required so the graph will tell you the interval. The same predictions would be used as it is on the same day/tide and the same Percentage Springs value would be found as a result.

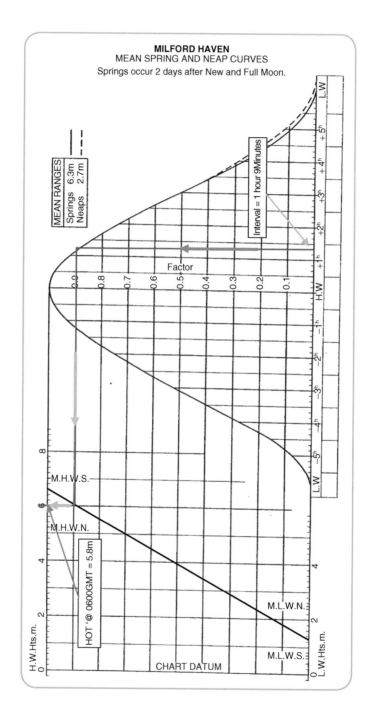

MILFORD HAVEN
MEAN SPRING AND NEAP CURVES
Springs occur 2 days after New and Full Moon.

MEAN RANGES
Springs 6.3m
Neaps 2.7m

Interval = 1 hour 9Minutes

Factor

M.H.W.S.

M.H.W.N.

HOT ' @ 0600GMT = 5.8m

M.L.W.N.

M.L.W.S.

CHART DATUM

H.W.Hts.m.

L.W.Hts.m.

L.W

H.W

L.W

FIGURE 4.3 Entering the Milford Haven tidal curves with the Time Interval to find HOT – example 1

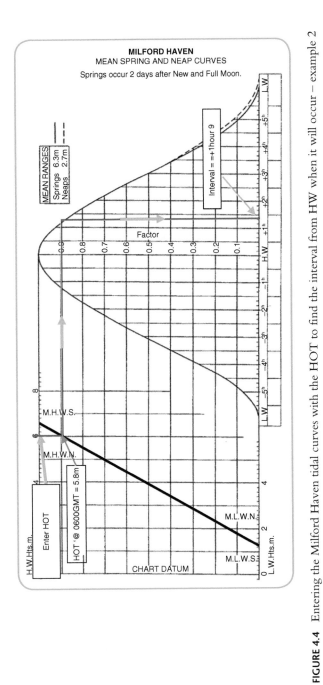

FIGURE 4.4 Entering the Milford Haven tidal curves with the HOT to find the interval from HW when it will occur – example 2

FIGURE 4.5   What you need to find before entering Milford Haven curves – example 2

The appropriate predicted tides are still:

04 : 51 GMT   6.4m ⎫
??????   5.8m ⎬ Slopy Line values
11 : 09 GMT   1.1m ⎭

This time, after plotting the 'Slopy Line', the graph is entered with the height of tide (HOT) and a horizontal line drawn from where this line intersects the 'Slopy Line' across to the curve. At this point a vertical is dropped to read off the interval. In this case the interval is read after High Water as the tide is ebbing, which gives the interval as + 1 hour 9 minutes after High Water.

Time of High Water Milford Haven is 04:51 + 1hr 09minutes makes HOT = 5.8m @ 06:00 UT/ GMT (See Fig 4.6.)

4.3 Interpolating between the curves

In most cases the value of the Percentage Springs will be somewhere between 0% and 100% of the Spring difference. If the Percentage Springs are 30%, for example, the rule to follow is that the 30% of the distance on the graph between the Spring and Neap curves, is always measured *from* the **Neap** curve *towards* the **Spring** curve as shown in Figure 4.6.

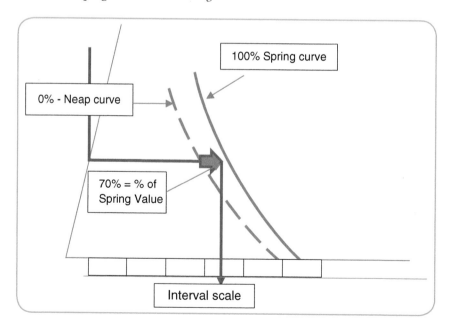

FIGURE 4.6   Interpolating the curves for percentage springs between 0% and 100%

4.4 Avonmouth – example 3

Find the earliest time when the height of tide will be sufficient for a cargo vessel with a departure draft of 4.7m and requiring a 1.5m UKC to sail from Avonmouth during daylight in the morning of the 15th August. The vessel will be required to have sufficient water to cross over a shallow patch charted at 3.0m.

There are several important things to spot from the question:

1) Earliest means probably the morning tide but the question stipulates that it must be *daylight*.
2) The question in this case clarifies that it is the morning required but you should always check the wording.
3) The expression *earliest* also confirms that as the vessel needs to cross a bank of Charted Depth 3.0 m so we must use a rising tide.
4) If it was a falling tide required, the question would have asked for the *latest (Last)*. The question asks for a *time* so the graph must be entered with the appropriate HOT required.

In order to find required HOT, a diagram is required. Once the HOT required is ascertained from the diagram, extract the appropriate morning set of tides on the day from the tide tables, which must be the morning flood tide.

Tides on the morning flood at Avonmouth on the 15th August

$$
\left.\begin{matrix} 05{:}27 - & 1.8\text{m} \\ 11{:}06 - & 12.1\text{m} \end{matrix}\right\} \text{Slopy Line}
$$

Predicted Range (PR) = 10.3m

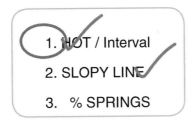

1. HOT / Interval
2. SLOPY LINE
3. % SPRINGS

FIGURE 4.7   What you need to find before entering Avonmouth curves – example 3

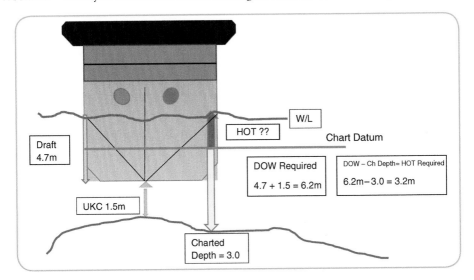

W/L

HOT ??

Chart Datum

Draft
4.7m

DOW Required

4.7 + 1.5 = 6.2m

DOW – Ch Depth= HOT Required

6.2m−3.0 = 3.2m

UKC 1.5m

Charted
Depth = 3.0

FIGURE 4.8   Determining the HOT required at Avonmouth – example 3

All that is now required is the value for % Springs

$$\%Sp = \frac{10.3 - 6.5}{12.3 - 6.5} \times 100 = \frac{3.8}{5.8} \times 100 = \boxed{65\%}$$

One final check to see what the tide is doing and we establish that it must be a flood tide, which means the plot must be on the left hand side of the tidal curves i.e. before High Water.

HW Avonmouth   = 11:06

Time interval for HOT 3.2m = −4:43

Earliest time to pass over shallow area is 06:23 Hrs on Monday 15th August.

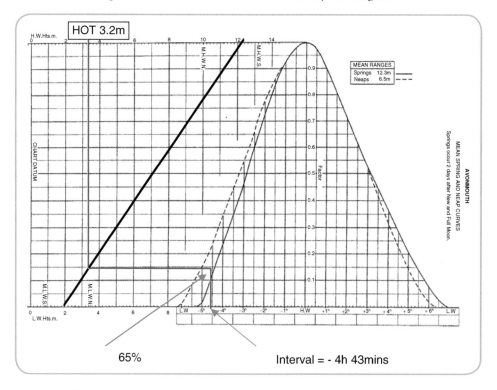

FIGURE 4.9 Use of the graph to determine earliest time for a given HOT at the Standard Port – Avonmouth – example 3

4.5 Southampton – example 4 (where the curve is referenced to Low Water)

Find the latest time for a vessel to berth at Southampton Container Port which is dredged to a charted depth of 14.9m during the afternoon of 6th November. The draft is 14.4m and the Port of Southampton require a UKC of 10% of the draft to be maintained until commence of discharge which will be upon arrival alongside.

In this example you will not only have to recognise the differences presented by the Southampton graph in terms of the LW reference from which the interval is taken, but you require the *latest* time so you need to be sure that you take care selecting the correct pair of tides. If the task is to find the latest time, this suggests that the tide *can* only be falling as any later, there would not be sufficient water to achieve it successfully. You will note that this question did not specify which tide, just that it would need to be in the afternoon.

Therefore the question is asking to find out when the HOT is 0.9m on the afternoon ebb tide at Southampton on the 6th November.

From the tide tables the correct pair of tides are:

$$\left.\begin{array}{ll} 10{:}53 & 4.7m \\ 16{:}45 & 0.5m \end{array}\right\} \text{SLOPY LINE}$$

In order to find the time you will again be required to enter the graph with the requisite HOT which requires a diagram (See Fig 4.11).

DOW required = Draft + 10% = 14.4 + 1.44 = 15.84m

Minimum HOT required = 0.9m

All that is now required is the value for % Springs.

$$\% \text{ Sp} = \frac{4.2-1.9}{4.0-1.9} \times 100 = \frac{2.3}{2.1} \times 100 = \boxed{110\%}$$

Remember that if the % Spring value is in excess of 100%, as discussed previously, you can only use the Spring curve.

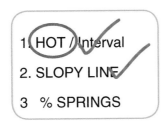

1. HOT / Interval

2. SLOPY LINE

3 % SPRINGS

FIGURE 4.10   What you need to find before entering the Southampton curves – example 4

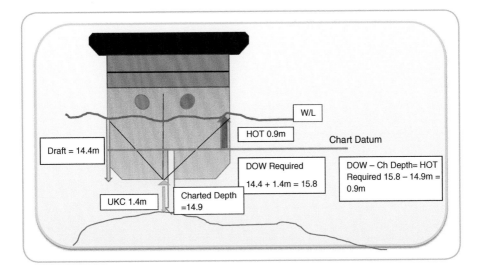

FIGURE 4.11   Determining the HOT required at Southampton – example 4

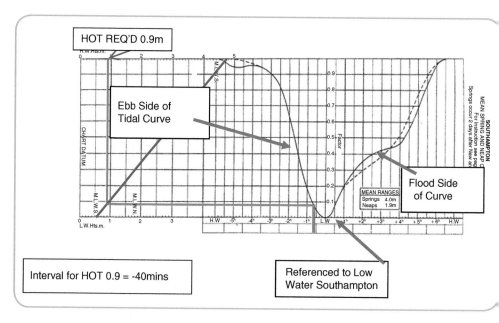

FIGURE 4.12 Determining the time interval before Low Water at Southampton – example 4

5

EUROPEAN SECONDARY PORTS TIDAL CALCULATIONS

Secondary Port calculations build upon the process which has been started in the Standard Port examples. The Standard Ports are main ports with their own graph and calendar of predictions for the whole year. Ports are paired together by the similarity of the characteristics of their tidal curves. This is not necessarily governed by geographical position, but by the movement of water which in turn is related by the shape of the seabed.

The list of Secondary Ports is found at the back of the ATT Part II where they are listed beneath the Standard Port to which their tidal curve is most closely aligned. Each port has a unique number and the ports are listed numerically to make it easier to locate and hence identify their appropriate Standard Port. The tables list the time differences and the height differences to be applied to the tide times and heights given on the Standard Port pages found in Part I.

By interpolating for the appropriate differences in time and heights, the values for the 'Slopy Line' and the time of High or Low Water can be determined to use on the Standard Port curves. These tidal curves can now be assumed to apply to the Secondary Port in question, with the tidal curve now effectively referenced to the time of HW at the Secondary Port. Any height or intervals determined from the curves, will now relate directly to the Secondary Port.

5.1 The Secondary Port time difference corrections

As mentioned, there are corrections for Secondary Ports found in Part II for time and height differences and in order to find the appropriate corrections to apply, interpolation of the data listed will be required. Interpolation for many is something that they have not done for a long time and some basic revision in the method may be required.

5.2 Interpolating the data

The layout of the information for Secondary Port corrections in Part II at the back of the tide tables is set out in columns, the order of which is standardised for ease of use throughout all of the Admiralty Tide Table volumes.

The first column lists the unique number assigned to the port and listings are in consecutive numerical order. This is followed by the position and name of the port – Standard Ports are listed in bold upper case, with the associated Secondary Ports listed below in lower case.

312				SCOTLAND, WEST COAST					
No.	PLACE	Lat. N.	Long. W.	— TIME DIFFERENCES (Zone G.M.T.)				HEIGHT DIFFERE MHWS MHWN	
				High Water		Low Water			
404	GREENOCK	(see page 86)		0000 and 1200	0600 and 1800	0000 and 1200	0600 and 1800	3·4	2·9
	Firth of Clyde								
391	Southend, Kintyre	55 19	5 38	−0020	−0040	−0040	+0035	−1·3	−1·
392	Sanda Island	55 17	5 35	−0040	−0040	☉	☉	−1·0	−0·
393	Campbeltown	55 25	5 36	+0010	+0005	+0005	+0020	−0·5	−0
393a	Loch Ranza	55 43	5 18	−0015	−0005	−0005	−0010	−0·4	−0
	Loch Fyne								
394	East Loch Tarbert	55 52	5 24	+0005	+0005	−0020	+0015	0·0	
395	Inverary	56 14	5 04	+0011	+0011	+0034	+0034	−0·1	+
	Kyles of Bute								
396	Rubha Bodach	55 55	5 09	−0020	−0010	−0007	−0007	−0·2	·
396a	Tighnabruich	55 55	5 13	+0007	−0010	−0002	−0015	0·0	
	Firth of Clyde (cont.)								
	Millport	55 45	4 56	−0005	−0025	−0025	−0005	0·0	
				−0020	−0015	−0010	−0002	+0·2	

FIGURE 5.1 Extract from ATT Volume 1 Part II (Secondary Port data)

(Courtesy of Hydrographer of the Navy)

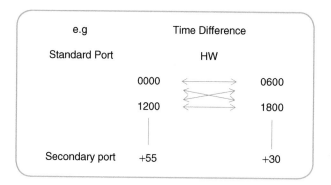

FIGURE 5.2 How the time difference corrections work

The next four columns are all headed under *Time Difference* and cover the 24 hour period in which any High or Low Water will occur. The first two of the four columns refer to **High** Water times at the Standard Port with the time difference at the Secondary Port listed directly below. The next two columns refer to the **Low** Water times at the Standard Port with similarly, the equivalent Secondary Port time differences listed in the columns below.

Figure 5.2 shows us that if the Time of HW at the Standard Port was either 00:00 hrs or 12:00 hrs, the difference at the Secondary Port would be +55 minutes later. If it was at 06:00 or 18:00, then the time of HW at the Secondary Port would be 30 minutes later. Invariably however, the time of the tide is not at any of these times but somewhere in between. This means that the time difference at the Secondary Port will also be somewhere in between. In order to find out *how far* in between, we need to interpolate between the figures.

5.3 Time difference calculation for the Secondary Port

Let us say that the High Water at the Standard Port is given as 16:40 hrs. 16:40hrs is between 12:00 and 18:00 hrs.

TOP TIP

It will be easier to interpolate the time difference if you convert the values into minutes (of time).

16:40 is four hours 40 minutes after 12:00 hrs and this equates to 280 minutes.
The difference between 12:00 and 18:00 hrs is six hours which is 360 minutes.
We need to find the equivalent of

$\dfrac{280}{360}$ th's of the difference between + 55 and + 30 minutes

$\dfrac{280}{360}$ × 25 minutes = 19 minutes.

19 minutes *is not the time difference*. 19 minutes is the same ratio of the difference between +55 and +30 minutes as 16:40 hrs, is between 12:00 and 18:00 hrs. The actual time difference if High Water at the Standard Port is 16:40 therefore is 55 less 19 minutes which is +36minutes. In this case the time of HW at Standard Port = 16:40 hrs

Time difference Secondary Port = + 36 minutes

The Time of HW at Secondary Port will be = 17 : 16 hrs

TOP TIP

In the majority of cases for European ports, the tidal diagrams are referenced to the time of HW so unless the Secondary Port is associated with Southampton, which is referenced to the LW time, just do the interpolation for the time of HW.

In an exam why waste time and risk making possible errors during the Low Water calculation if it is not needed. The opposite is valid for a Secondary Port associated with Southampton where in this case, the HW time would be irrelevent.

HEIGHT DIFFERENCES (IN METRES)				M.L.
MHWS	MHWN	MLWN	MLWS	Z_0 m.
3.4	2.9	1.0	0.4	
−1.3	−1.2	−0.5	−0.2	−1.17
−1.0	−0.9	⊙	⊙	⊙
−0.5	−0.3	+0.1	+0.2	1.80
−0.4	−0.3	−0.1	0.0	1.71
0.0	0.0	+0.1	−0.1	1.92
−0.1	+0.1	−0.5	−0.2	⊙

FIGURE 5.3 Secondary Port height differences

5.4 Height difference calculation for the Secondary Port

Once again there are four columns giving the mean High and Low Water levels at Spring and Neap Tides for the Standard Port. Directly below, the differences of height at the Secondary Ports are given in metres. The first two columns are for High Waters and the second two columns, for the Low Water height corrections. For height differences, unlike time, you will always be required to interpolate for both High and Low Water corrections as you need both values in order to construct the 'Slopy Line'.

The tidal information given in Fig 5.3, indicates that if the Height of Tide at HW was predicted to be 3.4m at the Standard Port (in this case still Greenock), the height difference at the Secondary Port would be 1.3m less. Similarly, if it was 2.9m, it would be 1.2m less at the Secondary Port below it. If the Height of Tide was predicted on a particular day to be somewhere in between, the difference would likewise be in between these two values.

If, on the day in question the predicted heights of HW and LW at Greenock were 3.1m and 0.6m respectively, the differences at the Secondary Port must lie between the tabulated corrections and these will be found by interpolation between the tabulated values.

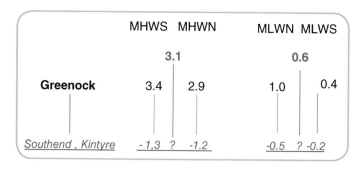

FIGURE 5.4 How the height difference corrections work

In this case the HW prediction is as follows:

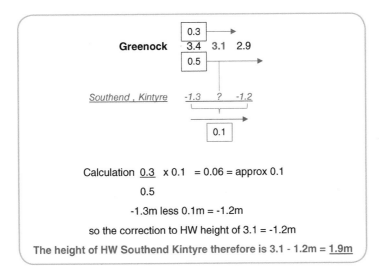

FIGURE 5.5   How to interpolate the HW Height difference

The same process is used to find the Low Water value as follows:

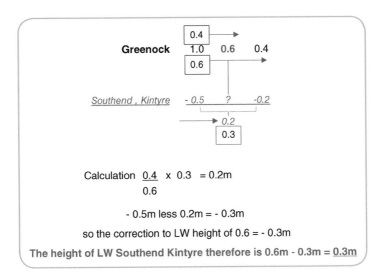

FIGURE 5.6   How to interpolate the LW height difference

The process of interpolation does not come naturally to everyone but there are other times that deck officers will have to interpolate, and exactly the same method is utilised. Interpolation is simply the method of finding equivalent ratios of a difference. For the tide height difference, if the values were laid out vertically, it would look like this:

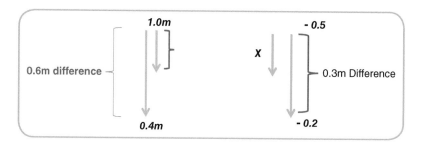

FIGURE 5.7 Alternative method of explanation of interpolation used to find the difference of height at the Secondary Port to apply to the Standard Port predicted value

Where

$$\frac{0.4}{0.6} = \frac{X}{0.3}$$

Re-arranged

$$\left[\frac{0.4 \times 0.3}{0.6}\right] = X = 0.2$$

0.2 is *not* the height difference to apply to the predicted HOT value at the Standard Port.
0.2m is the amount of the difference **between** −0.5 and −0.2 required to obtain the height difference to apply. It is important to make sure you apply the calculated difference in the correct direction. In this case (Fig 5.6), the difference X was 0.2m so the actual difference for the Secondary Port is: 0.5m less 0.2m. This makes the height difference at Southend equal to 0.3m.

HOT Greenock (Standard Port)	= 0.6m
Difference for Southend	= −0.3m
Ht of LW@ Southend (Secondary Port)	= 0.3m

If this is still difficult to understand, try following a simpler example where interpolation is more commonly used by deck officers but doesn't seem to cause such confusion.
   Ship's Head by Magnetic Compass Deviation

$$\begin{array}{cc}
000° & 1\frac{1}{2}°\,E \\
10 \left\{ \begin{array}{l} 020° \\ 030° \\ 040° \end{array} \right\} 20 \quad & \left. \begin{array}{l} 2\frac{1}{2}°E \\ 3\frac{1}{2}°E \end{array} \right\} 1° \\
060° & 5°\,E
\end{array}$$

If the Magnetic Compass heading was 030°. What would be the deviation for this heading?
   The amount to apply to $2\frac{1}{2}°E$ would be 10/20ths of 1° which is ½°.
   The Deviation therefore = $(2\frac{1}{2}°E + \frac{1}{2}°E) = 3°E.$
   The layout used in the previous tide examples have been arranged to interpolate horizontally because the information given for time and height differences in the tide tables is presented this way. If it makes it easier, then the same information may be laid out vertically but whichever you choose, it will produce exactly the same answer.

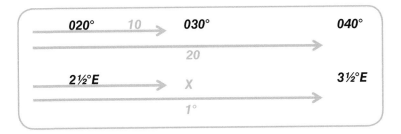

FIGURE 5.8   Alternative explanation of simple ratio using interpolation to find the equivalent difference to apply (vertical layout)

$$\frac{10°}{20°} = \frac{X}{1°}$$

Re-arrange the formula as follows:

$$\left[\frac{\cancel{10°}}{\cancel{20°}}\right] = \frac{1}{2} \times 1° = X = \tfrac{1}{2}° \text{ Deviation for 030° Magnetic} \left(2\tfrac{1}{2}° + \tfrac{1}{2}°E\right) = 3°E$$

5.5 Seasonal Correction

So far we have discussed the corrections for time and height difference in order to find the High and Low Water values at the Secondary Port. There is one final set of corrections to be applied and these are also tabulated in Part II of the ATT at the bottom of each page. This correction is described as the Seasonal Correction in Mean Level but is generally referred to as the Seasonal Correction (S.Corr).

This correction refers to the different seasonal increase or decrease in mean sea levels throughout the different months of the year. This correction adjusts for changes to the amount of predicted rainfall during wet or dry seasons at the various Secondary Ports during the year.

The seasonal rainfall has not hitherto required a separate correction to the predicted heights given on the daily Standard Port pages so did not appear in the Standard Port calculation. It is however, now necessary to *include* this correction to all Secondary Port calculations. The reason it was not required for Standard Ports is that Standard Ports have their own pages of predictions given for every month of the year and the Admiralty have already applied the correction on our behalf for each day of each month. This has avoided the need to look up the Seasonal Corrections in Part II of the tables. In short, this means that the predicted values on the Standard Port pages can be used directly on the adjacent tidal curves as the values for the 'Slopy Line' without further corrections. These predicted heights of tides therefore can be described as *CORRECTED* heights of tide, as the Seasonal Correction has already been applied.

In the case of Secondary Ports however, these corrections must be deduced from the tables and applied in order to give the correct HOT at the Secondary Ports. The tables give the monthly correction required for every port which are listed on the left hand side of the table and identified by their unique port identification number. The values given are listed for the first of the month and where there is a different value between consecutive months, providing the difference is only small, perhaps only one centimetre, the '15 day rule' if applied is usually considered sufficient.

5.6 The '15 day rule'

In order to avoid calculating and plotting heights of tides on the tidal curve to more accuracy than the nearest 0.1m, which on the scale provided would be impossible and unnecessary for navigational needs, the 15 day rule suffices to avoid this.

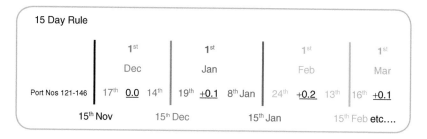

FIGURE 5.9 Use of the 15 day rule to find the Seasonal Correction

The table is set out in Part II for the first day of each of the 12 months of the year with the monthly seasonal correction directly below.

In the case where there is a different correction value in consecutive months, the correction is deemed pertinent to the 15 days before and 15 days after the first of that month. In other words if it is the 1st January and the correction is +0.1m, this is valid from the 15th December to the 15th January. If the date in question was the 16th January, the correction to apply would be the February figure, as that would cover the period from the 15th January through to the 15th February.

This will be acceptable providing the differences in the corrections between consecutive months are only 0.1 different from each other, because if your date was the 15th technically you could use either month's correction. If this was the case then the final HOT calculated would be correct as it will give an answer which falls within acceptable accuracy demanded by the Maritime & Coastguard Agency (MCA) of +/− 0.1m.

If however the difference in consecutive monthly corrections is greater than 0.1 difference, then the correction would be found by interpolation for the day of the month assuming a 30 day calendar month.

Extract from Instructions for the use of the tables in ATT states the following:

> It MUST be noted that the predictions for Standard Ports include the seasonal change for the Standard Port which may be different from the Secondary Port. The Seasonal Change tabulated in Part II and (Part III) are shown for the 1st day of each month. They should be suitably interpolated for the day of the Prediction . . .
> (Hydrographer of the Navy, 2017)

The reason for this becomes more obvious when you look at the monthly differences in the corrections at somewhere like Nantes (1667), where these are significant. As this is a large value, it could, if applied incorrectly, make a dangerous difference to the calculated HOT and hence the accurate assessment of UKC at any time.

Under the '15 day rule', if the date in question was the 5th January, the correct value of the correction would normally have been +0.1m because it fell before the 15th. Likewise, if the date was the 22nd January, you would normally use the February correction +0.8m. This however is a nonsense and it would be clearly dangerous to assume that before the 15th it was +0.1m and on the 16th it changed to +0.8m. This is a difference of 0.7m which will make a large difference to any

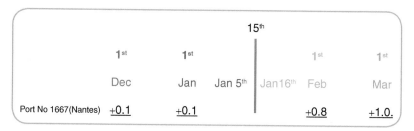

FIGURE 5.10 Example where it is unacceptable to use the '15 day rule' to find the Seasonal Correction

eventual calculated HOT. The effect the gradient of the 'Slopy Line' will make any readings taken of the curve incorrect, potentially putting the vessel in danger.

The correct way to find the right value to apply is to interpolate for the specific day, so that for the 5th January the interpolation would be:

> 5/30 x (difference between corrections)
>
> 5/30 x 0.7m = 0.1
>
> {+0.1(correction for 1st Jan) + 0.1} = correction for 5th Jan
>
> S. Corr for Nantes 5th Jan = +0.2m

FIGURE 5.11 To find the Seasonal Correction where monthly differences make the '15 day rule' unacceptable (5th January)

> 16/30 x (difference between corrections)
>
> 16/30 x 0.7m = 0.374m = approx 0.4m
>
> {+0.1(correction for 1st Jan) + 0.4} = correction for 16th Jan
>
> S. Corr for Nantes 16th Jan = +0.5m

FIGURE 5.12 To find the Seasonal Correction where monthly differences make the '15 day rule' unnacceptable to use (16th January)

5.7 The Secondary Port tidal calculation

We have now dealt with the three corrections to apply to the predicted HOT at the Standard Port in order to find the corrected HOT at the Secondary Port. Once again, there are only two questions in tidal calculations and both will require the 'Slopy Line' values for High and Low Water at the Secondary Port. To make it easy to remember and do away with the reliance on a grid, it will be useful to use the same layout for all Secondary Port problems, including those for Pacific Ports. This avoids any *'new'* learning. In a fairly short period, with practice, this layout becomes part of a process which you will be able to replicate quickly and easily in an examination or where required in the future.

As explained in sub-paragraph 5.6, the calculation will start with the appropriate set of tides which will dictate the *corrected* values of High and Low Water. However, this correction is appropriate to the Standard Port and the ship may be at a Secondary Port hundreds of miles away where the correction for that month is different. This requires you to find out the value of the seasonal correction which has been applied (using the tables in Part II of the ATT volume) and take it off. What you will have now is the Heights of Tides at the Standard Port minus the Seasonal Correction leaving an *UNCORRECTED* Value for Heights of Tide at the Standard Port.

> For Example:
>
> Hts of Tides Standard Port - Avonmouth (Corrected) 11.0m
>
> Seasonal Correction (S.Correction) (+0.1)   -0.1m
>
> Hts of Tides Standard Port - Avonmouth (Uncorrected) 10.9m

FIGURE 5.13 Removal of Seasonal Correction made to the Standard Port height prediction to produce the UNCORRECTED value

Note: REVERSE the sign of the correction

Hts of Tides Standard Port - Avonmouth (Uncorrected)	10.9m
Ht Difference Secondary Port – Newport	-2.6m
Hts of Tides Secondary Port – Newport (Uncorrected)	8.3m
Seasonal Correction Secondary Port – Newport	(+0.1)
Hts of Tides Secondary Port – Newport Corrected)	8.4m

FIGURE 5.14 Application of the height differences and Seasonal Corrections at the Secondary Port

Note: this time, the sign of the correction has been applied AS TABULATED because this is the seasonal correction to apply for Newport, (Secondary Port).

Hts of Tides Standard Port - Avonmouth (Corrected)	11.0m	
Seasonal Correction (S.Correction)	(+0.1)	
Hts of Tides Standard Port - Avonmouth (Uncorrected)	10.9m	
Ht Difference Secondary Port – Newport	-2.6m	
Hts of Tides Secondary Port – Newport (Uncorrected)	8.3m	
Seasonal Correction Secondary Port – Newport	(+0.1)	Correction applied as tabulated
Hts of Tides Secondary Port – Newport Corrected)	8.4m	

FIGURE 5.15 The standard layout for finding the height of HW and LW at a Secondary Port

Note: REVERSE the sign of the correction

(If the Seasonal Correction had been −0.1m then as we are taking the correction off, we would simply reverse the sign and apply + 0.1m making the uncorrected height at Avonmouth 11.1m).

Now that we have an uncorrected value for heights of HW and LW at the Standard Port, the next step is to interpolate the height differences for the Secondary Port in question using the method described in Fig 5.6 or Fig 5.7.

Applying the height differences found will give the heights of HW and LW at the Secondary Port but this will still be an *UNCORRECTED* value until the Seasonal Correction for the Secondary Port in question has been found and applied to these values.

The complete layout for a normal calculation to the 'Slopy Line' values of HW and LW at a Secondary Port would normally look as follows:

Now that the method to solve Secondary Port calculations has been explained, a typical example of the sort of examination question you will be expected to answer for HNC, HND and SQA papers would be as follows:

A vessel is on passage to Fishguard (490) and will be required to pass over a shoal with a Charted Depth of 2.5m during the afternoon flood tide of 7th October. The vessel has a draft of 6.0m and requires a UKC of 1.0m. Find the earliest time she can cross the shoal with the required clearance.

This question asks for a *time*, so the graph must be entered with the appropriate *HOT* required.

Required DOW	7.0m
Charted depth of shoal	2.5m
Req'd HOT	4.5m

FIGURE 5.16   Keeping on track of what remains still to be found before entering tidal curves

All that remains is to find the value of the Percentage Springs before plotting on the curve to produce a solution for the earliest time when the tide will be 4.5 metres providing a sufficient Depth of Water of 7.0 metres.

$$\%Sp = \frac{PR - Np(R)}{Sp(R) - Np(R)} \times 100$$

$$\frac{7.0 - 2.7}{6.3 - 2.7} \times 100 = 119\% \; \textit{Use Spring Curve only}$$

FIGURE 5.17   Keeping abreast of what is still left to find before entering tidal curves

For the Percentage Springs calculation, the ranges are found on the tidal curve for the Standard Port, and when calculated gives a value of 119%. This means that on this day, 7th October, it is at the top of the Spring Tide period, so only the Spring solid curve is used for the plot.

TOP TIP

For examinations it is recommended that you clearly state *why* the particular curve has been used so that it is not considered as a guess. Generally the Percentage Spring value will be somewhere between 0% and 100% and estimation between both curves is required. In this question, where it exceeds 100%, or where the value was a negative percentage, it must be clear that you understand how to use the curves, as incorrect use of the curves incur a 50% loss of marks.

All that remains is the calculation for the Heights of Tides at the Secondary Port, Fishguard, using the previously described layout and use of the corrections in Part II of the Tables. This will give the corrected HOT values for the 'Slopy Line'.

	LW	HW
Heights of Tides Milford Haven (Corrected)	0.5	7.5
Seasonal Correction Milford Haven (0.0)	0.0	0.0
Heights of Tides Milford Haven (Uncorrected)	0.5	7.5
Height differences at Fishguard	+0.2	−2.3
Heights of Tides at Fishguard (Uncorrected)	0.7	5.2
Seasonal Correction at Fishguard (0.0)	0.0	0.0
Heights of tides Fishguard (Corrected)	0.7	5.2

1. HOT / Interval
2. SLOPY LINE
3. % SPRINGS

FIGURE 5.18 Now all parameters have been calculated the 'Slopy Line' can be constructed

You must double check if the tide is an ebb or a flood tide to make sure that you plot on the correct side of the tidal curve.

TOP TIP

From the curve, if you want the time interval, after plotting the 'Slopy Line' 0.7m and 5.2m, enter the curve with a vertical line drawn down from 4.5m (HOT Req'd) to the 'Slopy Line'. From here, draw a horizontal line to the Spring curve and from where this meets drop a vertical line to the time interval scale (ensuring it is on the FLOOD side of the curve).

Reading off the scale, for this calculation, the time interval is one hour and 30 minutes **BEFORE** High Water (−1h 30m).
Therefore:

Time of HW at Fishguard	19:22 UT
Time interval for required HOT at Fishguard	−1:30
Earliest Time when HOT is 4.5m at Fishguard	17:52 UT

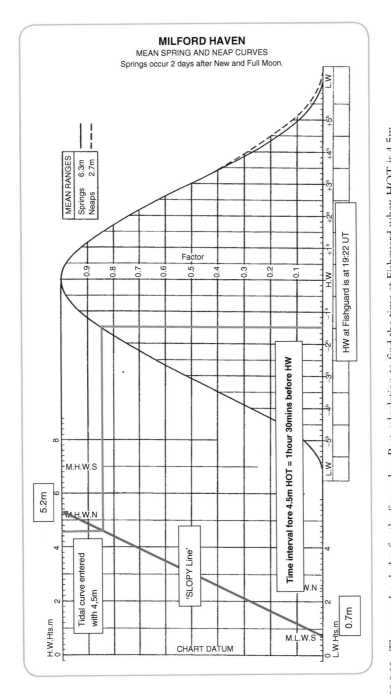

FIGURE 5.19 The completed plot for the Secondary Port calculation to find the time at Fishguard when HOT is 4.5m

6

PACIFIC STANDARD PORT TIDAL CALCULATIONS

Now that the process for calculating the tides has been established, there is little change in the technique required for calculating solutions to the 'what time?' and 'what height?' questions for Pacific tides. The time spent mastering the process involved in European tide calculations will for the most part be replicated and serve as revision with minor changes involved, when using the the Pacific tidal curves. The approach followed for Pacific port calculations will again mirror that which was followed for European ports. This chapter will look firstly at Standard Port calculations before subsequently progressing to Secondary Port calculations.

On inspection of the A.T.T. Volume 4 or Volume 6 covering the Pacific Ocean tides, the obvious difference is that the Standard Ports no longer have their own dedicated tidal curves. Instead, there is only one universal set of curves at the front of the volumes which is used in conjuction with the data from all the Standard Ports daily pages.

On further inspection the tidal curves show a similarity to those for European tides in as much as they still have an interval scale along the bottom of the curves although this time ranging between −7 hours before and +7 hours after the time of High Water. The HOT scales on the top and bottom of the left hand side of the curves will still be used to plot the 'Slopy Line' in exactly the same way as it was with European tide calculations.

6.1 Duration of tide curves

At first glance the most obvious difference from the European curves is the absence of the Spring and the Neap curves. These instead are replaced by three curves known as Duration of Tide (DOT) curves.

In European tides the concern was the difference in tidal range between High and Low Water but in Pacific tides, this is no longer of any significance as it is the *time difference* between High and Low Water, the duration of the tide (DOT), which is now key to using the curves.

In Pacific tide calculations, as there was with European tides, there are again three key parameters to obtain before entering the curves. With no Spring and Neap curves, we no longer have to calculate the Percentage Springs value which depended on mean ranges of the tides. For Pacific Ocean tides, now we are concerned about the time difference between the time of High and the time of Low Water – the *DURATION* of the tide (DOT). The difference between *the heights* of the predictions are now therefore irrelevent. The duration of the tides will fall between the five hour, six hour and the seven hour (See Fig 6.1) Duration of Tide curves.

As with the European curves, the same principals are used when using this tidal curve and it will follow the 'Slopy Line' method as previously described. Once again the graph is always referenced to

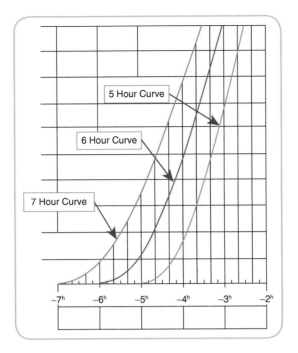

FIGURE 6.1 Duration of tide curves

the time of High Water at the port in question. The Time Interval scale has increased either side of High Water from −7 hours before, to +7 hours after.

The 'Slopy Line' is plotted in the same manner as previously done but on some curves but the scale may need to be doubled up. Care should be taken that if you do double the scales, you double BOTH the High and Low Water scales. More importantly, you must remember that you must double up the value of the HOT as well, when reading off the scales.

6.2 Use of the graphs

Once again before entering the curves you will need three essential pieces of information and this time they will be as follows:

> 1. HOT / Interval
>
> 2. SLOPY LINE
>
> 3. ***DURATION OF TIDE (DOT)***

FIGURE 6.2 What you need to find before entering Pacific tidal curves

The Duration of Tide for ports may or may not be in areas of the oceans where there is still a Semi-Diurnal Cycle of tides i.e. two High and Low Waters in a 24 hour period.

The graphs provide satisfactory results for heights and times of tides at intermediatory stages between High and Low Water providing that (i) the DOT lies within the scope of the curves – between five and seven hours and (ii) that there is no shallow Water correction. These corrections for the calculation required are found in Part III of the ATT Tables for Harmonic Constants using the Simplified Harmonic Method of Predictions N.p.159. (See Annex 1).

If *BOTH* these criteria are not satisfied then the curves will need to be plotted by hand for the required period by following the instructions given in the Preface of the ATT Tide Tables and which involve a long and convoluted calculation. Alternatively the GPS equipment contains the Harmonic Information and will automatically make the adjustments to the curves, providing acceptable predictions for these ports using the Total-Tide software which is pre-installed into GPS receivers.

As the method is already known and the same two questions remain to be solved, we can apply the same process to solve Standard Port calculations for Pacific ports.

6.3 Pacific Standard Ports example 1 – Vancouver

What will be the latest time a vessel could pass under a bridge at Vancouver (9133) on the 15th February on the morning flood tide? The vessel has an air draft of 27.0m and requires a clearance of 2.0m above the mast. The bridge has a charted height of 26.3m.

Once again the question asks for the time, so a diagram is required to resolve the HOT required to pass safely under the Bridge.

The first thing to address is: what does the question tell us about this bridge? It has a Charted Height of 26.3m which the chart indicates is measured above the reference level of Highest Astronomical Tide (HAT) at Vancouver. The Charted Height of the bridge is 26.3m above HAT and the height of HAT above Chart datum at Vancouver (as given in Part 1 of Table V in the relevant Volume of ATT's) is 5.1m. Theoretically the biggest gap between the waterline and the bridge will be when the HOT is zero and the water level is coincident with Chart Datum. At that point the maximum space available would be 31.4m. Out of this gap, how much space will the vessel require beneath the bridge?

Maximum space available = 31.4m
Space required will be the Air-draft + the clearance required = (27.0m + 2.0m) 29.0m

Maximum HOT require, in other words, the latest opportunity to pass under the bridge with the required 2.0m clearance will be when the Height of Tide reaches a height of (31.4 − 29.0) = **2.4m.**

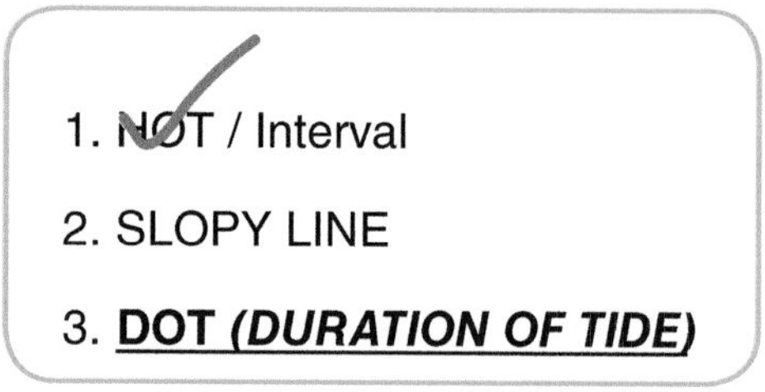

1. ~~NOT~~ / Interval

2. SLOPY LINE

3. **DOT _(DURATION OF TIDE)_**

FIGURE 6.3 What you need to find before entering Pacific tidal curves

The next thing to find out is the pair of relevant tides at Vancouver on the day so that the values for the 'slopy line' can be found.

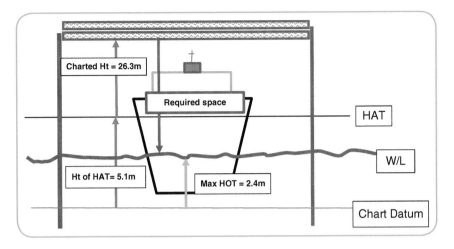

FIGURE 6.4 Example 1 – drawing the diagram to establish the required Height of Tide

On the date in question, the question requires the latest in the morning. This indicates that the morning flood tide on the 15th February is the only pair of tides that can be use. These are as follows:

LW 07 : 25 – **1.4m**
⎱ SLOPY LINE values
HW 13 : 20 – **3.5m**

1. N̶O̶T / Interval

2. SL̶O̶PY LINE

3. **DOT (DURATION OF TIDE)**

FIGURE 6.5 What you need to find before entering Pacific tidal curves

All that remains now as this is a Pacific tides problem, is to find the Duration of Tide in this case, which is simply the difference in the time between the Low Water and the High Water predictions.

LW   07 : 25
HW   13 : 20
DOT   5h 55m

1. N̶O̶T / Interval

2. SL̶O̶PY LINE

3. **DOT (DURATION OF TIDE)**

FIGURE 6.6 Everything needed has now been found to enter tidal curves

i. Plot the 'Slopy Line' LW & HW (1.4m and 3.5m). Enter the tidal curve at the required HOT which is 2.4m dropping a vertical line to the point where it intersects the 'Slopy Line'.

ii. At this intersection, draw a horizontal line across all three DOT curves.

iii. Estimate a DOT of 5 hours and 55minutes between the five hour and the six hour curves on the flood side of the diagram.

iv. At this point drop a vertical line down to the Interval Scale and read off the interval before the time of High Water at Vancouver.

In this case it is 2 hours and 58 minutes before HW at Vancouver which will give the latest time to pass under the Bridge at Vancouver on the morning flood of 15th February as:

HW Vancouver   13:20

Time interval   −2:55 from the curve

Latest time   10:25 hrs *ZONE TIME*

You will notice that I have accentuated that the answer is in Zone Time. This is not the first time that Zone Time has been mentioned but we have not paid much attention to it in previous European Port example questions. This is because Zone Time is the *Local* time which in most cases in UK was Greenwich Mean Time (Zone 0), so it didn't affect the outcomes.

6.4 Zone Time (Local Time)

In Pacific tide calculations, the Zone Time and the local time will not be UT (GMT) and it is important to get into the habit of automatically checking the Zone Time and appending it to the answer. The reason for this will be made abundantly clear when we look at Secondary Port calculations, where the Secondary Port may be hundreds if not thousands of miles away from its Standard Port, and in a completely different Time Zone.

So for the above example, HW Vancouver was 13:20 (Zone Time) and because of its position in Longitude 126° West.

The time at Vancouver can be referred to as:

13:20 (Zone +8)

13:20 Zone Time

Or   13:20 Local Time (in Vancouver)

When using the Admiralty Tide Tables for any port in the world, the predicted times of HW and LW are always given in Local Time at that place. In this case, HW Vancouver was given as 13:20hrs, so it will be at 13:20 local time on your watch. 13:20hrs will be the Zone Time which, if you needed to know, is given in the top left hand corner of each Standard Port page. In the case of Vancouver, the Time Zone happens to be (Z+8).

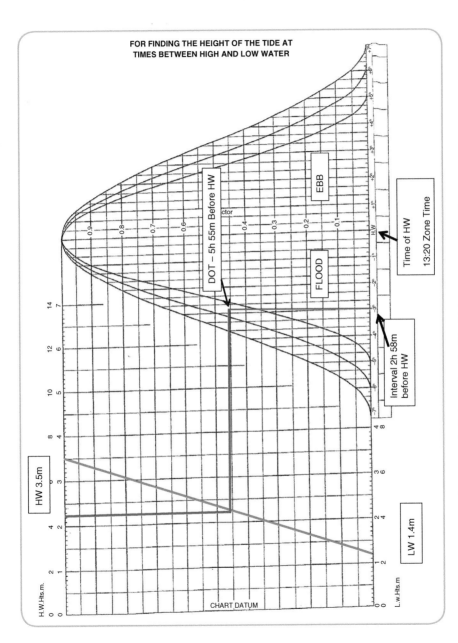

FIGURE 6.7 Worked solution for Secondary Port calculation using Pacific tidal curves – example 1

6.5 Conversion of Local Time to UT and From UT to equivalent Zone (Local) Time

(Z+8) simply tells us that if you needed to convert this time to UT(GMT) you would have to add eight hours to find the equivalent Time in UT at Greenwich (UK).

$$\text{Therefore HW Vancouver } 13:20 \, (Z+8)$$

$$\text{To convert to UT} \qquad +8:00 \, \text{Hrs}$$

$$\text{Equivalent time in } \mathbf{UK} \qquad \mathbf{21:20 \, hrs \, UT \, (GMT)}$$

Supposing the question was posed in a different way and you were required to find the height of tide in Vancouver at 21:20hrs **UT (GMT)** 15th February. If you went into the tide tables for Vancouver and looked for the pair of tides either side of 21:20hrs, this would be the wrong approach entirely as British Columbia, Canada, does not keep local time in UT (GMT).

British Columbia (B.C.) is in Zone Time (Z+8) so you would need to find the equivalent time in Vancouver and then find the pair of tides either side of that time locally.

For these questions it is easier if first you look up the Zone Time at the port in question. In this case we know that Vancouver is in Time Zone +8 (top left of the Vancouver Standard Port page). If you were in Vancouver, to get to UT you would simply add eight hours (Z+8). If you are already in UT (GMT) as in this case, the question is looking for a Height of Tide at 21:20hrs UT (GMT), to get to Vancouver local time, you would have to do the opposite. To get from UT to local (Zone) time 21:20hrs – eight hours gives the equivalent local time at Vancouver.

Vancouver at equivalent of	21:20 hrs UT (GMT)
Convert to Local time	−8:00 hrs
Equivalent Local Time @Vancouver	13:20 Zone Time (Z+8)

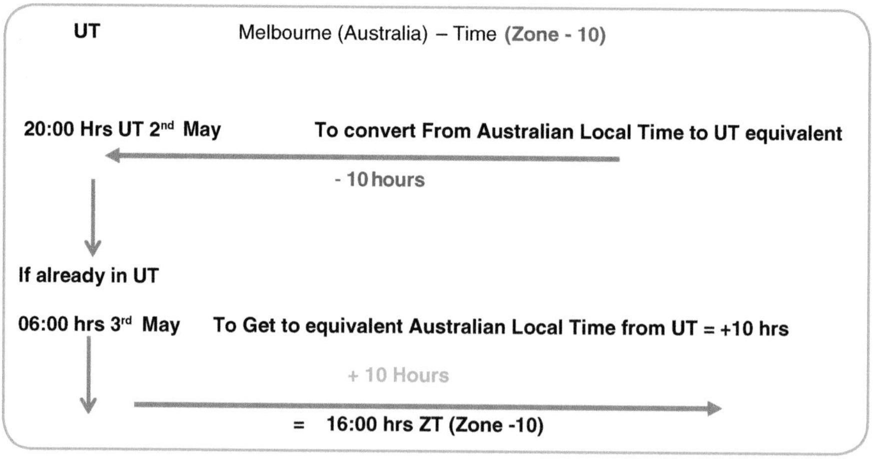

FIGURE 6.8 Converting UT (GMT) to the Local Time at the Standard Port in order to choose the correct pair of tides

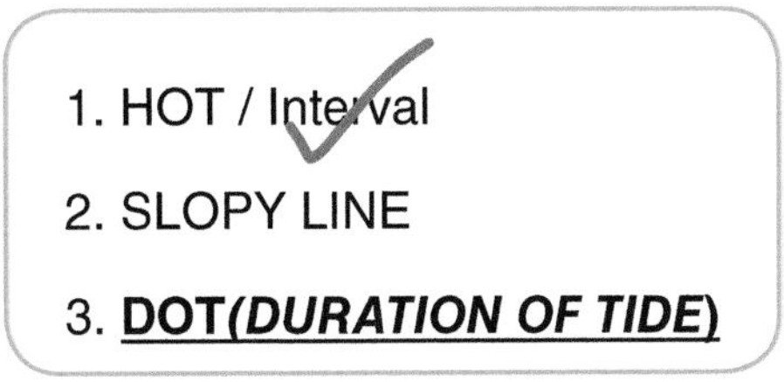

1. HOT / Interval

2. SLOPY LINE

3. **DOT(_DURATION OF TIDE_)**

FIGURE 6.9 What you need to find before entering Pacific tidal curves

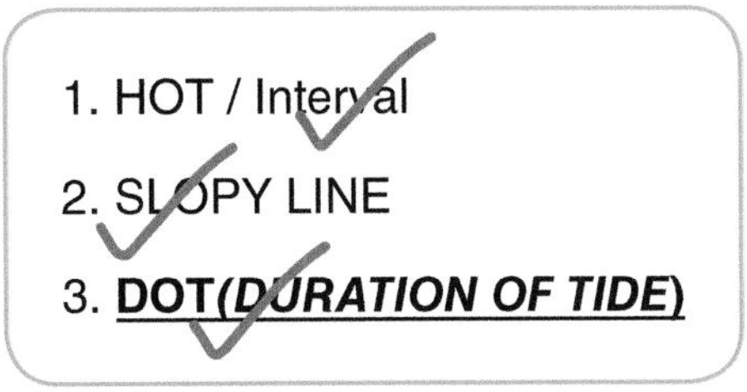

1. HOT / Interval

2. SLOPY LINE

3. **DOT(_DURATION OF TIDE_)**

FIGURE 6.10 Everything needed has now been found to enter tidal curves

6.6 Pacific Standard Ports example 2 – Melbourne

Find the Height of Tide at Melbourne at 06:00 UT (GMT) on 3rd May.

In order to start answering a question worded this way, it must be recognised that the pair of tides either side of 16:00 ZT (Local Time) at Melbourne need to be selected. To continue this question, the pair of tides either side of 16:00 hrs need to be selected. If the Height of Tide is required then this will allow the time interval to be calculated in order to enter the curves.

Time of HW $\quad 19:20\,(Z-10)$

Time Required $\;16:00\,(Z-10)$

Time Interval – 3hrs 20Minutes

$$\text{Required } 16:00 \text{ ZT}\,(Z-10) \left.\begin{array}{l} \text{LW } 12:25 \;\textbf{0.1m} \\ \\ \text{HW } 19:20 \;\textbf{0.9m} \end{array}\right\} \textbf{SLOPY LINE}$$

DOT = – 6hrs 55 Min

All that is required is to check that the plot is on the correct side of the curves i.e flood side, and then plot on the curve to find HOT 3hrs 20minutes before HW at Melbourne. The HOT at this time from the curves (Fig 6.11) is 0.6m.

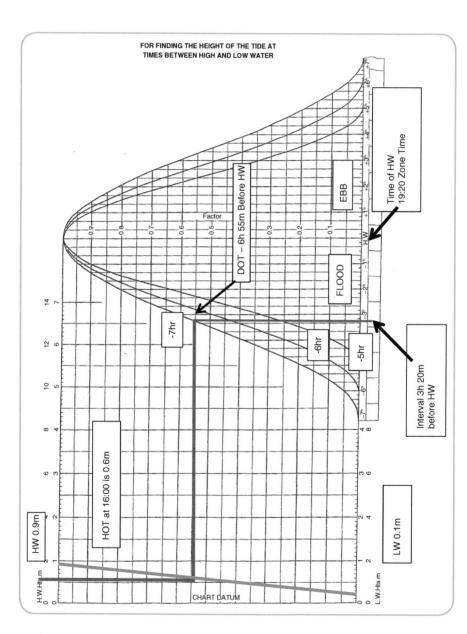

FIGURE 6.11 Worked solution for Secondary Port calculation using Pacific tidal curves –example 2

6.7 Pacific Standard Ports example 3 – Darwin

Find the earliest Zone Time and UT for a vessel to pass over a shoal charted at 4.8m on the approaches to Darwin during the afternoon on the 22nd January with a minimum UKC of 2.0m. The vessel has a draft 7.0m.

DOW Required = 7.0 + 2.0 = 9.0m

DOW – Charted Depth = Required HOT

9.0 – 4.8 = 4.2m HOT

In this case the earliest time required suggests that the tide MUST be a flood tide and the question states that it will be in the afternoon. The correct pair of tides to choose will therefore be as follows:

LW 15 : 39 ZT 2.6m ⎫
                      ⎬ SLOPY LINE VALUES
HW 22 : 29 ZT 6.4m ⎭

DOT   6 Hours 50 Minutes

TOP TIP

If the question asks for Zone Time and UT just work out the answer in Local (Zone Time) and then on the last line apply the Time Zone difference to obtain UT (GMT). Do not try to convert to GMT at the beginning because this may cause you to select the wrong pair of tides and will incur heavy loss of marks as your HW time and any results from the curves may be incorrect.

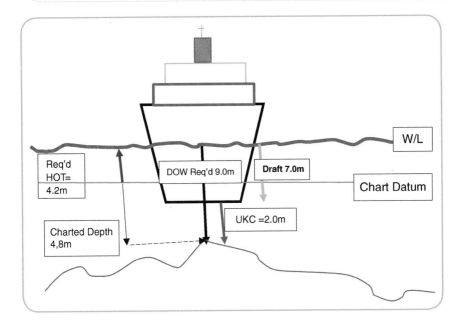

FIGURE 6.12 Diagram to determine HOT required – example 3

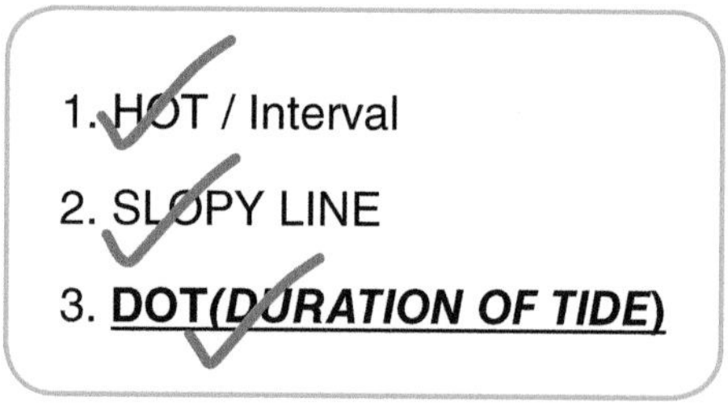

```
1. HOT / Interval ✓

2. SLOPY LINE ✓

3. DOT(DURATION OF TIDE) ✓
```

FIGURE 6.13 Everything needed has now been found to enter tidal curves

From the tidal curve (Fig 6.14) the interval before HW when the HOT will be 4.2m is 3 hours and 50 minutes before HW Darwin.

HW Darwin   22:29 ZT

Interval From Curves   −3hours 50mins

Earliest time to cross the shoal will be at 18:39 ZT (local time).

The question requires the answer in Zone Time (ZT) and in UT(GMT).

In order to convert the time to GMT, the Time Zone found at top left of the Standard Port page for Darwin is (Z − 8) so to convert to UT(GMT):

Earliest Time will be $18:39$ ZT $(Z-8)$

Convert to $UT(GMT) - 8:00$

Earliest Time equivalent is $10:39$

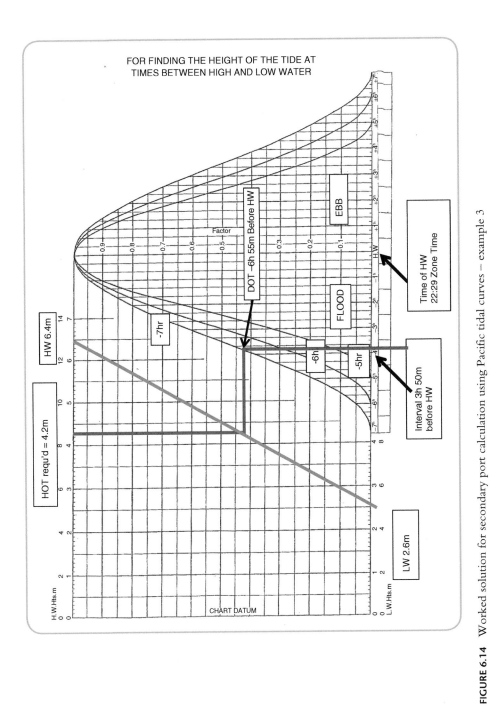

FIGURE 6.14 Worked solution for secondary port calculation using Pacific tidal curves – example 3

7

PACIFIC SECONDARY PORTS TIDAL CALCULATIONS

7.1 Pacific Secondary Port calculations

Now that the use of the Duration of Tides curves is better understood, Secondary Ports should be much easier to master. All the difficult learning of the process has already been achieved whilst learning how to calculate the European Secondary Port calculations in the initial chapters.

The same process that was used for European Secondary Port calculations, and the table that was created, will remain unchanged and be perfectly suited for use with Pacific calculations. We will use the same corrections, in the same order and derived from the same tables as previously and in fact, some corrections are made easier. The time difference corrections for the Secondary Ports in Part II have only a single correction for High and Low Water which eliminates the need for interpolation for the time differences to be applied to the predicted times.

Once again by applying the Secondary Port height difference corrections, the 'Slopy Line' for the heights of High and Low Water at the Secondary Port will be plotted on the curves.

The Duration of Tide (DOT) will be the DOT at the *Secondary Port* after the time differences have been applied. Once again this leaves either the Height of Tide, or the Time Interval to be derived, depending on the nature of the problem to be solved.

7.2 Pacific Time Zones

The only difference which needs to be appreciated is that in the Pacific Ocean, Secondary Ports may have their Standard Port not just hundreds of miles away, but in some cases thousands of miles apart. The connection between the two is not geographical but the characteristic of the tidal flow. It will now require more care to ensure the significance of the differences in the Time Zones between the two ports, when choosing the appropriate pair of tides when commencing the calculation.

When choosing the appropriate pair of tides to start any calculation, if there is a significant time difference to apply at the Secondary Port, it should be borne in mind that it is the date and time of day at the Secondary Port which is relevant. In other words you must check that after the time difference has been applied, giving the times of the tides at the Secondary Port, they are on correct date or time of day *at the Secondary Port* where *your ship*, bridge, sandbank or *other danger* exists to cause concern.

The ATT Tide Tables will show the Time Zone at the top left hand side of each Standard Port page but it is now far more possible that the Time Zone at the Secondary Port may be different. It is therefore *even more* important in Pacific tidal calculations to check by running your eye up the page for the Secondary Port in question in Part II of the tables, and confirming whether the Secondary Port is in the same Time Zone or if it has changed.

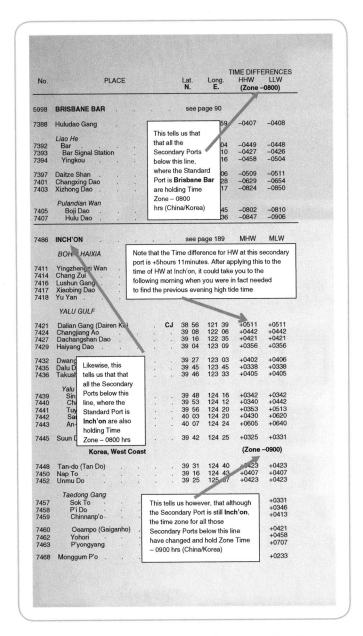

				TIME DIFFERENCES	
No.	PLACE	Lat. N.	Long. E.	HHW (Zone −0800)	LLW
5998	**BRISBANE BAR**	see page 90			
7388	Huludao Gang		59	−0407	−0408
	Liao He				
7392	Bar		04	−0449	−0448
7393	Bar Signal Station		10	−0427	−0426
7394	Yingkou		16	−0458	−0504
7397	Daitze Shan		06	−0509	−0511
7401	Changxing Dao		28	−0629	−0654
7403	Xizhong Dao		17	−0824	−0850
	Pulandian Wan				
7405	Boji Dao		45	−0802	−0810
7407	Hulu Dao		36	−0847	−0906
7486	**INCH'ON**	see page 189		MHW	MLW
	BOHAI HAIXIA				
7411	Yingzhengzi Wan				
7414	Chang Zui				
7416	Lushun Gang				
7417	Xiaobing Dao				
7418	Yu Yan				
	YALU GULF				
7421	Dalian Gang (Dairen K.)	CJ	38 56	121 39	+0511	+0511
7424	Changjiang Ao		39 08	122 06	+0442	+0442
7427	Dachangshan Dao		39 16	122 35	+0421	+0421
7429	Haiyang Dao		39 04	123 09	+0356	+0356
7432	Dwang		39 27	123 03	+0402	+0406
7435	Dalu D		39 45	123 45	+0338	+0338
7436	Takush		39 46	123 33	+0405	+0405
7439	Sin		39 48	124 16	+0342	+0342
7440	Ch		39 53	124 12	+0340	+0442
7441	Tu		39 56	124 20	+0353	+0513
7442	Sa		40 03	124 20	+0430	+0620
7443	An-		40 07	124 24	+0605	+0640
7445	Suun D		39 42	124 25	+0325	+0331
	Korea, West Coast			(Zone −0900)	
7448	Tan-do (Tan Do)		39 31	124 40	+0423	+0423
7450	Nap To		39 16	124 43	+0407	+0407
7452	Unmu Do		39 25	125 07	+0423	+0423
	Taedong Gang				
7457	Sok To				+0331
7458	P'i Do				+0346
7459	Chinnanp'o				+0413
7460	Oeampo (Gaiganho)				+0421
7462	Yohori				+0458
7463	P'yongyang				+0707
7468	Monggum P'o				+0233

Callout boxes within the table:

- This tells us that that all the Secondary Ports below this line, where the Standard Port is **Brisbane Bar** are holding Time Zone – 0800 hrs (China/Korea)
- Note that the Time difference for HW at this secondary port is +5hours 11minutes. After applying this to the time of HW at Inch'on, it could take you to the following morning when you were in fact needed to find the previous evening high tide time
- Likewise, this tells us that that all the Secondary Ports below this line, where the Standard Port is **Inch'on** are also holding Time Zone – 0800 hrs
- This tells us however, that although the Secondary Port is still **Inch'on**, the time zone for all those Secondary Ports below this line have changed and hold Zone Time – 0900 hrs (China/Korea)

FIGURE 7.1   Changes in Time Zones between Standard and Secondary Ports

Most Secondary Ports will operate in the same Time Zones as their parent Standard Port and after applying time corrections both will remain on the same date, or time of day. Examples where this will *not* be the case will occur if the two ports are separated by several Time Zones and are consequently a considerable distance removed from one another.

Examples where this is the case will be dealt with later in the chapter illustrating how best to ensure the correct pair of tides are selected initially. By dealing with this situation separately it will make it easier to follow and avoid potentially dangerous mistakes affecting heights and times plotted on the tidal curves.

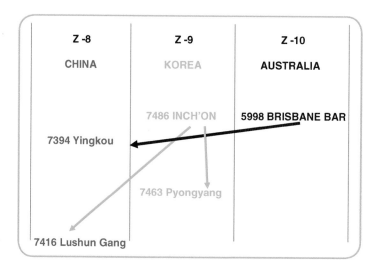

FIGURE 7.2 Inconsistency in Time Zones between both Standard and Secondary Ports and national Time Zones

As you will notice from the diagram, the Standard Port for the first nine Secondary Ports on this page in Part II of the Tide Tables is Brisbane Bar which is in Australia in the Southern Hemisphere. The Port No 7394 Yingkou however, a Secondary Port whose Standard Port is listed as being in Australia, is located in China – far to the North.

Always double check to see whether there has been a Time Zone change and when the time difference for High Water and Low Water is applied, the resultant time will be changed to the new time zone without any further alteration.

The listed time difference has already taken into account the difference in distance between the two ports, which may have spanned over several time zones and be in a totally different longitude. Inconsistencies litter the Tide Tables with regards to Time Zones so care needs to be taken. Take a look at the page of the Tidal Information in Figure 7.1, the Standard Port for the Secondary Port of Pyongyang (7463) in Korea, is Inch'on (also in Korea) and both with a Time Zone of (Z − 9). Inch'on is also the Standard Port for the Secondary Port of Lushun Gang (7416) but Inch'on is in Time Zone (Z − 9) and Lushun Gan, which is in China, is in Time Zone (Z − 8). Yingkou is a Secondary Port in China but the Standard Port for Yingkou is at Brisbane Bar Australia, where the Zone Time is (Z − 10).

TOP TIP

ALWAYS CHECK THE TIME ZONES BETWEEN THE TWO PORTS AS THEY MAY WELL BE DIFFERENT.

7.3 Pacific tidal curves – adapting HOT Scales

Adapting the HOT scales on the tidal curves to suit the predicted values for the 'Slopy Line' for unusually high or low Heights of Tides may be necessary but care needs to be taken to achieve correct readings of the scales.

LW Shimonoseki   02:56 (Z – 8)   −0.3m

HW Shimonoseki   09:54 (Z – 8)   2.2m

The Low Water height is −0.3m which should not cause concern as this only indicates that at Low Water the seabed is exposed and dries out 0.3m above Chart Datum. Although the scales on the curves start at zero, there are several ways to plot this value.

1. Extend the scale to the left measuring 0.3m and using a pair of dividers, plot 0.3m off the scale to the left of the origin.
2. On the Low Water scale at the bottom of the curves, create a false Zero. Where the scale is marked as 1.0m, change this to the new Zero and then plot 0.3m to the left of this point. It is important now that the High Water scale Zero is also moved so that the gradient of the 'Slopy Line' is not affected. Even more importantly, when reading off the resulting Height of Tide, make sure that the height is read using the modified scale and not the original.
3. If the Height of the Tides to be plotted extends above the scale provided, then the scale can again either be extended if there is room to the left of the HW point on the curves, or halved. This must be done to both the Low Water and the High Water scales and the answer must be read off the modified scale. This may sound like it is stating the obvious but this is yet another common area of examination mistakes which are very costly. This is another example where a second glance to make sure that the correct scale has been used is warranted which will give rise to an incorrect Height of Tide.

7.4 Pacific Secondary Ports example 1 – Nagasaki

Find the Height of Tide at 10:00 Zone Time (ZT) at Nagasaki (7666) on the 7th January.

As the question requires the HOT, the curves will provide the answer so it will be required to enter the curves with the Time Interval. The first thing to check will be the Standard Port for Nagasaki which is in this case Shimonoseki in Japan.

The pair of tides from the Tide tables will be as follows:

Shimonoseki $08:03\left(Z-8\right)$ 0.5m

Nagasaki $\Rightarrow$ **10 : 00 ZT**   **???**

Shimonoseki $14:23\left(Z-8\right)$ 2.2m

In order to find the Time Interval, the time of the tides at Nagasaki must be found. This in turn will lead to the Duration of the Tides; leaving just the table of corrections of the predicted Heights at Shimonoseki to provide the 'Slopy Line' to be used.

	LW	*HW*
Times of Tides Shimonoseki	08:03 **(Z–8)**	14:23 **(Z–8)**
Time Differences Nagasaki	−1:04	−0:57
Times of Tides Nagasaki	06:59 **(Z–9)**	13:26 **(Z–9)**
Time Required		10:00hrs

TOP TIP

Take note of the change in Time Zone which automatically occurs when the time difference is applied – **YOU JUST NEED TO NOTICE THAT IT HAPPENED.**

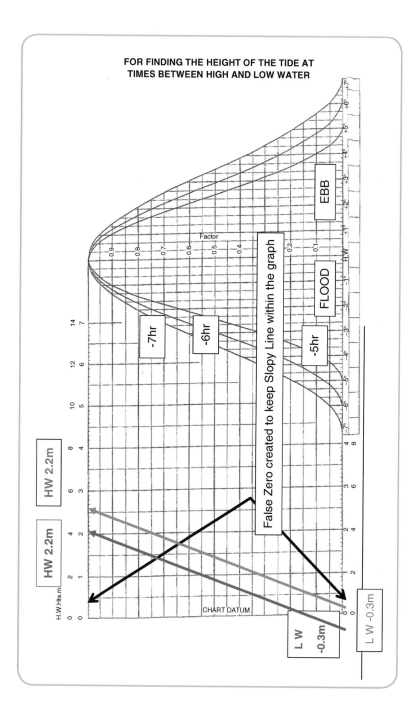

FIGURE 7.3 Plotting Drying Heights and creating a false Zero to make better use of the scale

1. HOT / Interval

2. SLOPY LINE

3. **DOT *(DURATION OF TIDE)***

FIGURE 7.4   What you need to find before entering Pacific tidal curves – example 1

Times of Tides	Nagasaki	06:59 **(Z–9)** LW
		13:26 **(Z–9)** HW

The Time Interval to enter the curves at 10:00 Hrs ZT = –3hrs 26minutes before HW

1. HOT / Interval ✓

2. SLOPY LINE

3. **DOT *(DURATION OF TIDE)*** ✓

FIGURE 7.5   What you need to find before entering Pacific tidal curves – example 1

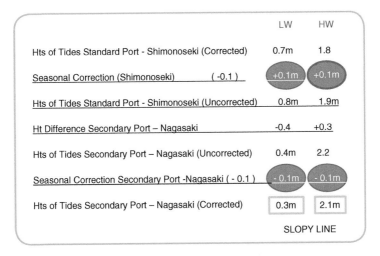

	LW	HW
Hts of Tides Standard Port - Shimonoseki (Corrected)	0.7m	1.8
Seasonal Correction (Shimonoseki) (-0.1)	+0.1m	+0.1m
Hts of Tides Standard Port - Shimonoseki (Uncorrected)	0.8m	1.9m
Ht Difference Secondary Port – Nagasaki	-0.4	+0.3
Hts of Tides Secondary Port – Nagasaki (Uncorrected)	0.4m	2.2
Seasonal Correction Secondary Port -Nagasaki (- 0.1)	- 0.1m	- 0.1m
Hts of Tides Secondary Port – Nagasaki (Corrected)	0.3m	2.1m
		SLOPY LINE

FIGURE 7.6   Correcting the Standard Port heights to find the values for the 'Slopy Line' at Nagasaki – example 1

All that remains is to calculate the values of the HW and LW at Nagasaki in order to plot the 'Slopy Line'.

7.5 Interpolation for height differences

Now that all the components have been found all that is required is to confirm what it is you are trying to find by referring to the question and double check you are plotting on the correct side of the graph.

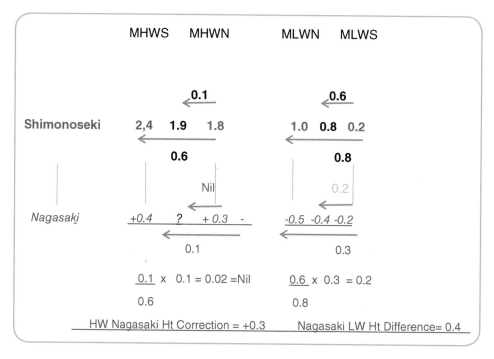

FIGURE 7.7 Interpolation for the height differences at Nagasaki – example 1

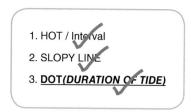

FIGURE 7.8 What you need to find before entering Pacific tidal curves – example 1 Nagasaki

7.6 Constructing the plot using the Pacific tidal curves to derive the answer

Initially construct the 'Slopy Line' using the Low and High Waters values derived for the Secondary Port i.e. Nagasaki, and enter the curves at a time interval of 3 hours and 26 minutes BEFORE High Water. From this point, take a vertical line which will cross all the DOT curves. The Duration of Tide (DOT) calculated previously was –6 hours and 27 minutes. Estimate where this occurs between the –6 hour and the –7 hour curves and take a horizontal line across to the 'Slopy Line'. From the intersection point take a second vertical line up to the HOT scale and the answer given for the Height of Tide at Nagasaki at 10:00 hrs is in this case 1.1m.

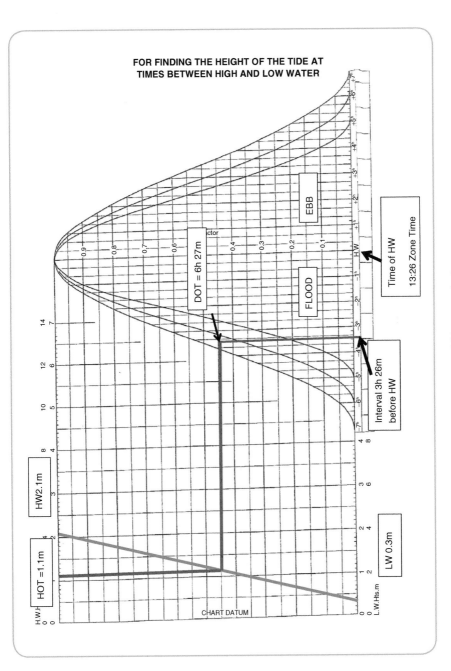

FIGURE 7.9 Plot to determine the Height of Tide at 10:00 ZT at Nagasaki in example 1

7.7 Pacific Secondary Port example 2 – Guayaquil

Find the latest time during the morning tide when a vessel could pass under a temporary overhead cable at Guayaquil, Ecuador (9544) on 19th October. The cable has a charted height of 15.0m. The vessel has an air draft of 15.0m and a minimum clearance below the cable of 2.0m is required. In this question, in order to find the latest time to pass under the temporary cable, it will be necessary to calculate the maximum Height of Tide required to allow a 2.0m minimum clearance.

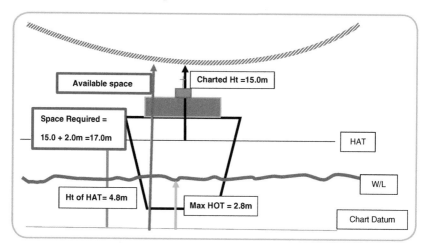

FIGURE 7.10 Example 2 Guayaquil. Drawing the diagram to establish the required Height of Tide

In choosing the correct pair of tides, care must be taken to consider what the tide must be doing in order for it to be the *latest* time.

Space required = 15.0m (Airdraft) + 2.0m (Clearance) = 17.0m

Total space available = 4.8m (HAT) = 15.0m (Charted Ht) = 19.8m

Maximum allowable HOT = 2.8m

In this case it is clear that the tide *must* be rising and the question asks for the morning at Guayaquil so you will need the pair of tides at Balboa which, when the time difference is applied, gives the times of flood tide in the morning 19th October at Guayaquil.

		LW	HW
Times of Tides	Balboa	19:03 (Z + 5) **18th Oct**	01:27 (Z + 5) **19th Oct**
Time Differences	Guayaquil	+5:11	+4:23
Times of Tides	Guayaquil	00:14 (Z + 5) 19th Oct	05:50 (Z + 5) 19th Oct

LW Guayaquil 00:14

HW Guayaquil 05:50

DOT = 5hours 6minutes.

7.8 Extracting the correct pair of tides when correcting for time differences between ports

**Notice that this time there have been no Time Zone changes but we have needed to use the tide on the previous evening (18th October) at Balboa. Looking at the time differences for Guayaquil, if the morning flood had been chosen, after the application of the time differences, times of the tides at Guayaquil would have been in the afternoon. This is the one of the main difference between European and Pacific Secondary Port calculations and requires great care in order to ensure that the calculation starts with the correct set of tides.

Time differences between Pacific ports are often now considerably larger than with most European Secondary Ports. Incorrectly applying these to tides at the Standard Port may result in giving Secondary Port High and Low Water times which can be at the wrong time of day or even the wrong date required. It is at the Secondary Port where your ship and associated navigational hazards exist so make sure you have the appropriate information for the Secondary Port. For example, in this question the morning tides at Balboa on the 19th October are as follows:

		LW	*HW*
Times of Tides	Balboa 19th	07:35 (Z + 5) 19th Oct	13:43 (Z+5) 19th Oct
Time Differences	Guayaquil	+5:11	+4:23
Times of Tides	Guayaquil 19th	12:45 **(Z + 5)**	18:06 **(Z+5)**

This incorrect choice of tides at Balboa results in an ***afternoon*** tide at Guayaquil when the question clearly required the ***morning*** at Guayaquil. All that remains is to calculate the values of the HW and LW at Guayaquil in order to plot the 'Slopy Line'.

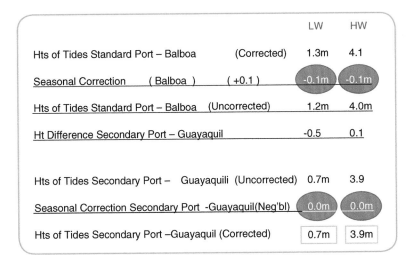

FIGURE 7.11   Correcting the Standard Port heights to find the values for the 'Slopy Line' values at Guayaquil – example 2

7.9 Extrapolating correction values for height differences at the Secondary Port

In this question, you will notice that where we are used to calculating the differences for the Secondary Port by interpolating between the tabulated values, here the value for the Low Water fell outside the MLWS and MHWN values given for Balboa in the Part II extract. This is common and in cases like this, where interpolation is not possible, ***Extrapolation*** will be required.

This is a valid method if it is assumed that the rate of change of the height correction used against the difference in Height of Tide is linear. This allows us to assume that the difference in height we would apply if we had interpolated will be the same outside the tabulated values. It is as a result of the daily predicted Heights of Tides being either slightly greater or smaller on that day than the average MHW or MLW levels.

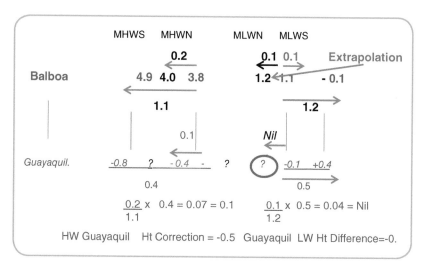

FIGURE 7.12 Interpolation and extrapolation for the height differences at Guayaquil – example 2

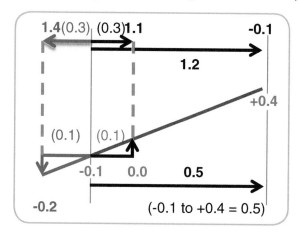

FIGURE 7.13 Extrapolation for the height differences are firstly interpolated and the difference applied outside the tabulated values

7.10 Extrapolation explained

The above figure (Fig 7.13) shows an extract from the height difference corrections from the Admiralty Tide Tables. If the uncorrected height of Low Water at the Standard Port was 1.4m, it becomes clear that this value lies outside the tabulated values for MLWN and MLWS which makes interpolating impossible. What is done to overcome this problem is simple – we have to cheat!

We assume that *the rate of change* of the height difference at the Secondary Port follows a linear relationship with the height of tide at the Standard Port.

This means that the correction will reduce or increase at the same rate **beyond** the tabulated values, as they do **between** them. In the example above the Standard Port value for LW to be corrected is 1.4m. 1.4m is not contained within the tabulated values and is 0.3m beyond the tabulated MLWN value.

If interpolating, this 0.1 would normally be applied to the tabulated value of −0.1m giving a correction to apply to the Standard Port HOT of Nil. However as this required an **extrapolation** and

TOP TIP

Interpolate as if it was 0.3m in the opposite direction as if in fact you needed the correction for 0.8m and interpolate as normal. You need to find the equivalent ratio between the two tabulated corrections

$$\frac{0.3}{1.2} \times 0.5 \left(-0.1\,\text{to} +0.4 \right) = 0.125 = 0.1\text{m}$$

we assume the rate of change will be the same in the opposite direction, the *actual* correction will be 0.1m applied outside the tabulated value which in this case, by inspection, will decrease in a negative direction of change. Therefore the equivalent correction to apply to the Standard Port Value of 1.4m will be:

–0.2 giving a HOT (Uncorrected) at the Secondary Port of 1.4–0.2 = 1.2m.

7.11 Pacific Secondary Ports example 3 – Quingdao China

Find the earliest time on the morning of the 16th May when a container vessel with a draft of 9.0m could safely cross over an obstruction in the channel charted at 9.6m at Quingdao (7318). The vessel must clear the obstruction with a minimum Under-Keel Clearance (UKC) of 2.0m.

Draft 9.0 +2.0 Req'd DOW   = 11.0m

Charted Depth Obstruction   = 9.6m

Minimum Required HOT   = 1.4m

Times of Tides	Shimonoseki	04:30 (Z–9) 16th April	10:09 (Z–9) 16th April
Corrections	Quingdao	+8:08	+7:55
Times of Tides	Quingdao	12:38 (Z–8)	18:04 (Z–8) 16th April

DOT 6h:34mins

At this point everything looks fine except that on closer inspection, the flood tide at the port where the obstruction is – Quingdao – does not start until after midday so it will be well into the afternoon before there is sufficient water to clear the obstruction. It has to be the flood tide in the morning at Quingdao, so clearly the wrong pair of tides at Shimonoseki have been selected.

In this example, the pair chosen at the Standard Port gave times in the afternoon at the Secondary Port. In order to get the correct times for the morning flood therefore, the previous pair of flood tides at Shimonoseki will be required.

This requires taking the pair of evening flood tides from the 15th April at Shimonoseki which, when the time difference is then applied, will result in times for a flood tide in the morning both on the correct date and the right time of the day

Times of Tides	Shimonoseki	16:09 (Z–9) 15th April	22:38 (Z–9) 15th April
Corrections	Quingdao	+8:08	+7:55
Times of Tides	Quingdao	00:17 (Z–8) 16th April	06:33 (Z–8) 16th April

Now that the correct tides are being used, the Slopy Line calculation will be initiated with the tidal height predictions from the flood tides the previous evening at the Standard Port.

LW Quingdao.	00:17 ZT (Z–8) 16th April
HW Quingdao.	06:33 ZT (Z–8) 16th April
Duration of Tide (DOT)	6hrs 16mins

HOT Required = 1.4m

Interval from Graph – 4hours 10mins before HW

Time HW Quingdao	06:33 ZT
Interval	–4:10
Earliest time to cross obstruction in channel	02:23 ZT (Z–8) 16th April

In this question the interpolation for the correction to the High Water value at the Secondary Port required no interpolation as a Height of Tide of 2.4m correolated to the Spring value tabulated and was solved by eye. In an exam it is always advisable to indicate why there is no evidence of workings. A simple note to say that the answer was found by inspection is better than no explanation. Remember that you need to demonstrate that you know what you are doing using the correct methods not simply by guessing. Even if the guess turns out to be correct do not give the marker the opportunity to deduct precious marks as you did not explain your workings.

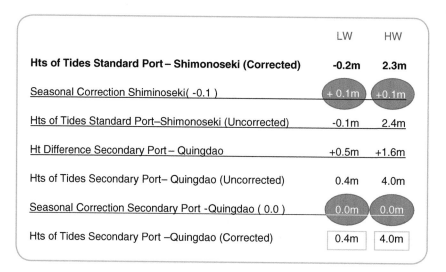

	LW	HW
Hts of Tides Standard Port – Shimonoseki (Corrected)	-0.2m	2.3m
Seasonal Correction Shiminoseki(-0.1)	+ 0.1m	+0.1m
Hts of Tides Standard Port–Shimonoseki (Uncorrected)	-0.1m	2.4m
Ht Difference Secondary Port – Quingdao	+0.5m	+1.6m
Hts of Tides Secondary Port– Quingdao (Uncorrected)	0.4m	4.0m
Seasonal Correction Secondary Port -Quingdao (0.0)	0.0m	0.0m
Hts of Tides Secondary Port –Quingdao (Corrected)	0.4m	4.0m

FIGURE 7.14 Correcting the Standard Port heights to find the values for the 'Slopy Line' values at Quingdao – example 3

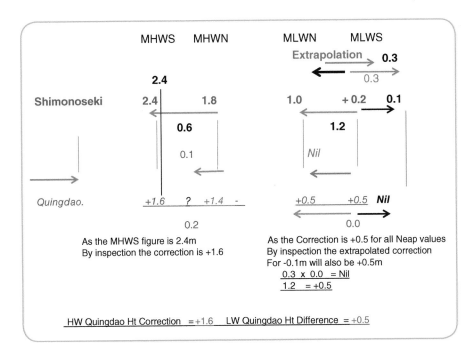

FIGURE 7.15 Interpolation and extrapolation for the height differences at Quingdao – example 3

In the case of the Low Water difference, the LW to be corrected for the Secondary Port of Quingdao was outside the tabulated value and therefore required an extrapolation to find the difference to apply. Once again on inspection the difference was +0.5 for both the Neap and Spring LW heights tabulated so even though it was an extrapolation, it was clear that the correction was a constant 0.5m difference at the Secondary Port, whatever the height at the Standard Port.

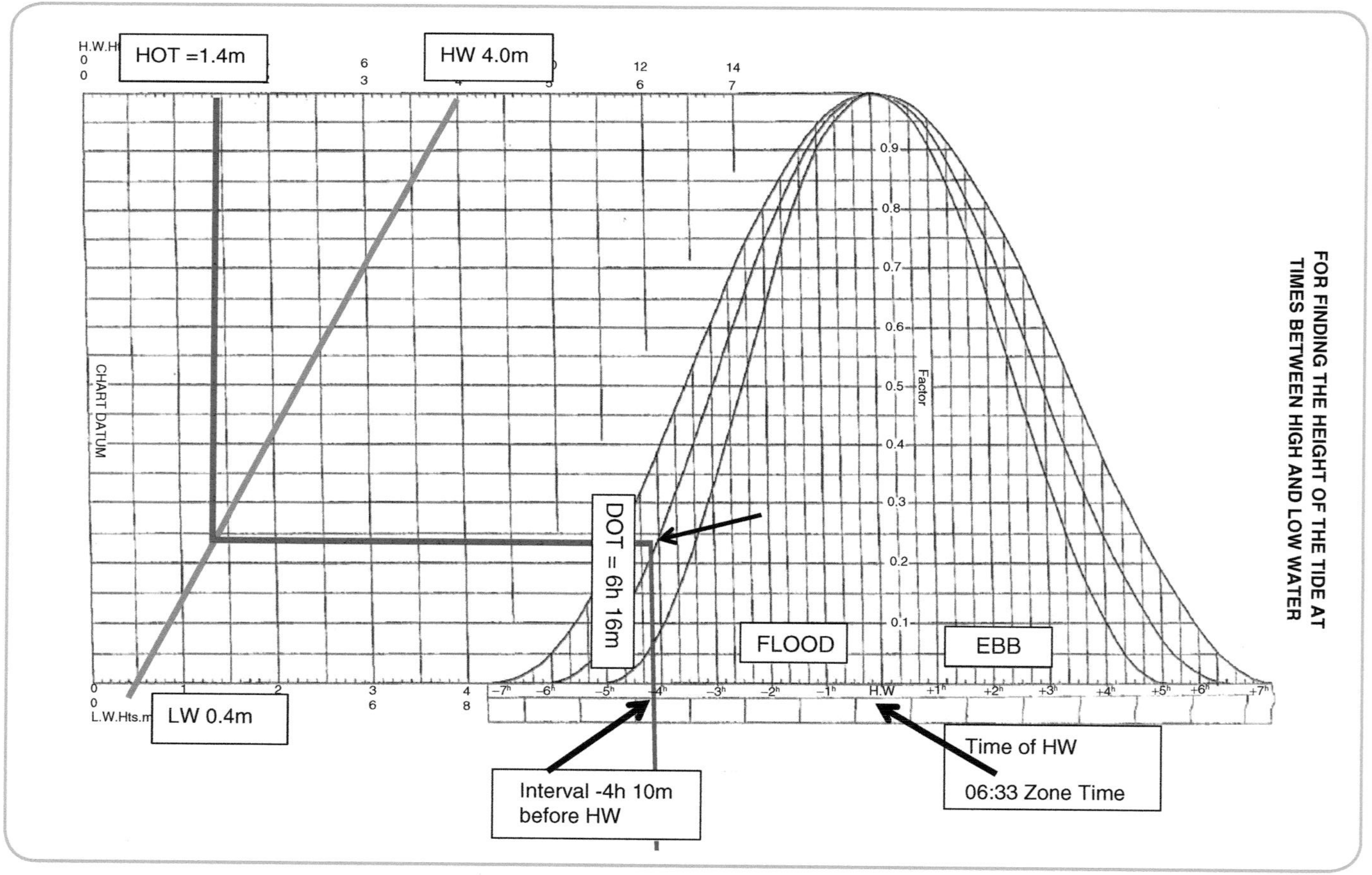
FOR FINDING THE HEIGHT OF THE TIDE AT
TIMES BETWEEN HIGH AND LOW WATER
HOT =1.4m
HW 4.0m
LW 0.4m
DOT = 6h 16m
FLOOD
EBB
Factor
CHART DATUM
H.W.Hts.m
L.W.Hts.m
H.W
Interval -4h 10m
before HW
Time of HW
06:33 Zone Time

FIGURE 7.16 Plot to determine the earliest time to cross an obstruction at Quingdao in example 3

8

VERTICAL SEXTANT ANGLES, TIDAL STREAMS AND CO-TIDAL DATA

8.1 Use of the sextant for tidal height calculations

The sextant is a marine instrument designed to measure angles, generally vertical angles in conjunction with astronomical bodies in celestial navigation, to ascertain position lines and determine a ship's position. It can be used both vertically to measure altitude above the horizon, and horizontally to determine the angle between terrestrial marks such as prominent lights or features marked on a chart used for navigational position fixing.

When using the sextant to obtain Vertical Sextant Angles (VSAs), the height of an object such as a lighthouse will be marked alongside it, given on the chart as a Charted Height. This may be given in feet, but more generally in metres, which will need to be checked as it will make a difference to any calculation to find the distance off the land. The Charted Height measurement does not indicate the height which the object is above Sea Level, it is the height above a reference datum, which may change depending on the location. The reference datum used will be marked in the Legend of the chart in use and will normally be one of the accepted known levels such as Mean High Water Springs (MHWS), Mean Sea Level (MSL) or Highest Astronomical Tide (HAT). This will fix its height above the reference level used and the height at which the reference level, measured above Chart Datum, can be found in the Tide Tables. For information of this kind, the heights of these levels are given in *Table V*, in the front of the Admiralty Tide Tables.

As the Height of Tide rises and falls then the actual height of the object above the sea level at any time will vary. In extreme cases the range of the tidal heights between High and Low Water can vary up to 14 metres. As the prime use of a Vertical Sextant Angle is to determine distance off the land using trigonometry, the Height of the Tide at the time of the sextant shot must firstly be calculated. This will enable the correct value for the height of the object above sea level to be determined, for use in the calculation required to obtain a position line.

By finding the distance off a terrestrial landmark then, whilst an accurate bearing may not be possible, a position circle can be drawn to seaward of the mark. If several distances off the land can be found at the same time, each will produce a position circle in the same way. When all the circles are plotted to seaward from the conspicuous terrestrial marks, a positional fix can be obtained where the arcs of each radius circle intersects. Whilst this is an outdated method of finding the ships position nowadays, with so many alternative methods available to the mariner, this nonetheless is still a viable method of fixing the ship's position should all else fail providing the height of tide can be found.

It follows that a sextant angle for navigational accuracy is of no real value without an accurate time noted at the time of its use. Equally when using a sextant to determine the distance off the shore, the

Height of Tide calculation is dependent on knowing the time. When the Height of Tide is known, using the Charted Height information given on the charts and basic trigonometry, it is possible to calculate the distance off the land. A series of Vertical Sextant Angles can be used to obtain a fix as follows:

Find the distance off a lighthouse at Peterhead at 10:30 BST hrs on the 6th May if a Vertical Sextant angle (VSA) of 01° 19' was taken using a light house with a charted height of 30m.

Time HW Aberdeen	06:35 (GMT)
Time Difference Peterhead	−44
Time HW Peterhead	05:51 GMT
Convert to BST	+1:00
Time HW Peterhead	06:51 BST
Required Time	10:30 BST
Interval From graph	+3h 39mins

$$\% \text{ Springs} \frac{1.6 - 1.8}{3.7 - 1.8} \times 100 = -11\% \text{ Use Neap curve}$$

	HW	LW
HOT Aberdeen (Corr)	3.2	1.6
SC Aberdeen	+0.1	+0.1
HOT Aberdeen (Unc)	3.3	1.7
Ht Diff Peterhead	−0.3	−0.1
HOT Peterhead (Unc)	3.0	1.6
SC Peterhead	−0.1	−0.1
HOT Peterhead (Corr)	2.9	1.5
Interval Fm Graph + 3hs39 Minutes		
Gives HOT @ 10:30BST = 2.0m		

MHWS Aberdeen	4.3m
Difference at Peterhead	−0.5m
MHWS Peterhead	3.8m
Charted Ht Lighthouse	30.0m
Lighthouse is	33.8m above Chart Datum
HOT	is 2.0m
Lighthouse is 31.8m above Sea level (the horizon) at 10:30 BST	

Therefore if you want the ***distance off***

$$\text{Distance off} = \frac{\text{Opposite (Ht)}}{\text{Tan } \theta° \text{ (VSA)}} = \frac{31.8\text{m}}{\text{Tan } 0°19'} = 1383.6 \text{metres}$$

$$\frac{1383.6\text{m}}{1852\text{m}} = \textbf{0.75 Miles off}$$

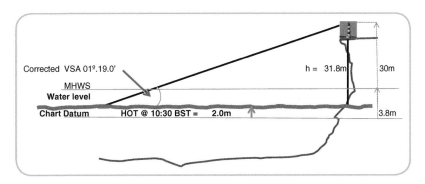

FIGURE 8.1 VSA diagram to determine distance off lighthouse at Peterhead

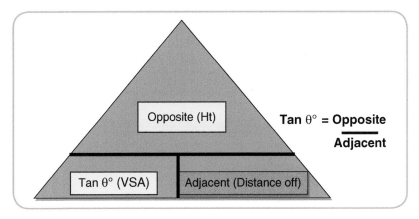

FIGURE 8.2 Use of the tangent formula to get the distance off the land

8.2 Tidal Stream Atlas

The Tidal Stream Atlases contain tidal stream information detailing the direction and strength of the tides for up to six hours before to six hours after High Water at key shipping areas around the UK such as the Dover Straits (Fig 8.4). These are produced by Admiralty publications (UKHO, 2003) and cover up to 20 key shipping areas around the UK and NW Europe. They pictorially describe the information found in the legend of a nautical chart, represented at reference positions and marked by tidal diamonds.

The arrows show the direction of the current and their thickness and intensity indicate where the strongest rates are to be expected. The figures alongside represent the mean Neap rate followed by the Spring rate in knots.

8.3 Procedure for determining tidal streams during periods between the Neap and the Spring Tides

1. Plot the position of the ship on the chart and determine the closest appropriate rates for the Spring and Neap tidal periods.
2. On the Computation of Rates graph in the front of the Tidal Stream Atlas, plot the Spring rate value on the axis labelled Springs on the left hand side of the graph, using the rate scale along the top of the graph.

NP 233
Edition 3 – 1995 (Reprinted 2003)

Tidal Stream Atlas
DOVER STRAIT

The Charts
This atlas contains a set of 13 charts showing tidal streams at hourly intervals commencing 6 hours before HW Dover and ending 6 hours after HW Dover. The times of HW Dover and other details of the predictions for this port are given in NP 201, Admiralty Tide Tables Vol 1, which is published annually. NP 201 also gives tidal predictions for a number of ports and tidal stations in the area covered by this atlas.

On the tidal stream charts the directions are shown by arrows which are graded in weight and, where possible, in length to indicate the approximate strength of the tidal stream. Thus ⟶ indicates a weak stream and ⟶ indicates a strong stream.

The figures against the arrows give the mean neap and spring rates in tenths of a knot, thus: 19,34 indicates a mean neap rate of 1·9 knots and a mean spring rate of 3·4 knots. The comma indicates the approximate position at which the observations were obtained.

Computation of Rates – Example .
Required to predict the rate of the tidal stream about 3 miles off Cap Gris—Nez at 0500 on a day for which the tidal prediction for Dover (extracted from NP 201) is:

0505	0·2m
0959	6·3m
1730	0·1m
2225	6·2m

Mean Range of tide at Dover for the day is 6·1m.

The appropriate chart in the atlas is that for *5 hours before HW Dover* and this gives mean neap and spring rates for the required position of 20,34 (2·0 kn, 3·4 kn).

Enter the diagram *Computation of Rates* opposite with these mean neap and spring rates, joining the dots representing them with a ruler. From the intersection of this line with the horizontal line representing the range at Dover (6·1) follow the line vertically to the scale of Tidal Srteam Rates (top or bottom) and read off the predicted rate— in this example 3·5 knots.

FIGURE 8.3 Extract from Tidal Stream Atlas NP 233 (UKHO, 2003)

3. Repeat the exercise for the Neap rate at the ship's position and plot the Neap rate value on the axis labelled Neap Rate on the left hand side of the graph using the rate scale along the bottom of the graph.
4. Join the two points marked with a diagonal line being sure to over extend it to cover the whole graph, top to bottom.

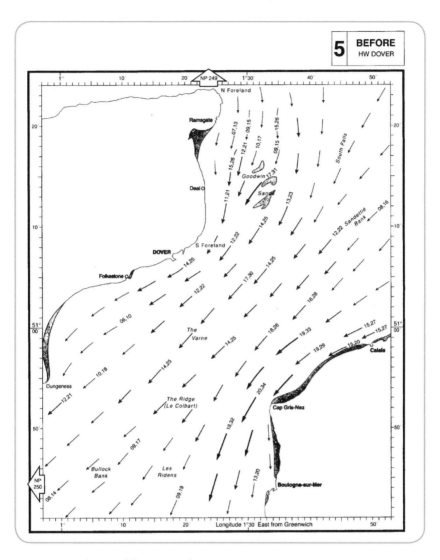

FIGURE 8.4 Extract from Tidal Stream Atlas NP 233 (UKHO, 2003)

5. Work out the predicted range for the appropriate pair of tides covering the time in question and using the Mean Range scale down the left hand side of the graph, draw a horizontal line which will intersect with the diagonal already plotted.

6. From the point where the two intersect, take a vertical line to either of the rate scales across the top, or at the bottom of the graph, and read of the rate corresponding to the predicted range on that day.

The direction of the stream can be determined from the directions of the arrows which will be shown on the chart.

Alternatively, the rate may be calculated by interpolation or extrapolation using the Neap and Spring ranges taken from the Computation of Rates diagram, or from the Tide Tables.

Rate = 2.0 + (6.1 − 3.2) ÷ (6.0 − 3.2) × (3.4 − 2.0)

Rate = 3.45

Rate = 3.5 knts

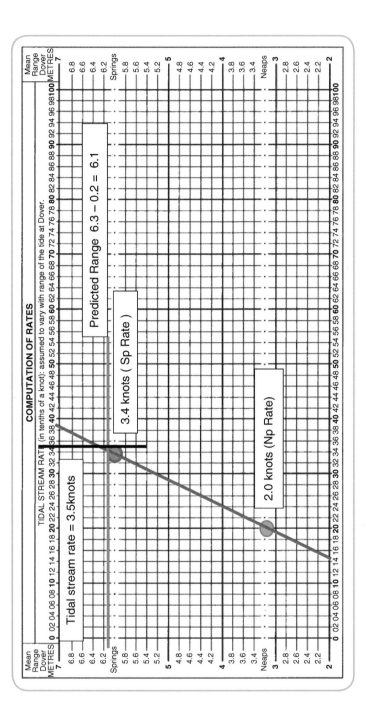

FIGURE 8.5 Use of computation of rates graph from Tidal Stream Atlas NP 233 (UKHO 2003)

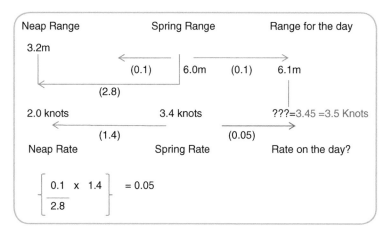

FIGURE 8.6 Interpolation of Mean Spring Rates using mean ranges given on computation of rates graph for the Standard Port

Hours	A 50 42'.3N 0 26'.5E			B 50 53'.0N 1 00',0E			C 51 01'.0N 1 10'.0E		
	Dir	Sp	Np	Dir	Sp	Np	Dir	Sp	Np
6 (Before HW)	248	0.8	0.4	213	1.6	0.9	224	0.9	0.5
5	067	0.5	0.3	214	2.1	1.2	239	1.0	0.6
4	068	1.9	1.0	215	1.8	1.1	235	1.1	0.6
3	071	2.6	1.5	213	0.9	0.5	242	0.6	0.4
2	069	2.3	1.3	S	/ a	c k	S	/ a	c k
1	0.68	1.2	0.6	033	0.8	0.5	052	0.6	0.3
HW	0.67	0.1	0.1	032	1.5	0.8	049	1.2	0.7
1 (After HW)	248	0.9	0.5	031	1.9	1.1	049	1.3	0.7
2	247	1.4	0.8	030	1.7	1.0	056	1.0	0.5
3	251	1.8	1.0	031	1.2	0.6	054	0.5	0.3
4	253	1.7	1.0	032	0.4	0.2	S	/ a	c k
5	250	1.6	0.9	211	0.4	0.2	219	0.4	0.2
6	249	1.2	0.7	212	1.3	0.7	217	0.8	0.4

FIGURE 8.7 Tidal stream information given in the Legend of a chart using tidal diamonds as the reference points

8.4 Tidal diamonds

An alternate source of tidal stream information can be found by consulting the legend of a chart where tidal diamonds at various positions on the chart have been recorded and the direction of set and the rates at Neap and Spring Tides tabulated. This is the same information which is used to represent the flow and strength of the tidal stream in a pictorial format for up to six hours either side of the HW at the given Standard Port in the Tidal Stream Atlas.

The rates of the tidal stream given are generally accepted as being effective up to half an hour before and half an hour after the hourly value. When using these diamonds then the nearest tidal diamond available to the ship's position or intended route should be used.

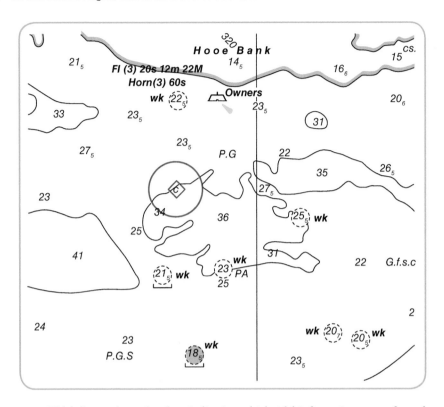

FIGURE 8.8 Tidal diamonds on the chart indicating which tidal information to use from the chart Legend

8.5 Neaping

The nature of Neap Tides is that they produce the highest Low Water and the lowest High Water thus the range is going to be much less than during Spring Tides. The High Water may not be high enough for a vessel to clear shallow areas or obstructions on the seabed with sufficient Under-Keel Clearance. In the case of passing below bridges or cables, there may not be sufficient clearance above the uppermost structure on the vessel to pass safely below them during Neap Tides. When this situation arises, the vessel is said to be 'NEAPED'. In order to find when the ranges will increase i.e. building towards the subsequent Spring Tide, then the next suitable tide must be determined.

8.5.1 'Neaping' – worked example

A vessel arrives off Dalu Dao (7435) on July 31st at 19:00 ZT with draught 11.0m Even Keel.

Find the earliest Zone Time that the vessel can enter and maintain an Under-Keel Clearance of 1.8m over a 7.0m shoal patch.

Calculate Height of Tide required.

Draught	11.0m
UKC	1.8m +
DOW Req'd	12.8m
Charted Depth	7.0m
HOT Required	5.8m at Dalu Dao (7435)

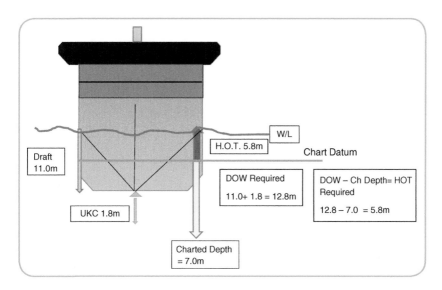

FIGURE 8.9 Determining the HOT required at Dalu Dao

Time of HW Inch'on	20:00 ZT
Time Difference Dalu Dao	−3:38
Time of HW Dalu Dao	16:22ZT

It can be seen that the HOT is less than the required 5.8m at High Water so there will be insufficient water for entry on this tide i.e. the vessel is neaped and the next suitable tide will be required to be calculated.

The required HOT at the Secondary Port, Dalu Dao, must be at least 5.8m. The predicted height at the Standard Port, Inch'on, which corresponds to the HOT of 5.8m at the Secondary Port, before correction, must be found.

From the tables find the HOT at the Secondary Port for MHWS and MHWN listed heights.

	MHWS	MHWN
Standard Port Inch'on (Uncorrected)	8.6m	6.5m
Ht Differences Dalu Dao	−3.1m	−2.1m
Dalu Dao Port (Uncorrected)	5.5m	4.4m
Seasonal Change Secondary Port	+0.2m	+0.2m
Secondary Port Dalu Dao Predicted (Corrected)	5.7m	4.6m

Before we can refer to the daily pages and find the earliest suitable tides, HW values at the Standard Port for the MHWS and MHWN levels need to be corrected, so that the equivalent height at the Standard Port for a corrected height of 5.8m at the Secondary Port can be interpolated. Interpolation is required to find Height of Tide at Standard Port corresponding to required Height of Tide at Secondary Port.

	MHWS	MHWN
Standard Port Inch'on (Uncorrected)	8.6m	6.5m
Seasonal Correction (Inch'on)	+0.1m	+0.1m
Standard Port Inch'on HW Predictions (Corrected)	8.7m	6.6m

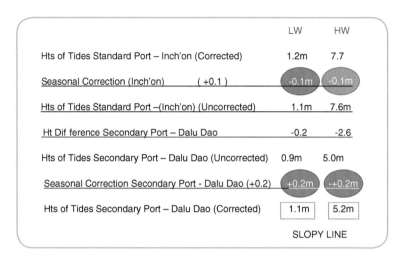

FIGURE 8.10 Correcting the Standard Port heights to find the Heights of Tide at Dalu Dao

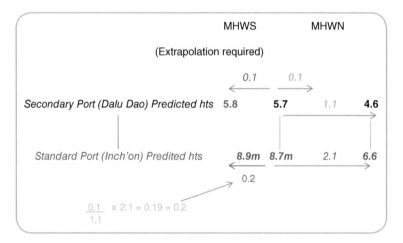

FIGURE 8.11 Interpolating to find the required minimum corrected HOT at the Standard Port Inch'on which will result in a 5.8m tide at Secondary Port, Dalu Dao

It can be seen that the minimum height of tide required at Inch'on must be at least 8.9m.

Referring back to the Tide Tables, the earliest date which shows a prediction of more that 8.9m will be on 9th August when the HW is predicted to be 9.1m at 05:01ZT.

8.6 Co-tidal data and the amphidromic point

1. AMPHIDROME: a point in the sea where there is zero tidal amplitude due to cancelling of tidal waves. CO-TIDAL LINES radiate from an amphidromic point and CO-RANGE LINES encircle it (Pugh, 2004).
2. The shape and depths of a particular area may cause the movement of the water within the UK and North Sea, Persian Gulf and the Malacca Straits to be rectilinear or rotational.
3. Co-tidal charts show lines of equal time differences of tides from a standard position.
4. Co-range charts show lines of equal tidal ranges. Co-tange diagrams of Mean Spring Range and Mean Neap Range.

5. Co-tidal data is used to find tidal information for positions offshore.
6. The principle application is to find the time of a particular Height of Tide or the Height of Tide at a particular time, at an offshore position.
7. They are published for areas around the United Kingdom, North Sea, Malacca Strait and Persian Gulf. Calculating Heights of Tides or the times they occur in these areas will be highly relevant to the passage planning for deep draft vessels as the rotational movement of the tide can result in water levels differing considerably from the Standard Port predictions.
8. The Persian Gulf charts depict the values of rate and direction for the four main harmonic constants.
9. In the Malacca Straits, the tide is predominantly diurnal, and one diagram with both co-tidal lines and co-range lines superimposed on the same chart is published.
10. In the North Sea the movement is rotational around amphidromic points where little movement occurs.
11. On Mean Interval Charts the lines are at half hour intervals, *except that between the 12:00 line and the 00:00 line there is a 00:25min interval.*

TOP TIP

The adopted time interval between the Moon's Meridian Passages is 24hrs and 50minutes.
One tide will run over half the interval, is 12hrs and 25minutes.
Therefore 12:25 of one cycle is 00:00 of the next cycle.
When the required position and the Standard Port lie on opposite sides of the 00:00 line, 12hrs and 25minutes must be added to the earlier time.

8.7 Use of co-tidal and co-range diagrams

To find the times and heights of High and Low Waters at a Position: 54°00'N 005°35'E for the 28th April.

i. Select a Standard Port near the ship's position preferably one of the five Standard Ports on the same amphidromic point diagrams – *use Helgoland.*
ii. Extract the High and Low Water heights and times for the Standard Port.
iii. Calculate the Mean Range, the difference between the predicted High and Low Waters on the appropriate tide.
iv. Extract the Mean Low Intervals for the Standard Port, Helgoland and the Position 54°00'N 005°35'E from the Mean Low Water Interval amphidromic chart.
v. Extract the Mean High Water Intervals for the Standard Port, Helgoland and the Position 54°00'N 005°35'E from the Mean High Water Interval amphidromic chart.
vi. Calculate the Differences between the Standard Port intervals at Helgoland and the ship's Position 54°00'N 005°35'E for High and Low Water.

The factor is the the ratio of the ranges at the position and the Standard Port, and is used to find the Height of Tide at the ship's position, knowing the Height of Tide at the Standard Port.

High Water interval at ship's position	8h:00m
High Water Interval at Helgoland	10h:35m
High Water Interval Difference	2h:35m before Helgoland

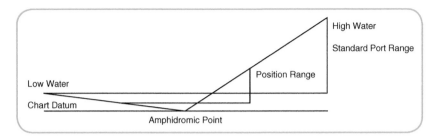

FIGURE 8.12 Differences in range at the Standard Port and at the ship's position required to determine the factor

Using the Low Water Interval Chart for the same ship's position,

Low Water interval at ship's position	2hrs:22minutes
Low Water Interval at Helgoland	4hrs:55minutes
Difference in the Low Water intervals	2hrs:33minutes before Helgoland

vii. Extract the Mean Spring and Neap Ranges at the Standard Port, Helgoland and the Position 54°00'N 005°35'E from the Mean Spring and Neap Range charts.

The Mean Spring Range at Helgoland is 2.8m, which is also the Mean Range predicted for the day. Using the diagram for Mean Spring Range, find the Mean Spring Range at the ship's position, which in this case is 1.45 metres.

viii. Calculate the Spring and Neap factors by dividing the Mean Range at the position by the Mean Range at the Standard Port.

8.8 Calculating the Factor

Daily Tides Predicted

	2345	2.7	Range --
Tides for the day at Helgoland 28th April	06:40	−0.2	Range 2.9
	12:11	2.5	Range 2.7
	18:56	−0.2	Range 2.7
29th April	00:26	2.7	Range 2.9
Mean range at Helgoland = 2.8m			

Mean range at Ship's Position	1.45m	
Mean Spring range at Helgoland– from the chart.	2.8m	
Factor = Mean Spring Range at Position	1.45	
Mean Spring Range at Standard Port	2.8	**FACTOR = 0.52**

ix. Apply the differences in time intervals found for High and Low Waters to Predicted Port times at Helgoland.

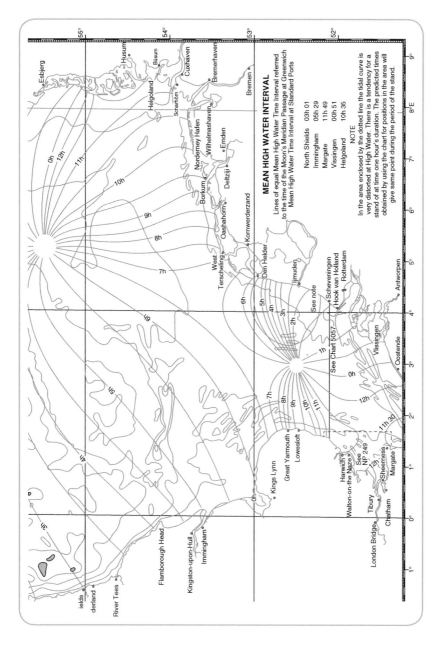

FIGURE 8.13 Co-tidal amphidromic diagram to determine time interval from HW Helgoland

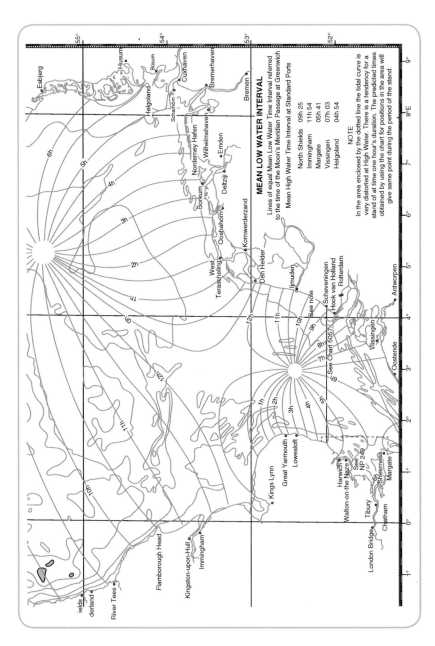

FIGURE 8.14 Co-tidal amphidromic diagram to determine time interval from LW Helgoland

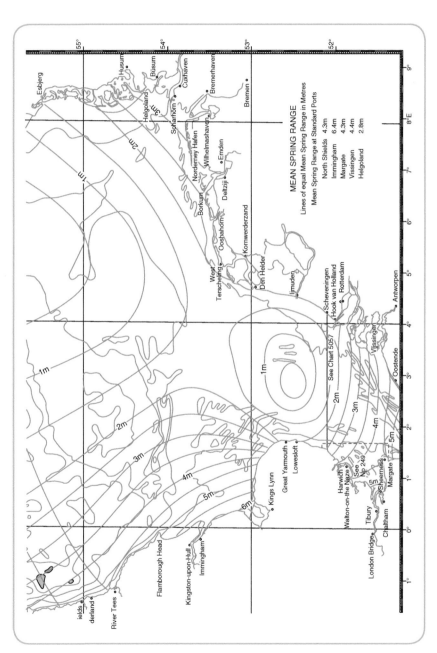

FIGURE 8.15 Co-range amphidromic diagram used to determine mean spring range at ship's position

	HW	LW	HW	LW	HW	
Helgoland Predicted Times	23:45	06:40	12:11	18:56	00:26	
Time Intervals	−2:35	−2:33	−2:35	−2:33	−2:35	Time Intervals from
Estimated Times at Ship's position	**21:10**	**04:07**	**09:36**	**16:23**	**21:51**	Amphidromic Point

x. Calculate the heights at the 54°00'N 005°35'E position by multiplying the Standard Port Heights at Helgoland, by the factor.

Predicted Hts	2.70	−0.20	2.50	−0.20	2.70
Multiplied by Factor	×0.52	×0.52	×0.52	×0.52	×0.52
Estimated High & Low Water Heights at Ship Position	**1.4m**	**−0.1m**	**1.3m**	**− 0.1m**	**1.4m**

xi. During periods when the daily ranges are between the Mean Neaps and Spring Range, interpolate (extrapolate) for the factor using Spring and Neap Ranges and the Spring and Neap Factors.

xii. To find the height at an intermediate time, or the time for an intermediate height, use the heights and times for the position using the tidal curves at the Standard Port using the HW and LW values and following the 'Slopy Line' and using the 'Percentage Spring' method in the normal way.

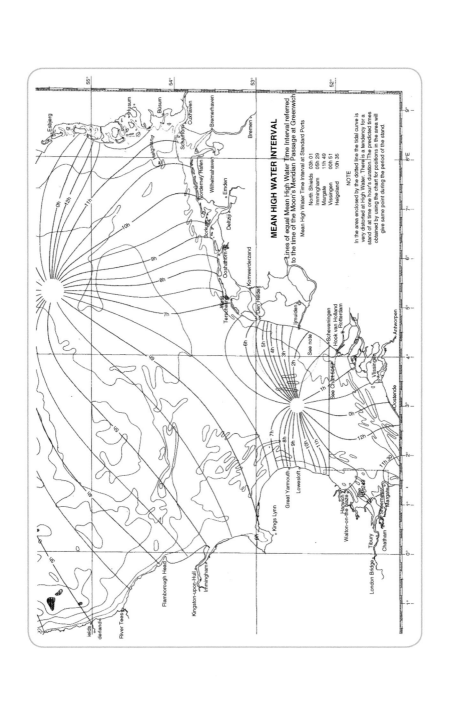

MEAN HIGH WATER INTERVAL

Lines of equal Mean High Water Time Interval referred to the time of the Moon's Meridian Passage at Greenwich

Mean High Water Time Interval at Standard Ports

North Shields	03h 01
Immingham	05h 29
Margate	11h 49
Vlissingen	00h 51
Helgoland	10h 35

NOTE

In the area enclosed by the dotted line the tidal curve is very distorted at High Water. There is a tendency for a stand of at time one hour's duration. The predicted times obtained by using the chart for positions in the area will give same point during the period of the stand.

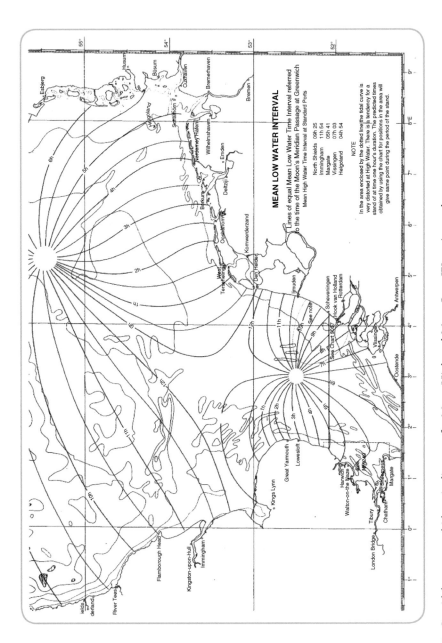

FIGURE 8.16 Co-tidal range amphidromic diagrams for Mean High and Mean Low Water intervals

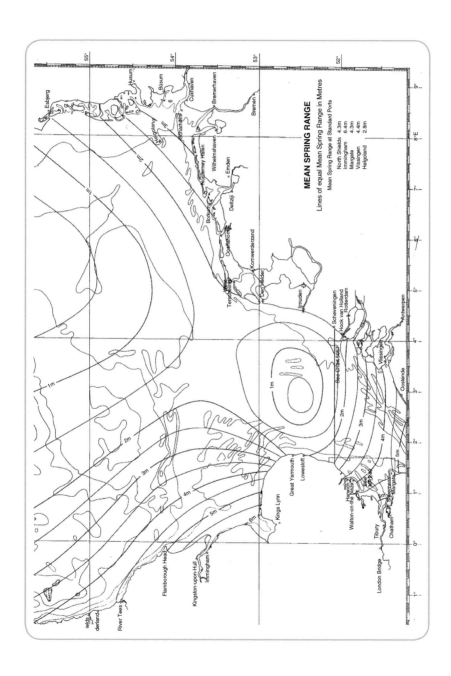

MEAN SPRING RANGE
Lines of equal Mean Spring Range in Metres

Mean Spring Range at Standard Ports

North Shields	4.3m
Immingham	6.4m
Margate	4.3m
Vlissingen	4.4m
Helgoland	2.8m

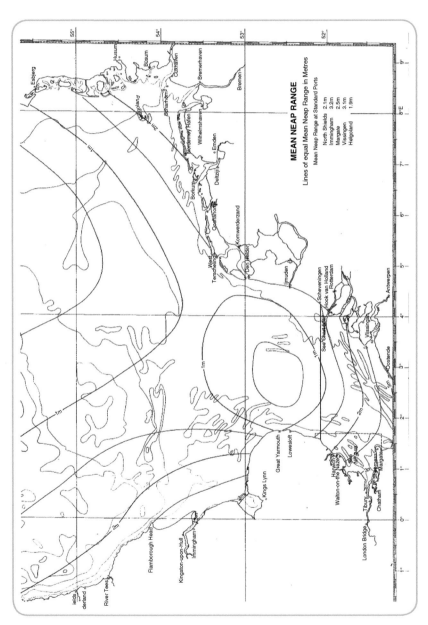

FIGURE 8.17 Co-range amphidromic diagrams for Mean Spring and Mean Neap tidal ranges

MEAN NEAP RANGE

Lines of equal Mean Neap Range in Metres

Mean Neap Range at Standard Ports

North Shields	2.1m
Immingham	3.2m
Margate	2.5m
Vlissingen	3.1m
Helgoland	1.9m

9

REVISION PAPERS WITH WORKED SOLUTIONS

9.1 Tides paper 1

1. European Standard
    Calculate the depth of water under the keel of a vessel with a draft of 8.1m arriving alongside a berth at Devonport Harbour (14) at 17:30 ZT on 25th January. The charted depth of the berth was given as 7.2m on the chart.

2. Pacific Secondary Port
    A vessel has to load bunker fuel-oil at Puerto Armuelles (9471) on the afternoon of 13th May. What is the earliest Zone Time on the mid-morning flood tide that she can pass over a shoal on the approaches to the bunker berth, which has a charted depth of 9.5m? Her draft is 10.4m and she requires a minimum of 1.5m UKC.

3. European Secondary Port
    A vessel on the approaches to Fishguard (490) is required to pass over a shoal with a charted depth of 2.5m during the afternoon flood tide on 7th October. The vessels' draft is 6.0m and she requires a UKC of 1.0m. Find the earliest time she can cross the shoal with the required clearance.

Answers to paper 1

No 1

Time of HW Devonport   12:55 UT.   Ht 4.5m

Time of LW Devonport   19:41 UT.   Ht 2.0m

Predicted Range = 2.5m

Time of HW   12:55 UT

Time Required 17:30 UT

Time Interval   +4h 35mins

$$\% \, \text{Springs} = \frac{2.5 - 2.2 \times 100}{4.7 - 2.2} = \frac{2.3 \times 100}{2.5} = 92\%$$

HOT from Tidal Curves   = 2.5m

Charted depth   = 7.2m

DOW   = 9.7m

UKC   = DOW − DRAFT = (9.7m − 8.1m)

Under-Keel Clearance at Berth = 1.6m

No 2

	LW	HW
Tide Times Balboa	09:37 (Z+5)	15:32 (Z+5)
Time Differences Puerto Armuelles	−10mins	−10mins
Tide Times Puerto Armuelles	09:27 (Z+5)	15:22 (Z+5)

Duration of Tide (DOT)	*5h 55mins*	
Tide Hts Balboa (corrected)	0.3	5.0
Seasonal. Correction Balboa	0.0	0.0
Hts Balboa (uncorrected)	0.3	5.0
Ht Differences Puerto Armuelles	0.0	−1.9
Hts Puerto Armuelles (uncorrected)	0.3	3.1
Seasonal Correction Puerto Armuelles	0.0	0.0
Hts Of Tide Puerto Armuelles	0.3m	3.1m SLOPY LINE

Draft −	10.4m
UKC Req	1.5m
DOW Req	11.9m
Charted depth	9.5m
HOT Req'd at Puerto Armuelles	2.4m

From Tidal Curve − for a HOT of 2.4m the Time Interval is −1hr 50mins

HW Puerto Armuelles	15:22
Interval from Graph	−1:50

Earliest Time to Pass Over Shoal 13:32 (Z+5)

No 3

Required DOW	7.0m
Charted depth of shoal at Fishguard	2.5m
Req'd HOT	4.5m

MH	HW	18:18 GMT
Time	diff	+1:04
Fishguard	HW	19:22 GMT
Int fm	graph	−1:3m
Time		17:52 GMT

Hts of Tides Milford Haven (corrected)	7.5	0.5
Seasonal Correction Milford Haven	0.0	0.0
Hts of Tides MH (uncorrected)	7.5	0.5
Ht Diffs Fishguard	−2.3	+0.2
Hts of Tides Fishguard (uncorrected)	5.2	0.7
Seasonal Correction Fishguard	0.0	0.0
Hts of Tides Fishguard (corrected)	5.2m	0.7m

$$\% \text{Springs} \frac{7.0 - 2.7}{6.3 - 2.7 \times 100} = 119\%$$

Use Spring curve only
The tide is a flood Tide

Interval from the tidal curve	1hr 30mins before HW
Time of HW Fishguard	19:22 GMT
Time interval from curve	1:30
Earliest time to cross Shoal	17:52 GMT

9.2 Tides paper 2

1. Pacific Standard Port
   Find the earliest time that a vessel with a draft of 11.0m can enter a dry dock for repairs in Inch'on with an Under-Keel Clearance of 1.0m on the afternoon flood tide of the 22nd December. The sill of the dry dock has a charted depth of 4.0m.
2. Pacific Secondary Port
   A bulk carrier is due to berth at Astoria (9228) on the afternoon flood tide on 6th January. The vessel must pass under a transporter bridge whose span rises to a maximum charted height of 16.0m. If the air draft is calculated to be 18.1m, find the latest time that the vessel can pass safely beneath the bridge with a minimum of 1.0m clearance.
3. European Standard Port
   Calculate the distance off a lighthouse a Vertical Sextant Angle of 0° 32.0' (I.E −2.0') is obtained at 15:30 Zone Time at Antwerp on 26th October. The chimney has a charted height of 31.5m.

Answers to paper 2

No 1

DOW required by Vessel		DRAFT + UKC = 11.0 + 1.0m = 12.0m
LW Inch'on	11:54	−0.2m
HW Inch'on	18:27	8.8m
Time Interval from Graph = −1hr 12mins		
		18:27
		1:12

Earliest time to enter Dry Dock 17:15 Zone Time

No 2

HAT Astoria	3.9m
Max space available = 16.5 + 3.9m	20.4m
Space required (18.1 + 1.0m)	19.1m
Max HOT	1.3m

	LW	HW
Tofino	11:55	17:40 (Z+8)
Diff Astiria	+1:00	+1:00
Tides Astoria	12:55	18:40 ZT (Zone Time +8)
		12:55

DOT 5hrs 45mins

	LW	HW
Hts of Tides Tofino (corrected)	1.5	2.9
Seasonal Correction Tofino (0.0)	0.0	0.0
Hts of Tides Tofino (uncorrected)	1.5	2.9
Ht Diffs @ Astoria	−0.7	−1.0
Hts of Tides Astoria (uncorrected)	0.8	1.9
Seasonal Correction @ Astoria (+0.1)	+0.1	+0.1
Hts of Tides Astoria (corrected)	0.9	2.0 = SLOPY LINE

Interval from Tidal Curves = −3hrs 20mins before HW

HW Astoria	18:40
Interval from Tidal Curves	−3hrs 20min

Latest time to pass below Bridge = 15:20 ZT (Local Time Astoria)

No 3

LW Antwerp	12:03	0.4m
HW Antwerp	17:39	5.7m
PR 5.3m		

$$\% \, Sp = \frac{5.3 - 3.2}{5.8 - 3.2} = \frac{2.1}{2.6} \times 100 = 81\%$$

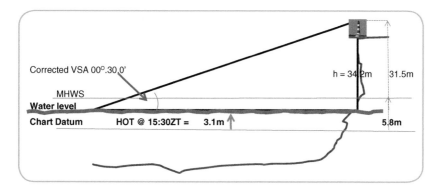

FIGURE 9.1 VSA diagram to determine distance off lighthouse in Antwerp

Interval for Tidal Curve

Time HW Antwerp	17:39
Time Required	−15:30
Time Interval to enter curve	−2hrs 09mins before HW
HOT for a time interval of	−2hrs 09mins = 3.1m

Charted Height of Lighthouse	31.5m + 5.8m (MHWS Antwerp) = 37.3m above CD
HOT at 15:30 when the VSA was taken	3.1m
At 15:30 Zone time the Beacon is	34.2m above W/line

VSA	0°32.0'
Index Error (I E)	−2.0
VSA (corrected)	0°30.0'

$$\text{Dist off} = 34.2 \, / \, \text{Tan } 0°\,32.0\,' = \frac{3919\text{m}}{1852\text{m}} = 2.12 \text{ Miles}$$

9.3 Tides paper 3

1. European Secondary Port

    A vessel is to pick up a pilot at 08:30 UT at Klaksvik (782) on the 1st March. If the Draft was 8.4m at the time what was the UKC if the Charted Depth was 12.5m at the pilot embarkation point.

2. Pacific Standard Port

    What is the earliest time (Zone Time and UT) on the morning flood tide of 19th November at Prince Rupert (8850) that a vessel with a draft of 3.5m can pass over an obstruction charted with a drying height of 0.5m. The vessel must have a 0.5m Under-Keel Clearance.

3. European Standard Port

    A lighthouse near Ullapool (334) has a charted height of 30.5m and subtended a Vertical Sextant Angle of 00.46.8 (I.E 2.0 on the arc) at 0800 Z on 1st November 1987. Find the distance from the lighthouse at this time.

Answers to paper 3

No 1

	LW	HW
Time of Tides @ Reykjavik	01:10 UT	07:23 UT
Time Difference Klaksvik		+3:45
Times of HW Klaksvik		11:08 UT
Time Required		08:30 UT
Time Interval before HW		−2hrs 38mins

	LW	HW
Hts of Tides Reykjavik (corrected)	0.1	4.5 PR 4.4m
Seasonal correction Reykjavik	−0.1	−0.1
Hts of Tides Reykjvik (uncorrected)	0.0	4.4
Ht differences Klalksvic	−0.1	−2.8
Hts of Tides Klaksvik (uncorrected)	−0.1	1.6
Seasonal correction Klaksvik	0.0	0.0
Hts of Tides Klaksvik (corrected)	−0.1	1.6 SLOPY LINE

$$\%\text{Springs} = \frac{4.4 - 1.6}{3.8 - 1.6} = \frac{2.8}{2.2} \times 100 = 127\%$$

CAN ONLY USE SPRING CURVE

HOT	0.9M
Draft	8.4m
Charted Depth	12.5m
DOW	12.5 + 0.9 = 13.4m
Draft	8.4m
UKC @ 08:30UT	5.0m

No 2

Draft	3.5m	Prince Rupert LW	05:35 (Z+8)
UKC required	0.5m	Prince Rupert HW	11:40 (Z+8)
DOW	4.0m	Duration of Tide 6hrs 05mins	
Charted Depth	−0.5m		
HOT Required	4.5m		

HW Prince Rupert	11:40 (Z+8)
Interval from graph	03:00
Earliest Time	0840 (Z+8) Zone Time
Convert to UT	+8:00
Earliest Time	16:40 UT

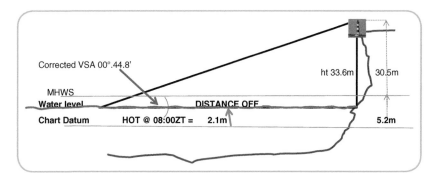

FIGURE 9.2 Diagram to determine distance off lighthouse at Ullapool

No 3

Sextant altitude VSA	0° 46.8'
Index Error	−2.0'
Corrected VSA	0° 44.8'

Charted Height of Lt House	30.5m
Ht MHWS Ullapool	5.2m
Height of Lt House above Chart Datum	35.7m

HW Ullapool	03:34	Ht 4.1m
LW Ullapool	09:40	Ht 1.7m
Predicted Range	2.4m	

$$\% \, Sp = \frac{2.4 - 1.8}{4.5 - 1.8} = \frac{0.6}{2.7} \times 100 = 22\%$$

$$Tan \, VSA = \frac{Opposite}{Adjacent} = \frac{Ht}{Distance \, off}$$

$$Dist \, off = \frac{33.6m}{Tan \, 0° \, 44.8'} = 2578m$$

$$= \frac{2578m}{1852m} = 1.4 \, Miles$$

9.4 Tides paper 4

1. Pacific Standard
   Find the distance from the shore for a vessel at anchor off Port Headland (6259) at 09:30 ZT on 21st February at which time a V.S.A of 0°.12.5'(I.E. 3.0' off the arc) was observed using a lighthouse with a Charted Height of 29.5m.

2. Pacific Secondary

A passenger vessel with a draft of 8.0m is intending to anchor off North Beru (6752) and is due to make a daylight arrival on the morning flood tide of the 25th of February. Access to the inner harbour anchorage is obstructed by a shallow patch with a charted depth of 8.0m and in order to cross this bank safely she must have a clearance under-keel of 1.0m. What is the earliest time she can cross this bank in order to enter the anchorage?

3. European Standard

What will be the latest time a vessel could pass under a bridge at Avonmouth (523) on the 5th May on the morning flood tide? The vessel has an air draft of 20.0m and will require a clearance of 2.0m above the mast. The bridge has a charted height of 15.0m.

Answers to paper 4

No 1

Sextant altitude VSA			0° 12.5'
Index Error			+3.0'
Corrected VSA			0° 15.5'
Charted Height of Chimney			29.5m
Ht MHWS Port Hedland			6.8m
Height of Chimney above Chart Datum			36.3m

HW Port Hedland	08:43	Ht 2.2m	14:54 ZT
LW Port Hedland	14:54	Ht 6.4m	09:30ZT
DOT	6 hrs 11mins	Interval	– 5hrs 24mins

HOT @ 09:30 ZT = 2.4m

$$\text{Tan VSA} = \frac{\text{OPPOSITE}}{\text{ADJACENT}} = \frac{\text{Ht}}{\text{Distance off}}$$

$$\text{Dist off} = \frac{33.9\text{m}}{\text{Tan } 0°15.5'} = 7518.6\text{m} = \frac{7518.6\text{m}}{1852} = 4.1\,\text{Miles}$$

Distance off Port Hedland at 09:30 ZT = 4.1 nautical miles

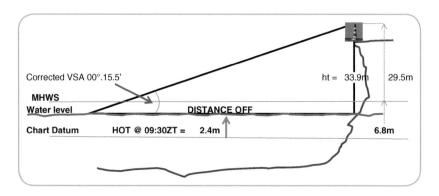

FIGURE 9.3 Diagram to determine distance off lighthouse at Ullapool

No 2

Draft 8.0 +1.0 Req'd DOW	9.0m
Charted Depth Shoal	8.0m
Required HOT	1.0m

	LW	*HW*
Times Puerto Montt	18:06 (Z+4) 25th Feb	00:24 (Z+4) 26th Feb
Corrections N Beru	−9:35	−9:35
Times N Beru	08:31 (Z−12)	14:49 (Z−12) 25th Feb

DOT 6hrs 18mins

	LW	*HW*
Hts of Tide P Montt (corrected)	1.3	6.6
Seasonal Correction P Montt	Negl'ble	Negl'ble
Hts of Tide P Montt (uncorrected)	1.3	6.6
Ht Differences N Beru	1.2	−5.3
Hts of Tides N Beru (uncorrected)	0.1	1.3
Seasonal Correction N Beru	Negl'ble	Negl'ble
Hts of Tides N Beru (corrected)	0.1	1.3 SLOPY LINE

HW N. Beru	14:49
Interval from Graph	−2:05

Latest Time 12:44 ZT (Z−12) 25th February

No 3

HAT Avonmouth	14.7m
Charted Height of Bridge	15.0m
Charted Ht of Bridge above Chart Datum	29.7m
Space required 20.0m + 2.0m Clearance	22.0m
Max HOT	7.7m

LW Avonmouth	05:11 UT	3.2m
HW Avonmouth	11:09 UT	9.4
		6.2m PR

$$\% \, Sp = \frac{6.2 - 6.5}{12.3 - 6.5} \times 100 = \frac{-0.3}{5.8} \times 100 = -5\%$$

CAN ONLY USE NEAP CURVE

Interval from Graph = 2hrs 20mins before HW Avonmouth	**11:09 UT**
	−2:20
Latest time to pass under @ Avonmouth will be at	**08:49 UTC**

9.5 Tides paper 5

1. Pacific Secondary Port
    A container ship ran aground in the entrance to Bligh Sound (6515), New Zealand at 19:30 UT on the 25th November. The vessel had a draft of 8.5m at the time. What was the Charted Depth of the seabed marked on the chart in that position?
2. European Secondary Port
    A VLCC Tanker is loading a full cargo of 300,000 tonnes of Brent crude oil at No 3 jetty at Sullom Voe oil terminal (293) in the Shetland Islands. She is due to complete loading and sail at 05:00 GMT on the 21st December. Find the earliest time that this vessel can pass over a shallow patch with a charted depth of 22.5m if her final loaded draft is 22.1m.

(She is required by the Port Authority to have a minimum 2.0m UKC at all stages of the pilotage.)

3. Pacific Secondary Port
    Calculate the earliest time when a vessel can pass under a bridge with a Charted Ht of 56.2m at Los Angeles (9351) on the 9th May. The vessel has an air draft is 55.5m above the waterline and a clearance of no less than 2.0m is required as she passed under the centre span of the bridge.

Answers to paper 5

No 1

Westport = Z−12 Therefore = 19:30 UT 25th Nov = 0:730 (Z−12) 26th at Bligh Sound

Times Westport	01:55 (Z−12)	08:16 (Z−12)
Corrections Bligh Sound	+1:00	+1:00
Times of Tides Bligh Sound	02:55 (Z−12)	09:16 (Z−12)
		02:55 (Z−12)
	DOT = 6hrs 21mins	

Times of Tides Bligh Sound	02:55 (Z−12)
Vessel Ran Aground	07:30 (Z−12)
Time Interval	**+4hrs 35mins after HW Bligh Sound**

	HW	*LW*
Hts of Tide Westport (corrected)	2.9	0.6
Seasonal Correction Westport	Negl'ble	Negl'ble
Hts of Tides Westport (uncorrected)	2.9	0.6
Ht Corrections Bligh Sound	−0.9	−0.2
Hts of Tides Bligh Sound (uncorrected)	2.0	0.4
Seasonal Correction Bligh Sound	Negl'ble	Negl'ble
HT LW Bligh Sound (Corr)	2.0	0.4 SLOPY LINE

From tidal curves HOT at	07:30 ZT at Bligh Sound = 0.7m
V/ls Draft	8.5m (DOW = 8.5m aground)
HOT	−0.7m

Charted Depth at the position of the grounding = 7.8m.

No 2

LW Lerwick	04:41 GMT	0.7m
HW Lerwick	10:58 GMT	2.3m

HW Lerwick	10:58 GMT
Time Diff Sullom Voe	1:33
Time HW Sullom Voe	09:25 GMT
DOW Required = Draft + UKC	(22.1m + 2.0m UKC) = 24.1m
Charted Depth of Shoal	22.5m
Minimum HOT Required	1.6m

	LW	*HW*
Hts of Tides Lerwick (corrected)	0.7	2.3 PR 1.6m
Seasonal correction Lerwick	−0.1	−0.1
Hts of Tides Lerwick (uncorrected)	0.6	2.2
Ht Diffs Sullom Voe	−0.2	+0.1
Hts of Tides Sullom Voe (uncorrected)	0.4	2.3
Seasonal correction Sullom Voe	+0.1	+0.1
HOT Sullom Voe (corrected)	0.5	2.4 SLOPY LINE

$$\% \text{ Springs} = \frac{1.6 - 0.7}{1.7 - 0.7} \times 100 = \frac{-0.9}{1.0} \times 100 = -90\%$$

Time HW Sullom Voe	09:25 UT
Time Difference for HOT 1.6m from tidal curves	−2:50

Earliest Time to cross shoal @ Sullom Voe 06:35 UT

No 3

Standard Port:	San Diego	Date: 9th May 1987
Secondary Port:	Los Angeles	Zone: +08:00

HAT Los Angeles	2.2m
Max Space Available	56.2 + 2.2m = 58.4m
Space Required	55.5 + 2.0m = 57.5m
Max HOT	0.9m

	HW	*LW*
San Diego	06:20	12:22
Differences	+00:13	+00:14
Secondary Port	06:33	12:36
		06:33 ZT
		12:36 ZT

Duration of Tide = 6hrs 03mins.

	HW	*LW*
Hts of Tides San Diego (corrected)	1.3	0.2
Seasonal Correction San Diego	Negl'ble	Negl'ble
Hts of Tides San Diego (uncorrected)	1.3	0.2
Ht Differences Los Angeles	−0.1	0.0
Hts of Tides Los Angeles (uncorrected)	1.2	0.2
Seasonal Correction Los Angeles	Negl'ble	Negl'ble
Hts of Tides Los Angeles (corrected)	1.2	0.2 SLOPY LINE

HW Los Angeles	06:33
Time interval fm graph for 0.9m	+2:12
Earliest time	08:45 ZT

9.6 Tides paper 6

1. Secondary Pacific Port
    At the time of low water on the same ebb tide at Bligh Sound (6515), a Vertical Sextant Angle of 0°25'.0 (I.E. 2.0' off the arc) was attained from a nearby lighthouse with a Charted Height of 26m. Calculate the distance-off this headland at Low Water.
2. Secondary Pacific Port
    On the 9th May, a Heavy Lift Vessel loaded a Jack up Oil Rig stowed on-deck, at the Port of Los Angeles (9351). During the outward transit through the port, the vessel must pass below a bridge which has a charted height of 56.2m, as they depart Los Angeles Harbour. The highest point on the oil rig was 54.5m above the waterline and a clearance of no less than 3.0m was required as she passed under the bridge
    Calculate the latest time during the afternoon when the vessel would be safe to pass under the bridge.

3. European Standard Port
    Find the earliest UT and BST time required for a vessel departing Portsmouth (65) to pass over a sandbank with a charted depth of 6.1m on the morning flood tide on the 17th November. The vessel with a draft of 7.0m requires a UKC of 2.0m.

Answers to paper 6

No 1

Sextant.Alt	0.25.0'
Index Error	+ 2.0'
Corrected VSA	.0.27.0'

MHWS Westport	=	3.2m
Diff Bligh Sound	=	− 1.1m
MHWS Bligh Sound	=	2.1m

Chart Datum to MHWS	**2.1m**
Charted Height of Lighthouse	**26.0m above MHWS**
Ht of Lighthouse above Chart Datum	**28.1m**
Height of Tide at Low Water	**0.4m**

Therefore Light House is (28.1m − 0.4m) = 27.7m above water level at LW

$$\text{Tan VSA} = \frac{\text{OPPOSITE}}{\text{ADJACENT}} = \frac{\text{Ht}}{\text{Distance off}}$$

$$\text{Dist off} = \frac{27.7\text{m}}{\text{Tan } 0°27.0'} = 3526\text{m} = \frac{3526\text{m}}{1852} = 1.9 \text{ Miles}$$

Distance off lighthouse @ LW Bligh Sound at 09:16 ZT = 1.9 nautical miles

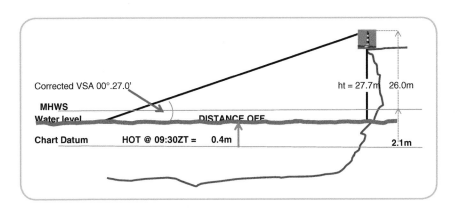

FIGURE 9.4 Diagram to determine distance off lighthouse at Bligh Sound

No 2

Standard Port	San Diego	Date: 9th May 1987
Secondary Port	Los Angeles	Zone: +08:00

Height Req' at
HAT Los Angeles 2.2m

Max Space Available	56.2 + 2.2m = 58.4m
Space Required	54.5 + 3.0m = 57.5m
Max HOT	0.9m

	LW	HW
San Diego	12:22	18:47
Differences	+00:14	+00:13
Secondary Port	12:36 ZT	19:00 ZT
		12:36 ZT

Duration of Tide = 6hrs 24mins

	LW	HW
Hts of Tides San Diego (corrected)	0.2	1.6
Seasonal Correction San Diego	Negl'ble	Negl'ble
Hts of Tides San Diego (uncorrected)	0.2	1.6
Ht Differences Los Angeles	0.0	−0.2
Hts of Tides Los Angeles (uncorrected)	0.2	1.4
Seasonal Correction Los Angeles	Negl'ble	Negl'ble
Hts of Tides Los Angeles (corrected)	0.2	1.4 SLOPY LINE

HW Los Angeles	19:00
Time interval fm graph for	0.9m − 4:20
Latest time to pass under Bridge	14:40 ZT

No 3

DOW Required	Draft + UKC = (7.0 + 2.0) = 9.0m
Charted Depth	6.1m
Minimum HOT required to pass overt sandbank	2.9m

LW	Portsmouth	01:17	1.7
HW	Portsmouth	08:23	4.1m

Predicted Range 2.4m

$$\% \text{ Springs} = \frac{2.4 - 2.0}{4.1 - 2.0} \times 100 = \frac{0.4}{2.1} \times 100 = 19\%$$

Time of High Water Portsmouth	**08:23 UT**
Time Interval from Graph	**−2:50**
Earliest time to cross Sandbank	**05:33 UT**
Convert to BST (+1 hr)	**+1:00**
Earliest time to cross Sandbank	**06:33 BST**

9.7 Tides paper 7

1. Pacific Secondary Port
    A vessel has a draught of 6.3m and is on passage to Kunsan (7504) in Korea. What will be the earliest time she can dock on October 19th during the morning flood tide if the charted depth alongside is 4.7m where the minimum UKC required is 2.0m? The berth will not be available until at least 10:00hrs on the morning of the 19th.
2. European Standard Port
    At 07:00 Zone Time on the morning of May 5th a cruise liner is on the outward pilotage transit from Esbjerg (1417) in Denmark, with a draft of 8.6m. Calculate the latest time by which she must have crossed the shallowest patch of the outward pilotage, which has a charted depth of 10.0m. The vessel will need an Under-Keel Clearance of 20m.
3. European Secondary Port
    Whilst en route towards the Pentland Firth, a vessel is forced to make an unscheduled port call at Peterhead (245). The Master is concerned about the berth allocated, which according to the Guide to Port entry has only 7.8m charted depth alongside. The vessels draft is currently 8.0m and she is due to berth alongside the South Breakwater Quay at 10:30 BST on 6th May. Find the HOT at 10:30 and state the UKC alongside at that time.

Answers to paper 7

No 1

	LW	*HW*
LW Inchon	09:17 (Z−9)	15:17 (Z−9)
Time Difference Kunsan	−1:30	−2:03
Times of Tides Kunsan (Outer Port)	07:47 (Z−9)	13:14 (Z−9)
		07:47 (Z−9)
Duration of Tide (DOT)		5hrs 27mins

DOW required Draft	**UKC = (6.3m + 2.0m) = 8.3m**
Charted Depth alongside	**4.7m**

Minimum HOT required 3.6m

	LW	*HW*
Hts of Tides Inch'on (corrected)	2.2	7.0
Seasonal Correction Inch'on (−0.1)	+0.1	+0.1
Hts of Tides Inch'on (uncorrected)	2.3	7.1
Ht Differences Kunsan	−0.3	−1.5
Hts of Tides Kunsan (uncorrected)	2.0	5.6
Seasonal Correction Kunsan (−0.1)	−0.1	−0.1
Hts of Tides Kunsan (corrected)	1.9m	5.5m SLOPY LINE

Time of HW Kunsan Outer Port	13:14 (Z−9)
Time Interval from Graph	− 2hrs 50mins before HW
Earliest time to dock at berth	10:24 (Z−9) 19th October

No 2

LW Esbjerg   06:49 (Z−1)   1.1m

HW Esbjerg   12:54 (Z−1)   0.2m

Predicted Range = 0.9m

$$\% \text{ Springs} = \frac{0.9 - 1.2}{1.7 - 1.2} \times 100 = -60\%$$

CAN ONLY USE NEAP CURVE

DOW Required	Draft = 8.6 + 2.0 UKC = 10.6m
Charted Depth of Shoal	10.0m
Minimum HOT Required	0.6m
Time HW Esbjerg	06:49 (Z−1)
Time Difference fm Graph	4hrs 40mins

Latest Time to pass shallow area must be before 11:29 (Z−1) Zone Time

No 3

Time HW Aberdeen	06:35 (GMT)
Time Difference Peterhead	−44
Time HW Peterhead	05:51 (GMT)
Convert to BST	+1:00
Time HW Peterhead	06:51 BST
Required Time	10:30 BST
Interval From graph	+3hrs 39mins

$$\% \text{ Springs} \frac{1.6 - 1.8}{3.7 - 1.8} \times 100 = -11\%$$

CAN ONLY USE NEAP CURVE

	HW	*LW*
Hts of Tides Aberdeen (corrected)	3.2	1.6
Seasonal Correction Aberdeen (−0.1)	+0.1	+0.1
Hts of Tides Aberdeen (uncorrected)	3.3	1.7
Ht Differences Peterhead	−0.3	−0.1
HOT Peterhead (uncorrected)	3.0	1.6
Seasonal Correction Peterhead (−0.1)	−0.1	−0.1
HOT Peterhead (corrected)	2.9	1.5 SLOPY LINE

Interval Fm Graph for + 3hrs 39mins gives a HOT	2.0m
Charted Depth alongside	7.8m
DOW alongside @10:30BST	9.8m
Draft	8.0m
UKC at Time of berthing	1.8m

9.8 Tides paper 8 Foundation Degree paper

The Gas Carrier M.V. Qatari Gold, with a Summer DWT of 105,000 Tonnes is on passage to Williamstown (6078), Melbourne, Australia. On her transit up the River Yarra, she will need to safely pass under the West Gate Suspension Bridge. The bridge has a charted height of 50.1m, measured to the underside of the maintenance gantry slung below the main span.

The overall height of the vessel from the keel to the top of the mast is 58.5m and the air-draft of the vessel is 48.5m.

i)   Calculate the latest time on the morning of 22nd February when the vessel could safely pass under the bridge with a clearance above the mast of 2.0m.

ii) Calculate the Even-Keel draft required to pass under the bridge with the required clearance for this time.

iii) The vessel will arrive at her berth *one hour* after passing the bridge. Calculate the Under-Keel Clearance at the dock for this time, using the draft calculated in part (ii), if the charted depth alongside the berth is 12.5m.

Answers to paper 8 (FD paper)

Fm MHWS to Maintenance Gantry	50.1m
Highest Astronomical Tide (HAT) −	1.1m
Gantry height above Chart Datum	51.2m

Air Draft	48.5m
Clearance	2.0m
Space required fm Gantry to W/l	50.5m

Available space	51.2m
Space required	50.5m
Max HOT	0.7m

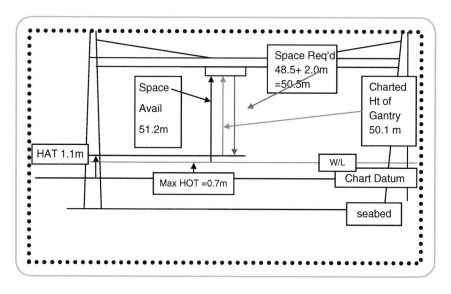

FIGURE 9.5 Diagram to determine max HOT to pass under West Gate Bridge, Melbourne

LW	0204	0.3m
HW	0748	0.8m
DOT 5hrs: 44min		

Fm Graph Interval	−1hr 50mins
High water is at 07:48 DOT	5hrs 44mins
Interval	−1hr 50mins

Latest Time = 05:58 ZT 22nd February
B) Ht from Keel to top of Mast = 58.5m
Air draft = 48.5m
DRAFT at time must be no less than **10.0m**
C) Latest time passing under bridge is 05:58ZT
Vessel will be in the turning basin one hour later at 06:58 ZT
From the graph, HOT @ 06:58 = Interval of −00hr 50mins

HOT @ 06:58 ZT	0.8m
Charted Depth in the turning basin alongside berth	12.5m
DOW alongside	13.3m
Draft	10.0m
UKC	3.3m

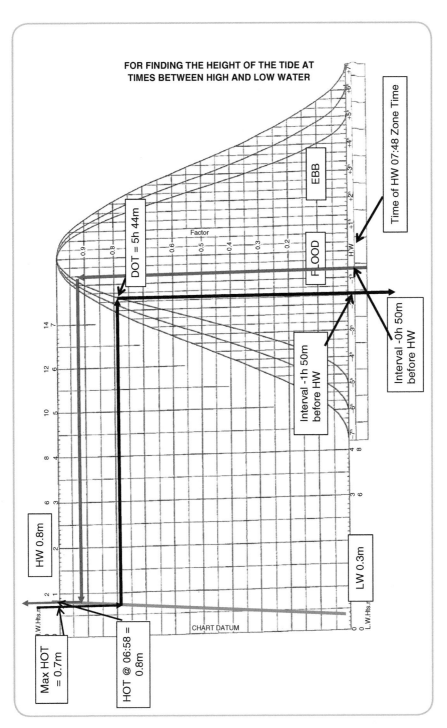

FIGURE 9.6 Plot for latest time passing West Gate Bridge, Melbourne

9.9 Tides paper 9 Foundation Degree paper

1. Pacific Secondary Port
    Find the earliest time that a small island freighter can pass over a reef at the harbour entrance at Nauru (6764) on morning flood of 27th April. The vessel has a draft of 3.5m and the reef has a charted depth of 3.7m. The vessel requires a UKC of 1.0m.

2. European Standard Port
    A vessel arriving at draft at Fawley, Refinery at Southampton (62) on the 12th October has a draft of 13.0m, what will be the latest time GMT and BST during the afternoon ebb that she could cross a bank in the channel which has a charted depth of 12.5m, if a UKC of 2.0m was required.

3. Pacific Secondary Port
    The Crane Barge 'Atlas Challenger' is currently settled aground alongside her berth at Puerto Armuelles (9471). She is to be towed to the adjacent Container terminal for a series of heavy lifts during the morning flood tide during **daylight** on the 13th May. Calculate the earliest time she will refloat with a UKC of 0.5m if the draft is 3.4m and the charted depth alongside is 1.5m.

Answers to paper 9 (FD paper)

No 1

	LW	*HW*
Times of Tides Puerto Montt (corrected)	19:36	01:36 (28th April)
Time Difference Nauru	−9:41	−9:38
Times of Tides Nauru	09:55	15:58
		09:55
Duration of Tide (DOT)		6hrs 03mins

DOW required Draft = UKC = (3.5m + 1.0m)	4.5m
Charted Depth at harbour entrance	3.7m
Minimum HOT required	0.8m

	LW	*HW*
Hts of Tides Puerto Montt (corrected)	0.5	6.4
Seasonal Correction Puerto Montt (−0.1)	Negl'ble	Negl'ble
Hts of Tides Puerto Montt (uncorrected)	0.5	6.4
Ht Differences Nauru:	−0.6	−4.6
Hts of Tides Nauru (uncorrected)	−0.1	1.8
Seasonal Correction Nauru (−0.1)	Negl'ble	Negl'ble
Hts of Tides Nauru (corrected)	−0.1m	1.8m SLOPY LINE

Time of HW Nauru	15:58 (Z−9)
Time Interval from Graph	−3hrs 05mins before HW
Earliest time to enter Nauru Harbour	12: 53 27th April

No 2

HW Southampton 12 Oct 13:47 = 4.1m   DOW Required 13.0 +2.0m = 15.0m
LW Southampton 12 Oct 19:23 = 1.2m   Charted depth of bank = 12.5m
PR 2.9m   **Min HOT = 2.5m**

$$\% \, Sp = \frac{2.9 - 1.9}{4.0 - 1.9} \times 100 = \frac{1.0}{2.1} \times 100 = 48\%$$

Interval from graph = −1hr 35mins before LW

LW Southampton = 19 : 23

                                   − 1 : 35 mins

Latest time   17 : 48 GMT

Convert to BST   + 1 : 00

Latest Time   = 1848 BST

No 3

	LW	*HW*
Tide Times Balboa	09:37 (Z+5)	15:32 (Z+5)
Time Diff Puerto Armuelles	−10mins	−10mins
Tide Times Puerto Armuelles	09:27	15:22 (Z+5)
Duration of Tide	5hrs 55mins	

Hts of Tides Balboa (corrected)	0.3	5.0
Seasonal Correction Balboa	0.0	0.0
Hts Balboa (uncorrected)	0.3	5.0
Ht Diffs Puerto Armuelles	0.0	−1.9
Hts Puerto Armuelles (uncorrected)	0.3	3.1
Seasonal Correction Puerto Armuelles	0.0	0.0
Hts of Tide Puerto Armuelles (corrected)	0.3m	3.1m

Draft	3.4m
UKC Req'd	0.5m
DOW Req'd	3.9m
Charted Depth	1.5m
HOT Req'd	2.4m

From Graph the Interval = −1hr 50mins before HW

HW Puerto Armuelles	15:22
Interval from Graph	−1:50
Earliest Time to Refloat With 0.5 UKC	13:32 (Z+5)

9.10 Tides paper 10 Foundation Degree paper

European Secondary port

On the 11th August the semi-submersible heavy lift ship was chartered to carry the completed hull of newly built warship from the shipbuilders at El Ferrol de Cuillado (1716) in Spain, to Melbourne, Australia. The semi-submersible vessel needed to submerge to a draft of 20.7m, in order to allow the warship to be floated-in sufficiently clear of all blocks and sea fastenings on her main deck.

The lifting operation was scheduled to take place just off the shipyard at El Ferrol de Cuillado, where the charted depth was 20.2m at the commencement of the lift a UKC of 3.0m was required below the Semi Submersible's hull.

Given the above information find the earliest time during when there would have been sufficient depth of water to commence the loading operation during the afternoon flood tide on the 11th August. At which time would the operation need to have been completed on the next ebb tide?

Answers to paper 10 (FD paper)

DOW Required	20.7m + 3.0 = 23.7m
Charted Depth of Seabed	20.2
Minimum HOT	3.5m

HW Pointe De Grave	18:39 (Z−1)
Time Diffs El Ferrol de Cuillado	−51
Time of HW El Ferrol de Cuillado	17:48 (Z−1)

	LW	HW
Hts P De Grave (corrected)	0.7	5.7 PR 5.0m
Seasonal Correction P De Grave	0.0	0.0
Hts Pte De Grave (uncorrected)	0.7	5.7
Ht Diffs El Ferrol De Cuillado (uncorrected)	−0.3	−1.7
Hts of Tides El Ferrol De Cuillado (uncorrected)	0.4	4.0
Seasonal Correction El Ferrol De Cuillado	0.0	0.0
Hts of Tides El Ferrol De Cuillado Ht (corrected)	0.4	4.0 SLOPY LINE

$$\% \, Sp = \frac{PR - Np(R)}{Sp(R) - Np(R)} \times 100 = \frac{5.0 - 2.2}{4.3 - 2.2} = \frac{2.8}{2.1} = 133\%$$

CAN ONLY USE SPRING CURVE

Time of HW El Ferrol de Cuillado	17:48 (Z−1)
Interval from graph	−1hr 50mins = 15:58 ZT
Earliest Time	15:58 (Z+1) August 11th

HW Pointe De Grave	18:39 (Z–1)
Time Diffs El Ferrol de Cuillado	−51
Time of HW El Ferrol de Cuillado	17:48 (Z–1)

	LW	HW
Hts P De Grave (corrected)	0.6	5.7 PR 5.0m
Sesonal Correction P De Grave	0.0	0.0
Hts Pte De Grave (uncorrected)	0.6	5.7
Ht Diffs El Ferrol De Cuillado (uncorrected)	−0.3	−1.7
Hts of Tides El Ferrol De Cuillado (uncorrected)	0.3	4.0
Seasonal Correction El Ferrol De Cuillado	0.0	0.0
Hts of Tides El Ferrol De Cuillado Ht (corrected)	0.3	4.0 SLOPY LINE

$$\% \, Sp = \frac{PR - Np(R)}{Sp(R) - Np(R)} \times 100 = \frac{5.1 - 2.2}{4.3 - 2.2} = \frac{2.9}{2.2} = 132\%$$

CAN ONLY USE SPRING CURVE

Time of HW El Ferrol de Cuillado	17:48 (Z–1)
Interval from graph	−1hr 20 mins = 15:58 ZT
Latest Time	19:08 (Z–1) August 11th

9.11 Tides paper 11

1. Explain with the aid of a diagram describe how the relative positions of the Earth, Sun and Moon influence the tidal ranges experienced over a period one Lunar Month?
2. Explain in detail what meteorological conditions may affect the Heights of Tides and hence Pilots' decisions as to time of entry into a port or channel?
3. Explain the reliability of the tidal information contained in the Admiralty Tide Tables?
4. Explain that when the Standard Port and Secondary Port are not in the same Time Zone, how is the time difference between the two countries accounted for within the Admiralty Tide Tables?
5. What are the two criteria which must both be satisfied in order to use the Duration of Tide Curves? What other indication in the ATTs is available to the mariner, to indicate that the curves may not be used without the use of Harmonic Constants?
6. Using The Tidal Stream Atlas chartlet provided, work out the tidal rate for a vessel in position off Dungeness at **50 53 N 001 01 E** at 15:30 hours (5 hours before HW Dover) using data provided as follows:

   02:30   0.4m
   08:17   5.1m
   14:13   0.3m
   20:28   5.1m

7. What is the meaning of a Semi-Diurnal Tidal Cycle and why is this pattern of tides so different in some South East Asian ports?
8. Explain the meanings of the following terms:

   Drying Height
   HAT
   Charted Depth
   MLWS
   Chart Datum

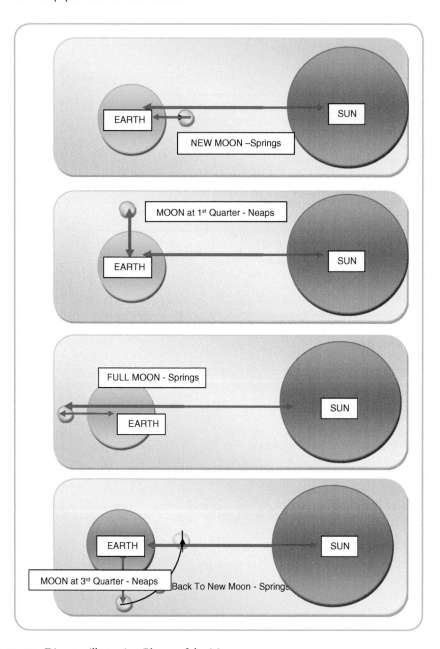

FIGURE 9.7 Diagram illustrating Phases of the Moon

Answers to paper 11

No 1

Explain with the aid of a diagram describe how the relative positions of the Earth, Sun and Moon influence the tidal ranges experienced over a period one Lunar Month?

Phase of the Moon and their effects on tidal height and ranges

On the first day of the Lunar cycle, the moon is not visible as the rays of the sun are shining on the face of the back of the moon and reflected away from an observer on Earth. At this phase, the NEW MOON, the Earth, Moon and Sun are in the same vertical plane and the gravitational forces between them are acting in the same direction combining TO PRODUCE A MAXIMUM FORCE and attempting to pull the earth towards them. This force affects the liquid oceans and creates a 'bulge' drawing water towards them in equatorial latitudes. The resultant of this is to produce the *HIGHEST* High Waters (HHW) and, as the Earth rotates six hours later, the observer will be in a position 90° away from the strongest effects of the 'bulge'. Here at Low Water, an observer will experience the *LOWEST* Low Waters (LLW). This situation is known as the SPRING TIDE and will produce *THE MAXIMUM RANGES OF TIDES* during this period. In the subsequent days the planets become misaligned and a Crescent moon becomes visible and the combination of the gravitation forces begins to reduce. When the planets are displaced by 90°, which takes about seven days from New Moon, the combined gravitational forces are at their minimum and the effect on the surface of the oceans is at its least. In terms of the Heights of Tides. This produces the *LOWEST* High Water (LHW) and the *HIGHEST* Low Water (HLW) and thus the *MINIMUM RANGES OF TIDES* called NEAP TIDES during this Phase of the Moon referred to as the First Quarter (Half Moon) the cycle continues but as the Moon continues orbiting the Earth daily, the planets now re-align themselves and as the Gibbous moon grows, the combined gravitational increases until once again they will be in a line at FULL MOON when again there will be a Spring Tide giving HHW and LLW and maximum range of tides. From this point on once again the planets get out of alignment and the Moon continues its orbit until once again the Earth, Moon and Sun are 90° to each other. As the combined gravitational forces are once again at their weakest, Neap Tides occur again producing the LHW and HLW and at this point of the cycle which is referred to as the Third Quarter, the Minimum Ranges of tides are also experienced.

From Third Quarter (Half Moon) the planets once again re-align with gravitational forces building until again all are in the same vertical plane and as the Moon reduces from a Half Moon to an ever reducing crescent, it disappears from sight when once again the phase returns to New Moon and repeats itself. The time taken for a complete cycle varies slightly between 27 1/3 and 29 days depending on the orbit of the Earth around the Sun which is elliptical and one full cycle from New Moon to New Moon is known as a LUNAR MONTH.

No 2

Explain in detail what meteorological conditions may affect the Heights of Tides and hence Pilots decisions as to the time of entry into a port or channel?

The Tide Tables are published annually, all the tidal information they contain must be completed before the 1st of January with predictions for the full year ahead. The predictions are made on the basis of decades or longer of continual observations resulting in huge amounts of data stored recording the heights and times of every High and Low Water at major ports throughout the world. At the same time, other records such as atmospheric pressure, wind speed and direction, rainfall and unseasonal storms have been collected and the averages for each of these meteorological phenomenon calculated.

Let us say that for the port of Southampton, the average atmospheric pressure is 1012 millibars and the annual average wind speed and direction works out to be South Westerly force 3/4. Together with the averages of all the tidal height data, these figures are all entered into a computer which is programmed to take all these different factors into account. The results will produce the predicted values used for compilation of the tables providing that the actual features of the weather, are the same as the average conditions expected in that area.

If, however, any of the factors are different, for instance the pressure is higher, or the wind direction is from the opposite direction, the actual values may be slightly higher or lower than the predictions given for the daily heights of tide in the Tide Table for the days. This will have a direct bearing on the

Under-Keel Clearance. The pilot will look at the predictions and the weather conditions on the day and make a decision whether perhaps to anchor and await changes in the weather or, if there is likely to be more water than predicted, get to the berth earlier than originally planned.

Factors effecting Height of Tides and hence Depth of Water.

1. Atmospheric pressure
    HIGHER than average pressure will cause actual levels to be LOWER than predicted.
    LOWER than average pressure will cause actual levels to be HIGHER than predicted.
2. Wind direction and force
    ONSHORE and stronger than average force winds will result in HIGHER Levels than predicted.
    OFFSHORE will result in actual levels of tides to be LOWER than predicted.
3. Rainfall
    FLOODING – actual levels HIGHER than predictions.
    DROUGHT – actual levels LOWER than predicted values.
4. Storm Surges
    Storm surges will give HIGHER than predicted Heights of Tides.
    Negative storm surges will take waterway from the coasts and will give levels LOWER than predicted heights.
5. Seiches
    Seiches will cause HIGHER levels in the downwind end of a bay or inlet. These will cause greatly reduced levels towards the windward end of the bay or inlet.

No 3

Explain the reliability of the tidal information contained in the Admiralty Tide Tables.

Reliability is a measure of how much trust you would place in the tidal predictions given in the Tide Tables.

As with the reliability of any product, you would look at the quality and the quantity of the product sold. If you take a popular car, it sells in large numbers because of a reliable quality of build. It is the same for tidal data. If the infrastructure at a port allows for regular readings of the Heights of Tides of every tide over a period of decades, then the averages used are fine-tuned and can be relied upon. If the port is brand new, or there are changes such as new buildings such as piers and finger berths, then there will be much less quantity of data and the averages used may be less accurate.

The Admiralty Hydrographic Department do not claim to work out all the tidal information found in the Tide Tables themselves which is received from National Hydrographic Authorities but do undertake sample checks to ensure as far as possible that the data is accurate. Whilst that random check on the figures is conducted, they do say that they are confident of the accuracy and if there are differences found, they are only very small and may be due to the method of calculation or computer algorithms used by different Hydrographic Authorities give slightly different predictions.

No 4

Explain that when the Standard Port and Secondary Port are not in the same Time Zone, how is the time difference between the two countries accounted for within the Admiralty Tide Tables?

The Zone Time of the Standard Port is given at the top left hand side of the pages in Part 1 tidal information. The time differences given in the data in Part II of the Tables can be interpolated to give the time differences to apply to determine the times of tides at an associated Secondary Port. For ports in the UK it is likely that both the Standard and the Secondary port are in the same time zone and the time difference will be quite small. In order to check that there has been no Time Zone change between the two, by running your eye up to the top of the page, it will indicate which Time Zone those ports directly below here will be holding.

When the Standard Port and the Secondary Port have a large Difference of Longitude between them, they will often be in different Time Zones. The appropriate Time Zones will be indicated part way down the pages showing the Time Zone held by all the ports listed below these entries.

The time differences which can be calculated have already taken the distance and hence difference of longitude between the two ports into account, and by applying these (highlighted in brackets), the Zone Time will automatically change to the Time Zone being held at the Secondary Port listed. YOU MUST ALWAYS MAKE A POINT OF CHECKING IF THERE HAS BEEN A TIME ZONE CHANGE – AND simply changing the time found to the new Time Zone applicable.

The Time Zone for port 6771 = Zone −12 which by applying −1 hr 17mins to the time of HW at the Standard Port Davao, changes the Time Zone from (Zone −8) at Davao to (Zone −12) at Ailinglapalap Atoll.

The Time Zone for port 6795 after applying the Time Difference to the times listed at Davao will automatically change the Time Zone to (Zone −11) at Ponape Island but you must always check.

The Time Zone for port 6797 Oroluk Island is (Zone −11) but the Standard Port, Valparaiso, holds a Standard Time of (Zone +4). By applying a time difference of −7hrs 37mins to the time of HW at Valparaiso, the time difference caused by the distance that the two ports are apart, will take you to the correct Zone Time at Oroluk Island which you can confirm by running your eye up the page to the Bracketed indication which gives the new time zone applicable for this port

No 5

What are the two criteria which must both be satisfied in order to use the Duration of Tide Curves? What other indication in the ATTs is available to the mariner, to indicate that the curves may not be used without the use of Simplified Harmonic Constants?

No.	PLACE	Lat. N.	Long. E.	TIME DIFFERENCES MHW (Zone −1200)	MLW	HEIGHT DIFFERENCES (IN METRES) MHWS	MHWN	MLWN	MLWS	M.L. Z_0 m.
5062	**DAVAO**	see page 24				1.6	1.0	0.5	−0.2	
6771	Ailinglapalap Atoll	7 17	168 45	−0117	−0117	+0.1	+0.1	+0.2	+0.4	0.92
6772	Maloelap Atoll	8 43	171 14	−0129	−0129	+0.1	+0.2	+0.2	+0.4	0.90
6775	Wotje Atoll	9 28	170 14	−0135	−0135	0.0	+0.1	+0.1	+0.4	0.85
6776	Kwajalein Atoll U	8 44	167 44	−0130	−0130	+0.1	+0.1	+0.2	+0.3	0.91
6776a	Nimuru To	9 27	167 29	−0132	−0132	+0.1	+0.2	+0.2	+0.4	0.92
6777	Likiep Atoll	9 49	169 17	−0125	−0125	0.0	+0.1	+0.2	+0.4	0.92
6782	Rongerik Atoll	11 23	167 31	−0138	−0138	0.0	+0.2	+0.2	+0.4	0.90
6783	Rongelap Atoll	11 09	166 54	−0131	−0131	0.0	+0.1	−0.2	+0.4	0.86
6786	Bikini Atoll	11 36	165 33	−0142	−0142	0.0	+0.1	0.3	+0.4	0.92
6787	Eniwetok Atoll	11 26	162 23	−0128	−0129	−0.3	−0.1	+0.1	+0.4	0.79
6787a	Runit Island	11 33	162 21	−0133	−0133	−0.2	0.0	+0.1	+0.4	0.79
6788	Ujelang Atoll	9 46	160 58	−0121	−0121	−0.2	−0.1	+0.1	+0.4	0.80
6790	Wake Island	19 17	166 37	−0200	−0200	−0.7				
	Caroline Islands									
6792	Kusaie Island	5 20	163 01	−0125	−0125	0.0				
				(Zone −1100)						
6795	Ponape Island U	6 59	158 13	−0243	−0243	−0.4				
6795a	Metaranimo Ko	6 52	158 23	−0239	−0240	−0.2				
						MHHW	MHLW	MHLW	MLLW	
9644	**VALPARAISO**	see page 231		HHW	LLW	1.5	1.2	0.5	0.4	
6797	Oroluk Island	7 40	155 10	−0737	−0707	−0.6	−0.6	−0.1	0.0	0.59
6798	Nomoi Islands	5 20	153 44	p	p	−0.5	−0.6	−0.1	0.0	0.62
6798a	Moro Tu	5 29	153 33	p	p	−0.5	−0.6	−0.1	0.0	0.61
6800	Hall Islands	8 36	152 15	p	p	−0.7	−0.7	−0.1	0.0	0.51
				(Zone −1000)						
6800a	Nomuin To	8 27	151 47	p	p	−0.8	−0.8	−0.1	−0.1	0.47
5599	**DREGER HARBOUR**	see page 63				1.5	Δ	Δ	0.9	

Changes in Zone indicated directly above those Secondary Ports which, by applying the Time Differences, will be automatically adjusted to the Zone time indicated in the brackets

FIGURE 9.8 Changes of Zone Times within Part 11 of Admiralty Tide Tables

The Use of the Duration of Tides Curves for Pacific Tides to determine intermediary times and Heights of Tides can only be used if the following criteria are BOTH satisfied:

i)   The Duration of Tide must lie within the scope of the curves
ii) There must be no shallow water correction (Part 111)

If either of these two criteria is not met, intermediate heights must be predicted by a Simplified Harmonic Method of Tidal Prediction.

   Where there are ports which have diurnal tidal cycles i.e. only one High water and one Low Water over a period of 24 hours the Duration of Tide will often exceed the time, either less than five hours or

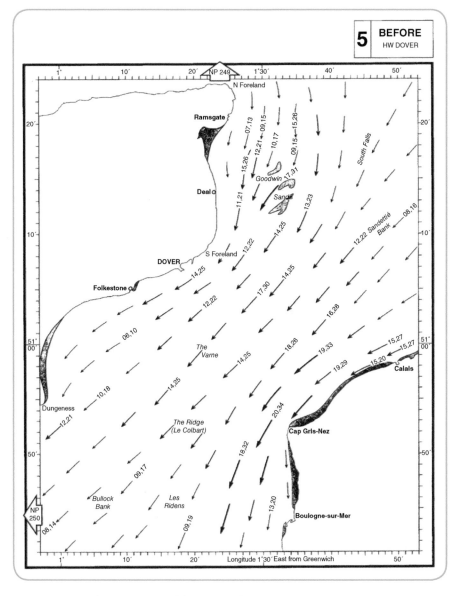

FIGURE 9.9   Extract from NP 233 Tidal Stream Atlas (UKHO, 2003)

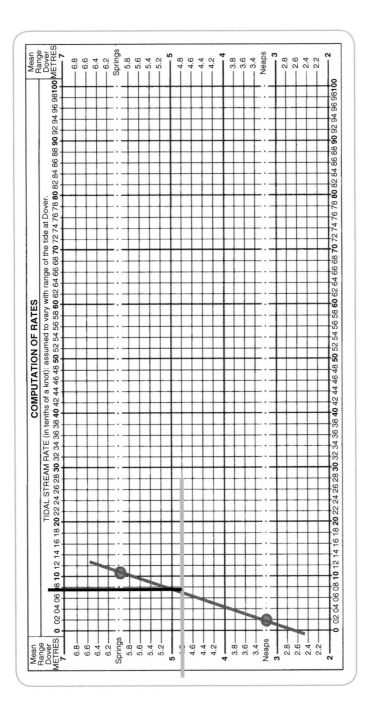

FIGURE 9.10 The rate for a position off Dungeness at 15:30 is 1.7 knots

more than seven hours, and in order to draw the mariner's attention to the fact that Harmonic Method of Calculation will be required, there is a warning at the bottom of those Standard Ports which may be affected.

No 6

Using the Tidal Stream Atlas chartlet provided, work out the tidal rate for a vessel in position off Dungeness at **50°52' N. 001° 00' E** *at 15:30 hours (five hours before HW Dover) using data provided as follows:*

Daily Tidal information

LW Dover

02:30	0.4m
08:17	5.1m
14:13	0.3m **Time required 15:30 = 5 Hours before HW Dover at 20:28**
20:28	5.1m **Predicted range 4.8m**

Spring rate and Neap rates from the chartlet is 1.2 Neap and 2.1 knots Spring Rate.

No 7

What is the meaning of a Semi-Diurnal Tidal Cycle and why is this pattern of tides so different in some south East Asian ports?

Semi-Diurnal Tidal Cycles produce two consecutive cycles of High Water followed by Low Water over a period of 24 hours. At the Equator when the planets are in alignment, the gravitational effect is at its maximum and levels of water under the direct influence of the force will have extra High Waters. Six hours later, the observer will have spun 90 degrees and as there is only a finite amount of water on the Earth, if it is being 'dragged' higher at the equator then the levels will be reduced somewhere else closer to the Poles. The water supplied to Northern Europe originates from the Atlantic Ocean, which has a fairly unobstructed flow in and out every six hours.

No 8

Explain the meanings of the following terms:

DRYING HEIGHT: The height of the seabed exposed above Chart Datum at LAT.

HAT: The highest ever expected levels of High Water given the worst combination of meteorological conditions and Spring Tides.

CHARTED DEPTH: The depths marked on a chart to indicate the depth from Chart Datum down to the seabed.

MLWS: The average of the lowest Low Waters recorded during Spring Tides.

CHART DATUM: The point at which the level of the water level will not normally be expected to drop below. It is normally, but not always, coincident with the level of Lowest Astronomical Tide (LAT). It is the reference level for Heights of Tide, Charted Depths and other levels such as MHWS, MLWS. MLWN, MHWN and Mean Sea Level (MSL).

APPENDIX 1
European waters

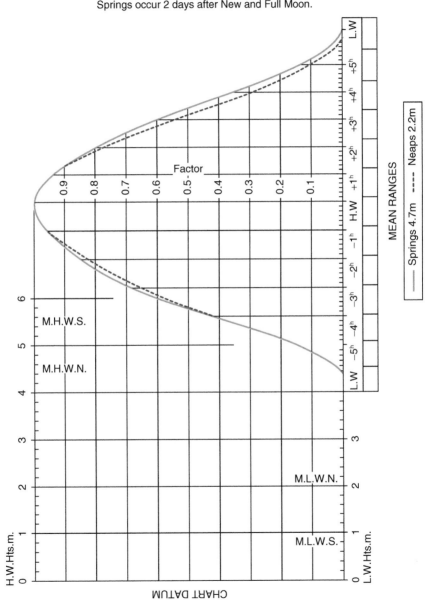

DEVONPORT
MEAN SPRING AND NEAP CURVES
Springs occur 2 days after New and Full Moon.

MEAN RANGES

—— Springs 4.7m   ----- Neaps 2.2m

ENGLAND, SOUTH COAST – PLYMOUTH (DEVONPORT)

LAT 50°22'N LONG 4°11'W

TIME ZONE GMT TIMES AND HEIGHTS OF HIGH AND LOW WATERS YEAR 1987

JANUARY

Day		TIME	M	TIME	M	TIME	M	TIME	M
1	TH	0014	0.9	0634	5.6	1242	0.7	1908	5.4
2	F	0103	0.9	0727	5.7	1331	0.7	2000	5.4
3	SA	0149	0.9	0815	5.6	1417	0.8	2047	5.2
4	SU	0234	1.0	0859	5.5	1501	1.0	2128	5.1
5	M	0317	1.3	0939	5.3	1545	1.3	2207	4.9
6	TU	0401	1.5	1017	5.1	1630	1.5	2246	4.7
7	W	0448	1.8	1058	4.8	1719	1.8	2332	4.6
8	TH	0542	2.0	1148	4.6	1816	2.0		
9	F	0031	4.5	0645	2.1	1252	4.5	1922	2.0
10	SA	0142	4.5	0756	2.1	1407	4.5	2030	2.0
11	SU	0249	4.7	0903	2.0	1513	4.6	2131	1.8
12	M	0346	4.8	1001	1.8	1609	4.8	2223	1.6
13	TU	0434	5.0	1050	1.6	1657	4.9	2309	1.5
14	W	0518	5.1	1134	1.4	1741	5.0	2349	1.4
15	TH ○	0557	5.2	1212	1.3	1820	5.0		
16	F	0025	1.3	0633	5.3	1247	1.2	1855	5.0
17	SA	0058	1.3	0705	5.3	1319	1.2	1926	5.0
18	SU	0130	1.3	0734	5.3	1350	1.2	1953	5.0
19	M	0201	1.3	0759	5.2	1422	1.3	2018	5.0
20	TU	0233	1.4	0823	5.2	1455	1.4	2044	4.9
21	W	0308	1.5	0852	5.1	1532	1.5	2117	4.8
22	TH	0348	1.7	0930	5.0	1615	1.7	2202	4.7
23	F	0436	1.9	1020	4.8	1709	1.8	2302	4.6
24	SA	0537	2.0	1128	4.6	1817	2.0		
25	SU	0019	4.5	0656	2.1	1255	4.5	1941	2.0
26	M	0153	4.6	0825	1.9	1436	4.6	2106	1.8
27	TU	0321	4.8	0945	1.6	1602	4.8	2216	1.5
28	W	0432	5.1	1049	1.2	1710	5.1	2314	1.1
29	TH ●	0535	5.4	1145	0.8	1810	5.3		
30	F	0006	0.8	0630	5.7	1235	0.6	1903	5.4
31	SA	0053	0.7	0720	5.8	1320	0.5	1949	5.5

FEBRUARY

Day		TIME	M	TIME	M	TIME	M	TIME	M
1	SU	0136	0.6	0803	5.8	1401	0.6	2028	5.4
2	M	0216	0.7	0840	5.6	1439	0.8	2101	5.3
3	TU	0253	0.9	0909	5.4	1515	1.0	2128	5.1
4	W	0328	1.2	0935	5.2	1549	1.4	2154	4.9
5	TH	0404	1.6	1002	4.9	1626	1.7	2228	4.7
6	F	0446	1.9	1042	4.6	1712	2.0	2319	4.4
7	SA	0540	2.2	1144	4.3	1815	2.2		
8	SU	0035	4.3	0701	2.4	1312	4.2	1946	2.3
9	M	0206	4.3	0837	2.3	1446	4.2	2110	2.1
10	TU	0321	4.5	0947	2.0	1553	4.4	2208	1.9
11	W	0416	4.7	1037	1.7	1643	4.6	2253	1.6
12	TH	0500	5.0	1118	1.4	1725	4.8	2331	1.4
13	F ●	0539	5.2	1153	1.2	1802	5.0		
14	SA	0005	1.2	0614	5.3	1226	1.1	1836	5.1
15	SU	0038	1.1	0646	5.4	1258	1.0	1907	5.2
16	M	0111	1.0	0716	5.4	1330	0.9	1936	5.2
17	TU	0143	1.0	0743	5.4	1402	0.9	2001	5.2
18	W	0215	1.1	0806	5.3	1435	1.1	2024	5.1
19	TH	0249	1.2	0830	5.2	1509	1.2	2050	5.0
20	F	0325	1.4	0902	5.0	1547	1.5	2127	4.8
21	SA	0408	1.7	0947	4.8	1634	1.8	2222	4.6
22	SU	0503	1.9	1053	4.5	1738	2.1	2342	4.4
23	M	0624	2.2	1233	4.2	1914	2.2		
24	TU	0133	4.4	0815	2.1	1437	4.3	2057	2.0
25	W	0316	4.7	0941	1.6	1604	4.6	2208	1.5
26	TH	0429	5.1	1042	1.1	1708	5.0	2304	1.1
27	F	0528	5.5	1133	0.7	1801	5.3	2352	0.7
28	SA ●	0619	5.7	1219	0.5	1847	5.5		

MARCH

Day		TIME	M	TIME	M	TIME	M	TIME	M
1	SU	0035	0.5	0702	5.8	1259	0.4	1926	5.6
2	M	0115	0.5	0739	5.8	1336	0.5	1959	5.5
3	TU	0150	0.6	0809	5.6	1410	0.7	2024	5.4
4	W	0223	0.8	0831	5.4	1440	1.0	2044	5.2
5	TH	0253	1.1	0850	5.2	1509	1.3	2105	5.0
6	F	0323	1.4	0915	4.9	1540	1.6	2136	4.7
7	SA	0358	1.8	0953	4.5	1616	2.0	2224	4.4
8	SU	0443	2.2	1052	4.2	1710	2.4	2337	4.2
9	M	0600	2.5	1223	3.9	1851	2.6		
10	TU	0115	4.1	0809	2.5	1412	3.9	2044	2.4
11	W	0246	4.3	0926	2.1	1528	4.2	2143	2.0
12	TH	0346	4.6	1012	1.8	1618	4.5	2225	1.7
13	F	0431	4.9	1049	1.5	1657	4.8	2302	1.4
14	SA	0509	5.1	1124	1.2	1733	5.1	2337	1.1
15	SU ○	0545	5.3	1157	0.9	1808	5.3		
16	M	0012	0.9	0620	5.5	1232	0.8	1843	5.4
17	TU	0046	0.8	0654	5.5	1306	0.7	1915	5.4
18	W	0121	0.8	0724	5.5	1340	0.8	1944	5.3
19	TH	0155	0.9	0751	5.4	1414	0.9	2008	5.2
20	F	0230	1.0	0816	5.2	1449	1.1	2033	5.1
21	SA	0307	1.2	0847	4.9	1526	1.5	2108	4.8
22	SU	0349	1.6	0933	4.6	1612	1.8	2204	4.6
23	M	0446	1.9	1048	4.3	1718	2.2	2333	4.4
24	TU	0616	2.2	1242	4.1	1907	2.3		
25	W	0129	4.4	0812	2.0	1441	4.3	2048	2.0
26	TH	0309	4.7	0930	1.5	1557	4.7	2153	1.5
27	F	0417	5.1	1025	1.1	1652	5.1	2244	1.0
28	SA	0510	5.5	1112	0.7	1738	5.4	2329	0.7
29	SU ●	0554	5.7	1154	0.5	1818	5.6		
30	M	0010	0.5	0633	5.7	1231	0.5	1853	5.6
31	TU	0047	0.5	0706	5.6	1306	0.6	1921	5.5

APRIL

Day		TIME	M	TIME	M	TIME	M	TIME	M
1	W	0121	0.7	0732	5.5	1337	0.8	1944	5.4
2	TH	0152	0.9	0753	5.3	1407	1.0	2004	5.2
3	F	0221	1.1	0814	5.1	1434	1.3	2028	5.0
4	SA	0250	1.4	0842	4.8	1503	1.6	2101	4.8
5	SU	0323	1.7	0921	4.5	1536	2.0	2147	4.5
6	M	0404	2.1	1018	4.1	1621	2.4	2252	4.2
7	TU	0510	2.4	1138	3.8	1744	2.6		
8	W	0017	4.1	0711	2.5	1315	3.8	1949	2.5
9	TH	0145	4.2	0840	2.3	1438	4.1	2057	2.2
10	F	0255	4.4	0928	1.9	1532	4.4	2143	1.8
11	SA	0345	4.8	1008	1.5	1615	4.8	2224	1.5
12	SU	0428	5.1	1046	1.2	1655	5.1	2203	1.1
13	M	0508	5.3	1124	0.9	1735	5.3	2342	0.9
14	TU	0549	5.5	1203	0.7	1814	5.5		
15	W ○	0020	0.7	0628	5.5	1241	0.7	1852	5.5
16	TH	0059	0.7	0707	5.5	1319	0.7	1927	5.4
17	F	0137	0.8	0743	5.3	1356	0.9	1959	5.3
18	SA	0215	0.9	0818	5.1	1434	1.2	2032	5.1
19	SU	0256	1.2	0859	4.8	1516	1.5	2115	4.8
20	M	0344	1.5	0954	4.5	1607	1.9	2217	4.6
21	TU	0448	1.8	1111	4.2	1719	2.2	2341	4.5
22	W	0619	2.0	1250	4.1	1859	2.2		
23	TH	0121	4.5	0756	1.8	1426	4.4	2025	1.9
24	F	0248	4.8	0904	1.5	1532	4.8	2126	1.4
25	SA	0350	5.1	0956	1.1	1622	5.1	2216	1.1
26	SU	0439	5.4	1041	0.8	1704	5.3	2300	0.8
27	M	0520	5.5	1122	0.7	1742	5.4	2340	0.7
28	TU	0556	5.5	1200	0.7	1815	5.5		
29	W	0018	0.8	0628	5.4	1235	0.8	1844	5.4
30	TH	0052	0.9	0657	5.3	1308	1.0	1911	5.3

PORTLAND

MEAN SPRING AND NEAP CURVES
Spring occurs 2 days after New and Full Moon

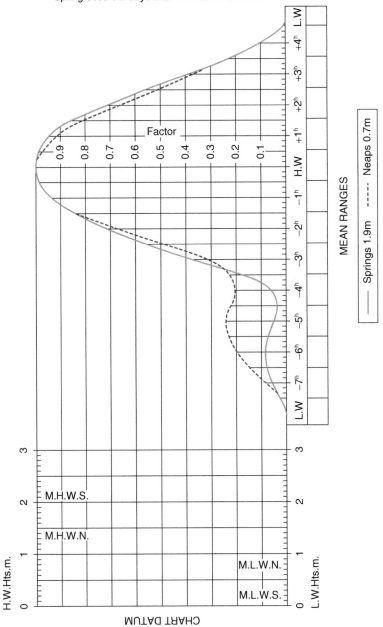

MEAN RANGES

——— Springs 1.9m   - - - - - Neaps 0.7m

TIME ZONE GMT   TIMES AND HEIGHTS OF HIGH AND LOW WATERS   YEAR 1987

JANUARY

Day	TIME	M	TIME	M		Day	TIME	M	TIME	M
1 TH	0741	2.4				16 F	0000	0.4		
	1213	0.4					0740	2.0		
	2013	2.2					1220	0.4		
							2010	1.8		
2 F	0030	0.4				17 SA	0030	0.4		
	0825	2.5					0808	2.0		
	1259	0.4					1250	0.4		
	2100	2.2					2044	1.8		
3 SA	0113	0.5				18 SU	0058	0.4		
	0905	2.5					0836	2.0		
	1345	0.4					1322	0.3		
	2141	2.1					2115	1.8		
4 SU	0156	0.5				19 M	0127	0.4		
	0943	2.3					0903	2.0		
	1432	0.3					1356	0.3		
	2222	1.9					2146	1.7		
5 M	0239	0.5				20 TU	0158	0.4		
	1023	2.2					0933	1.9		
	1520	0.4					1430	0.4		
	2301	1.7					2216	1.7		
6 TU	0321	0.6				21 W	0231	0.5		
	1105	1.9					1006	1.8		
	1607	0.4					1506	0.5		
	2343	1.6					2246	1.6		
7 W	0408	0.7				22 TH	0308	0.6		
	1150	1.7					1044	1.7		
	1657	0.5					1546	0.6		
							2324	1.5		
8 TH	0029	1.5				23 F	0354	0.7		
	0500	0.8					1131	1.6		
	1234	1.6					1637	0.7		
	1752	0.7								
9 F	0123	1.5				24 SA	0017	1.4		
	0606	0.8					0500	0.8		
	1346	1.5					1238	1.5		
	1852	0.7					1747	0.8		
10 SA	0226	1.5				25 SU	0133	1.4		
	0723	0.9					063.	0.9		
	1459	1.4					1409	1.4		
	1958	0.7					1913	0.8		
11 SU	0332	1.6				26 M	0302	1.5		
	0838	0.8					0804	0.8		
	1612	1.5					1546	1.5		
	2102	0.7					2037	0.7		
12 M	0435	1.7				27 TU	0425	1.7		
	0942	0.7					0920	0.6		
	1716	1.6					1708	1.7		
	2200	0.6					2148	0.5		
13 TU	0532	1.8				28 W	0537	2.0		
	1033	0.6					1023	0.5		
	1810	1.6					1818	1.9		
	2249	0.5					2248	0.4		
14 W	0624	1.9				29 TH	0640	2.2		
	1115	0.5					1115	0.3		
	1856	1.7				•	1916	2.0		
	2328	0.5					2337	0.3		
15 TH	0706	2.0				30 F	0731	2.4		
	1149	0.5					1202	0.2		
O	1935	1.8					2005	2.2		
						31 SA	0018	0.3		
							0813	2.5		
							1246	0.2		
							2045	2.2		

FEBRUARY

Day	TIME	M			Day	TIME	M
1 SU	0057	0.3		16 M	0039	0.2
	0849	2.5			0819	2.1
	1328	0.2			1300	0.2
	2121	2.2			2100	1.9
2 M	0134	0.3		17 TU	0107	0.2
	0923	2.4			0848	2.1
	1409	0.2			1334	0.2
	2153	2.1			2128	1.9
3 TU	0211	0.3		18 W	0138	0.2
	0959	2.2			0920	2.0
	1450	0.3			1407	0.2
	2224	1.9			2153	1.8
4 W	0248	0.4		19 TH	0211	0.3
	1033	1.9			0952	1.9
	1527	0.4			1439	0.3
	2252	1.7			2221	1.7
5 TH	0323	0.5		20 F	0242	0.5
	1105	1.7			1022	1.7
	1559	0.5			1512	0.4
	2319	1.5			2249	1.5
6 F	0356	0.6		21 SA	0319	0.5
	1136	1.5			1059	1.5
	1630	0.6			1551	0.6
	2351	1.4			2330	1.4
7 SA	0437	0.8		22 SU	0415	0.7
	1213	1.3			1158	1.4
	1711	0.8			1658	0.8
8 SU	0044	1.3		23 M	0050	1.3
	0553	0.9			0559	0.8
	1326	1.2			1349	1.3
	1829	0.8			1849	0.8
9 M	0224	1.3		24 TU	0243	1.4
	0800	0.9			0755	0.7
	1539	1.2			1545	1.4
	2022	0.8			2028	0.7
10 TU	0404	1.5		25 W	0413	1.7
	0927	0.8			0912	0.5
	1703	1.4			1705	1.6
	2141	0.7			2140	0.5
11 W	0512	1.7		26 TH	0524	1.9
	1019	0.6			1013	0.3
	1758	1.5			1811	1.8
	2234	0.5			2236	0.3
12 TH	0604	1.8		27 F	0625	2.2
	1058	0.5			1104	0.1
	1843	1.7			1906	2.0
	2315	0.4			2323	0.2
13 F	0648	2.0		28 SA	0715	2.4
	1131	0.3			1149	0.1
O	1923	1.8		•	1951	2.2
	2346	0.3				
14 SA	0722	2.0				
	1202	0.3				
	1958	1.9				
15 SU	0012	0.3				
	0751	2.1				
	1230	0.2				
	2030	1.9				

MARCH

Day	TIME	M			Day	TIME	M
1 SU	0002	0.2		16 M	0726	2.1
	0756	2.5			1205	0.1
	1227	0.0			2008	2.1
	2025	2.2				
2 M	0036	0.1		17 TU	0015	0.2
	0830	2.5			0759	2.2
	1304	0.0			1235	0.1
	2053	2.2			2037	2.1
3 TU	0109	0.2		18 W	0045	0.1
	0901	2.3			0833	2.1
	1340	0.1			1308	0.1
	2119	2.1			2105	2.0
4 W	0143	0.2		19 TH	0118	0.1
	0931	2.1			0907	2.0
	1416	0.2			1343	0.1
	2144	1.9			2134	1.9
5 TH	0215	0.3		20 F	0151	0.2
	0959	1.9			0939	1.9
	1445	0.3			1416	0.2
	2206	1.7			2201	1.7
6 F	0243	0.4		21 SA	0223	0.3
	1022	1.6			1008	1.7
	1507	0.4			1447	0.4
	2224	1.5			2229	1.6
7 SA	0306	0.5		22 SU	0301	0.5
	1039	1.4			1042	1.5
	1521	0.6			1542	0.5
	2241	1.4			2310	1.4
8 SU	0329	0.7		23 M	0401	0.6
	1053	1.2			1149	1.3
	1534	0.7			1637	0.7
	2306	1.3				
9 M	0419	0.8		24 TU	0037	1.3
	1141	1.1			0600	0.7
	1614	0.8			1404	1.2
					1848	0.8
10 TU	0043	1.2		25 W	0236	1.5
	0733	0.9			0747	0.6
	1545	1.1			1543	1.4
	2000	0.8			2019	0.6
11 W	0337	1.4		26 TH	0358	1.7
	0907	0.7			0857	0.4
	1650	1.3			1651	1.6
	2122	0.7			2123	0.5
12 TH	0443	1.6		27 F	0503	1.9
	0951	0.5			0955	0.2
	1734	1.5			1753	1.8
	2209	0.5			2217	0.3
13 F	0532	1.8		28 SA	0602	2.1
	1030	0.3			1047	0.0
	1819	1.7			1846	2.0
	2248	0.4			2303	0.2
14 SA	0614	2.0		29 SU	0652	2.3
	1105	0.2			1130	0.0
	1900	1.9		•	1928	2.1
	2320	0.3			2341	0.1
15 SU	0652	2.1		30 M	0733	2.3
	1136	0.2			1206	0.0
O	1936	2.0			1958	2.2
	2349	0.1				
				31 TU	0014	0.0
					0807	2.3
					1238	0.0
					2021	2.2

APRIL

Day	TIME	M			Day	TIME	M
1 W	0044	0.1		16 TH	0817	2.1
	0836	2.2			1244	0.2
	1308	0.1			2043	2.1
	2043	2.1				
2 TH	0115	0.1		17 F	0101	0.2
	0903	2.0			0855	2.0
	1339	0.2			1322	0.1
	2106	2.0			2116	2.0
3 F	0145	0.2		18 SA	0139	0.2
	0926	1.8			0931	1.8
	1404	0.3			1400	0.3
	2127	1.8			2150	1.8
4 SA	0210	0.4		19 SU	0216	0.3
	0944	1.6			1006	1.6
	1423	0.4			1437	0.4
	2146	1.6			2225	1.6
5 SU	0231	0.5		20 M	0303	0.4
	0956	1.3			1051	1.4
	1432	0.5			1523	0.5
	2200	1.4			2315	1.5
6 M	0253	0.7		21 TU	0414	0.5
	1009	1.2			1213	1.2
	1439	0.6			1647	0.7
	2221	1.3				
7 TU	0341	0.8		22 W	0041	1.4
	1104	1.0			0557	0.5
	1503	0.8			1404	1.3
	2339	1.2			1838	0.7
8 W	0644	0.8		23 TH	0217	1.5
	1519	1.1			0724	0.4
	1919	0.8			1521	1.4
					1956	0.6
9 TH	0245	1.3		24 F	0331	1.7
	0816	0.6			0831	0.3
	1608	1.3			1623	1.6
	2039	0.7			2057	0.5
10 F	0354	1.5		25 SA	0435	1.9
	0904	0.5			0930	0.2
	1653	1.5			1722	1.8
	2127	0.5			2150	0.3
11 SA	0444	1.7		26 SU	0533	2.0
	0948	0.3			1022	0.1
	1741	1.7			1815	1.9
	2209	0.4			2238	0.2
12 SU	0531	1.9		27 M	0625	2.0
	1029	0.2			1105	0.0
	1826	1.9			1857	2.1
	2246	0.3			2318	0.1
13 M	0615	2.0		28 TU	0707	2.1
	1104	0.2			1140	0.0
	1906	2.1		•	1928	2.1
	2302	0.2			2352	0.1
14 TU	0657	2.1		29 W	0742	2.1
	1136	0.1			1210	0.1
O	1940	2.1			1952	2.1
	2351	0.1				
15 W	0738	2.2		30 TH	0022	0.1
	1209	0.1			0810	2.0
	2011	2.2			1237	0.2
					2015	2.0

LOW WATERS – IMPORTANT NOTE. DOUBLE LOW WATERS OCCUR AT PORTLAND. THE PREDICTIONS ARE FOR THE FIRST LOW WATER.

SOUTHAMPTON

MEAN SPRING AND NEAP CURVES

For instructions see page XIV

Spring occurs 2 days after New and Full Moon

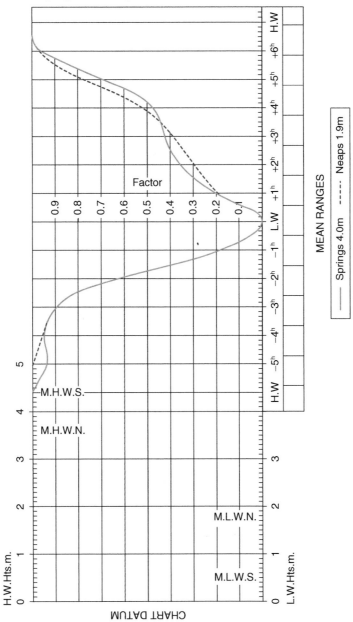

MEAN RANGES

——— Springs 4.0m   - - - - - Neaps 1.9m

JANUARY

Day	TIME	M	Day	TIME	M
1 TH	0504	0.6	16 F	0507	1.0
	1123	4.5		1138	4.3
	1728	0.4		1726	0.8
2 F	0000	4.6	17 SA	0010	4.3
	0555	0.6		0542	1.0
	1211	4.5		1212	4.3
	1817	0.3		1758	0.7
3 SA	0051	4.6	18 SU	0043	4.3
	0645	0.6		0616	0.9
	1301	4.5		1244	4.3
	1904	0.4		1832	0.7
4 SU	0142	4.6	19 M	0116	4.3
	0731	0.7		0650	1.0
	1352	4.4		1318	4.2
	1948	0.5		1904	0.8
5 M	0234	4.4	20 TU	0149	4.3
	0815	0.9		0724	1.0
	1445	4.2		1354	4.2
	2032	0.8		1937	0.9
6 TU	0328	4.3	21 W	0228	4.2
	0901	1.1		0800	1.2
	1541	4.0		1437	4.1
	2119	1.1		2014	1.1
7 W	0427	4.1	22 TH	0312	4.1
	0951	1.4		0841	1.4
	1642	3.8		1525	4.0
	2214	1.4		2059	1.4
8 TH	0530	3.9	23 F	0406	4.0
	1051	1.6		0934	1.6
	1750	3.7		1625	3.8
	2318	1.7		2159	1.6
9 F	0634	3.8	24 SA	0508	3.9
	1156	1.7		1043	1.7
	1902	3.7		1740	3.7
				2315	1.7
10 SA	0028	1.8	25 SU	0620	3.9
	0737	3.8		1204	1.7
	1303	1.7		1903	3.8
	2008	3.7			
11 SU	0133	1.8	26 M	0041	1.7
	0831	3.9		0734	4.0
	1403	1.6		1324	1.5
	2103	3.8		2019	3.9
12 M	0229	1.7	27 TU	0158	1.4
	0916	3.9		0838	4.1
	1455	1.4		1434	1.1
	2148	4.0		2121	4.2
13 TU	0317	1.5	28 W	0304	1.1
	0955	4.0		0935	4.3
	1539	1.2		1535	0.8
	2227	4.1		2216	4.4
14 W	0357	1.3	29 TH	0403	0.8
	1031	4.1		1025	4.4
	1617	1.0	●	1628	0.5
	2303	4.2		2305	4.6
15 TH	0434	1.2	30 F	0456	0.6
	1105	4.2		1114	4.5
O	1652	0.8		1719	0.3
	2339	4.2		2351	4.7
			31 SA	0544	0.4
				1159	4.6
				1805	0.2

FEBRUARY

Day	TIME	M	Day	TIME	M
1 SU	0036	4.7	16 M	0021	4.5
	0629	0.4		0600	0.7
	1245	4.6		1223	4.4
	1846	0.2		1815	0.5
2 M	0121	4.7	17 TU	0050	4.5
	0709	0.5		0633	0.6
	1329	4.5		1254	4.4
	1925	0.4		1845	0.5
3 TU	0204	4.6	18 W	0120	4.5
	0747	0.6		0703	0.7
	1413	4.3		1327	4.4
	2001	0.6		1915	0.7
4 W	0248	4.3	19 TH	0154	4.4
	0822	0.9		0733	0.9
	1500	4.1		1407	4.2
	2036	1.0		1946	0.9
5 TH	0334	4.1	20 F	0235	4.3
	0859	1.2		0809	1.1
	1550	3.9		1453	4.1
	2116	1.4		2027	1.2
6 F	0426	3.9	21 SA	0326	4.1
	0945	1.6		0855	1.4
	1649	3.6		1551	3.9
	2210	1.8		2122	1.6
7 SA	0527	3.7	22 SU	0428	3.9
	1028	1.9		1003	1.7
	1802	3.5		1711	3.7
	2324	2.1		2245	1.9
8 SU	0638	3.6	23 M	0550	3.8
	1208	2.0		1135	1.8
	1927	3.5		1851	3.7
9 M	0049	2.1	24 TU	0027	1.8
	0751	3.6		0718	3.8
	1329	1.9		1314	1.6
	2040	3.6		2017	3.9
10 TU	0203	2.0	25 W	0154	1.5
	0852	3.7		0832	4.0
	1435	1.6		1428	1.1
	2134	3.8		2118	4.2
11 W	0259	1.7	26 TH	0300	1.1
	0939	3.9		0928	4.2
	1523	1.3		1527	0.7
	2215	4.0		2207	4.4
12 TH	0342	1.4	27 F	0355	0.7
	1017	4.1		1016	4.4
	1602	1.0		1617	0.4
	2250	4.2		2252	4.6
13 F	0417	1.1	28 SA	0441	0.5
	1052	4.2		1100	4.5
O	1637	0.8	●	1701	0.2
	2322	4.3		2332	4.7
14 SA	0453	0.9			
	1124	4.3			
	1710	0.6			
15 SU	0527	0.7			
	1154	4.4			
	1743	0.5			

MARCH

Day	TIME	M	Day	TIME	M
1 SU	0524	0.3	16 M	0502	0.6
	1141	4.6		1130	4.4
	1744	0.1		1719	0.4
				2353	4.6
2 M	0012	4.8	17 TU	0536	0.5
	0605	0.3		1159	4.5
	1222	4.6		1751	0.4
	1822	0.2			
3 TU	0051	4.7	18 W	0021	4.6
	0641	0.3		0610	0.4
	1301	4.5		1231	4.5
	1856	0.3		1823	0.4
4 W	0128	4.5	19 TH	0052	4.6
	0713	0.5		0640	0.5
	1341	4.4		1304	4.4
	1926	0.6		1855	0.6
5 TH	0207	4.3	20 F	0127	4.5
	0741	0.8		0713	0.7
	1421	4.1		1345	4.3
	1953	1.0		1928	0.8
6 F	0245	4.1	21 SA	0209	4.3
	0810	1.2		0747	1.0
	1504	3.9		1434	4.1
	2026	1.4		2009	1.2
7 SA	0328	3.8	22 SU	0302	4.1
	0848	1.6		0836	1.3
	1556	3.6		1538	3.9
	2110	1.9		2109	1.6
8 SU	0421	3.6	23 M	0408	3.9
	0943	1.9		0947	1.7
	1705	3.4		1707	3.7
	2219	2.2		2240	1.9
9 M	0531	3.4	24 TU	0539	3.7
	1108	2.1		1128	1.7
	1841	3.4		1852	3.8
	2359	2.3			
10 TU	0704	3.4	25 W	0025	1.8
	1250	2.1		0713	3.8
	2012	3.5		1305	1.5
				2012	4.0
11 W	0130	2.1	26 TH	0147	1.5
	0822	3.6		0824	4.0
	1407	1.7		1416	1.1
	2110	3.7		2107	4.3
12 TH	0317	1.7	27 F	0247	1.0
	0914	3.8		0918	4.2
	1458	1.4		1510	0.7
	2151	4.0		2150	4.5
13 F	0317	1.3	28 SA	0335	0.7
	0954	4.0		1001	4.4
	1537	1.0		1556	0.4
	2224	4.2		2230	4.6
14 SA	0353	1.0	29 SU	0417	0.4
	1027	4.2		1040	4.5
	1611	0.7	●	1637	0.2
	2255	4.3		2308	4.7
15 SU	0427	0.7	30 M	0458	0.3
	1059	4.3		1120	4.6
O	1645	0.5		1716	0.2
	2323	4.5		2345	4.7
			31 TU	0534	0.3
				1156	4.5
				1752	0.3

APRIL

Day	TIME	M	Day	TIME	M
1 W	0018	4.6	16 TH	0546	0.4
	0608	0.4		1209	4.5
	1233	4.5		1803	0.4
	1823	0.5			
2 TH	0053	4.4	17 F	0028	4.6
	0638	0.5		0622	0.4
	1309	4.3		1248	4.4
	1850	0.7		1840	0.6
3 F	0126	4.2	18 SA	0107	4.4
	0704	0.8		0658	0.6
	1347	4.1		1335	4.3
	1917	1.1		1919	0.9
4 SA	0202	4.0	19 SU	0154	4.3
	0740	1.1		0738	0.9
	1428	3.9		1432	4.1
	1947	1.5		2007	1.2
5 SU	0243	3.8	20 M	0251	4.0
	0806	1.5		0832	1.2
	1518	3.6		1543	3.9
	2030	1.9		2113	1.6
6 M	0334	3.6	21 TU	0404	3.8
	0857	1.9		0947	1.5
	1625	3.4		1712	3.8
	2137	2.2		2243	1.8
7 TU	0442	3.4	22 W	0534	3.7
	1019	2.1		1121	1.6
	1757	3.4		1844	3.9
	2314	2.3			
8 W	0614	3.4	23 TH	0013	1.6
	1202	2.0		0702	3.8
	1928	3.5		1248	1.4
				1954	4.2
9 TH	0047	2.1	24 F	0126	1.3
	0738	3.5		0808	4.0
	1322	1.8		1352	1.0
	2028	3.7		2045	4.4
10 F	0153	1.7	25 SA	0221	1.0
	0836	3.7		0858	4.2
	1418	1.4		1444	0.7
	2110	4.0		2127	4.5
11 SA	0240	1.3	26 SU	0308	0.7
	0918	3.9		0939	4.3
	1459	1.0		1528	0.5
	2145	4.2		2204	4.6
12 SU	0317	0.9	27 M	0349	0.5
	0954	4.1		1018	4.4
	1535	0.7		1607	0.4
	2217	4.4		2240	4.6
13 M	0335	0.7	28 TU	0427	0.4
	1027	4.3		1056	4.4
	1610	0.5	●	1645	0.5
	2249	4.5		2315	4.5
14 TU	0430	0.5	29 W	0503	0.4
	1059	4.4		1132	4.4
O	1648	0.4		1719	0.6
				2348	4.4
15 W	0508	0.5	30 TH	0535	0.5
	1133	4.5		1208	4.3
	1726	0.4		1752	0.7
	2352	4.6			

HIGH WATERS – IMPORTANT NOTE. DOUBLE HIGH WATERS OCCUR AT SOUTHAMPTON. THE PREDICTIONS ARE FOR THE FIRST LOW WATER.

TIME ZONE GMT   TIMES AND HEIGHTS OF HIGH AND LOW WATERS   YEAR 1987

MAY

Day	Time	M
1 F	0021	4.3
	0605	0.7
	1243	4.2
	1819	0.9
2 SA	0054	4.1
	0632	0.9
	1321	4.0
	1848	1.2
3 SU	0129	4.0
	0703	1.1
	1404	3.8
	1922	1.5
4 M	0211	3.8
	0740	1.4
	1454	3.7
	2007	1.8
5 TU	0301	3.6
	0830	1.7
	1555	3.5
	2109	2.0
6 W	0405	3.5
	0941	1.9
	1731	3.5
	2231	2.1
7 TH	0525	3.4
	1106	1.9
	1831	3.6
	2353	2.0
8 F	0643	3.5
	1223	1.7
	1933	3.7
9 SA	0059	1.7
	0745	3.7
	1322	1.4
	2020	4.0
10 SU	0150	1.3
	0832	3.9
	1410	1.1
	2100	4.2
11 M	0234	1.0
	0913	4.1
	1451	0.8
	2136	4.3
12 TU	0315	0.7
	0953	4.2
	1533	0.7
	2212	4.5
13 W ○	0357	0.5
	1031	4.3
	1617	0.6
	2249	4.5
14 TH	0440	0.4
	1111	4.4
	1702	0.5
	2327	4.5
15 F	0525	0.4
	1154	4.4
	1747	0.6
16 SA	0009	4.5
	0608	0.4
	1241	4.4
	1832	0.7
17 SU	0055	4.4
	0652	0.6
	1334	4.3
	1919	0.9
18 M	0148	4.2
	0740	0.8
	1435	4.2
	2013	1.2
19 TU	0249	4.0
	0835	1.0
	1545	4.1
	2116	1.4
20 W	0400	3.9
	0943	1.2
	1702	4.0
	2231	1.5
21 TH	0519	3.8
	1101	1.3
	1818	4.1
	2345	1.5
22 F	0637	3.9
	1214	1.3
	1924	4.2
23 SA	0052	1.3
	0742	4.0
	1317	1.1
	2017	4.3
24 SU	0147	1.1
	0834	4.1
	1409	1.0
	2100	4.4
25 M	0235	0.9
	0918	4.2
	1457	0.9
	2139	4.4
26 TU	0318	0.7
	0958	4.2
	1538	0.8
	2215	4.4
27 W ●	0358	0.7
	1036	4.2
	1616	0.8
	2249	4.3
28 TH	0434	0.7
	1113	4.2
	1651	0.9
	2323	4.2
29 F	0509	0.7
	1150	4.2
	1725	1.0
	2356	4.2
30 SA	0541	0.8
	1226	4.1
	1757	1.1
31 SU	0030	4.1
	0612	0.9
	1305	4.0
	1830	1.3

JUNE

Day	Time	M
1 M	0107	4.0
	0645	1.1
	1347	3.9
	1907	1.4
2 TU	0148	3.8
	0723	1.3
	1433	3.8
	1950	1.6
3 W	0237	3.7
	0809	1.5
	1525	3.7
	2043	1.8
4 TH	0332	3.6
	0906	1.6
	1626	3.7
	2144	1.8
5 F	0436	3.6
	1010	1.7
	1731	3.7
	2253	1.8
6 SA	0544	3.6
	1118	1.6
	1831	3.8
	2359	1.6
7 SU	0649	3.7
	1223	1.5
	1926	4.0
8 M	0057	1.4
	0745	3.8
	1319	1.3
	2014	4.1
9 TU	0149	1.1
	0835	4.0
	1411	1.1
	2057	4.3
10 W	0240	0.9
	0922	4.1
	1502	0.9
	2141	4.4
11 TH ○	0329	0.7
	1009	4.3
	1553	0.8
	2224	4.4
12 F	0419	0.5
	1056	4.3
	1645	0.7
	2310	4.4
13 SA	0510	0.4
	1146	4.3
	1737	0.7
	2357	4.4
14 SU	0600	0.4
	1238	4.4
	1829	0.7
15 M	0048	4.4
	0650	0.5
	1332	4.4
	1919	0.8
16 TU	0143	4.3
	0738	0.6
	1429	4.3
	2010	0.9
17 W	0240	4.2
	0830	0.8
	1530	4.3
	2104	1.1
18 TH	0343	4.0
	0925	1.0
	1634	4.2
	2203	1.3
19 F	0451	3.9
	1026	1.2
	1742	4.1
	2306	1.4
20 SA	0610	3.9
	1132	1.3
	1846	4.1
21 SU	0009	1.4
	0707	3.9
	1236	1.3
	1942	4.1
22 M	0110	1.3
	0807	3.9
	1335	1.3
	2032	4.1
23 TU	0204	1.2
	0857	4.0
	1428	1.3
	2116	4.1
24 W	0253	1.1
	0941	4.1
	1514	1.2
	2154	4.1
25 TH	0335	1.0
	1022	4.1
	1554	1.2
	2229	4.1
26 F ●	0415	0.9
	1101	4.1
	1632	1.1
	2305	4.1
27 SA	0451	0.9
	1139	4.1
	1709	1.1
	2340	4.1
28 SU	0526	0.8
	1215	4.1
	1743	1.1
29 M	0015	4.1
	0600	0.9
	1251	4.1
	1817	1.2
30 TU	0050	4.0
	0633	0.9
	1327	4.1
	1853	1.2

JULY

Day	Time	M
1 W	0128	4.0
	0708	1.0
	1407	4.0
	1931	1.3
2 TH	0209	3.9
	0746	1.2
	1450	4.0
	2013	1.4
3 F	0255	3.8
	0828	1.3
	1537	3.9
	2101	1.6
4 SA	0347	3.8
	0918	1.5
	1632	3.9
	2155	1.7
5 SU	0448	3.7
	1017	1.6
	1732	3.9
	2259	1.7
6 M	0553	3.7
	1125	1.6
	1833	3.9
7 TU	0007	1.6
	0701	3.8
	1234	1.5
	1931	4.0
8 W	0112	1.4
	0805	3.9
	1339	1.4
	2027	4.2
9 TH	0214	1.1
	0902	4.1
	1441	1.1
	2118	4.3
10 F	0312	0.8
	0956	4.2
	1539	0.9
	2208	4.4
11 SA ○	0407	0.6
	1048	4.4
	1636	0.7
	2258	4.4
12 SU	0501	0.4
	1139	4.5
	1729	0.6
	2348	4.4
13 M	0553	0.3
	1228	4.5
	1821	0.6
14 TU	0038	4.4
	0642	0.3
	1319	4.6
	1909	0.6
15 W	0127	4.4
	0727	0.4
	1409	4.5
	1954	0.7
16 TH	0219	4.3
	0811	0.6
	1502	4.4
	2038	0.9
17 F	0314	4.2
	0855	0.8
	1556	4.3
	2125	1.1
18 SA	0413	4.0
	0944	1.2
	1656	4.1
	2219	1.4
19 SU	0518	3.8
	1043	1.5
	1800	3.9
	2322	1.6
20 M	0630	3.7
	1152	1.7
	1905	3.9
21 TU	0031	1.7
	0739	3.7
	1302	1.8
	2005	3.9
22 W	0136	1.6
	0842	3.8
	1405	1.7
	2056	3.9
23 TH	0234	1.4
	0932	3.9
	1458	1.6
	2139	4.0
24 F	0323	1.2
	1014	4.0
	1543	1.4
	2218	4.0
25 SA ●	0403	1.0
	1053	4.1
	1620	1.2
	2253	4.1
26 SU	0438	0.9
	1128	4.2
	1655	1.1
	2326	4.2
27 M	0512	0.8
	1200	4.2
	1728	1.0
	2358	4.2
28 TU	0545	0.7
	1232	4.3
	1802	1.0
29 W	0030	4.2
	0616	0.7
	1302	4.3
	1835	1.0
30 TH	0102	4.2
	0649	0.8
	1335	4.2
	1907	1.0
31 F	0138	4.1
	0720	0.9
	1411	4.2
	1941	1.1

AUGUST

Day	Time	M
1 SA	0217	4.0
	0753	1.1
	1451	4.1
	2019	1.3
2 SU	0303	3.9
	0834	1.3
	1540	4.0
	2106	1.6
3 M	0359	3.8
	0928	1.6
	1638	3.9
	2208	1.7
4 TU	0508	3.7
	1038	1.8
	1747	3.9
	2326	1.8
5 W	0631	3.7
	1203	1.8
	1901	3.9
6 TH	0049	1.6
	0750	3.8
	1325	1.6
	2010	4.0
7 F	0202	1.3
	0856	4.1
	1435	1.3
	2108	4.2
8 SA	0305	0.9
	0951	4.3
	1534	0.9
	2159	4.3
9 SU ○	0400	0.6
	1039	4.5
	1628	0.7
	2248	4.5
10 M	0451	0.3
	1126	4.6
	1718	0.5
	2334	4.6
11 TU	0540	0.2
	1210	4.7
	1805	0.4
12 W	0020	4.6
	0623	0.2
	1255	4.7
	1848	0.4
13 TH	0104	4.5
	0705	0.3
	1339	4.6
	1928	0.5
14 F	0150	4.3
	0742	0.5
	1424	4.5
	2004	0.7
15 SA	0238	4.2
	0819	0.8
	1513	4.2
	2043	1.1
16 SU	0330	4.0
	0900	1.3
	1606	4.0
	2128	1.5
17 M	0432	3.7
	0953	1.7
	1708	3.7
	2231	1.8
18 TU	0546	3.6
	1105	2.0
	1802	3.6
	2351	2.0
19 W	0713	3.5
	1231	2.2
	1937	3.6
20 TH	0113	1.9
	0829	3.6
	1348	2.0
	2040	3.7
21 F	0221	1.7
	0924	3.8
	1447	1.8
	2128	3.9
22 SA	0310	1.4
	1004	4.0
	1530	1.5
	2205	4.0
23 SU	0350	1.1
	1038	4.1
	1605	1.2
	2238	4.2
24 M ●	0423	0.8
	1107	4.3
	1637	1.0
	2308	4.3
25 TU	0454	0.7
	1137	4.4
	1708	0.8
	2337	4.3
26 W	0525	0.6
	1205	4.4
	1741	0.8
27 TH	0007	4.4
	0555	0.6
	1232	4.5
	1813	0.7
28 F	0036	4.4
	0626	0.6
	1302	4.4
	1843	0.8
29 SA	0108	4.3
	0655	0.7
	1334	4.4
	1913	0.9
30 SU	0144	4.2
	0725	1.0
	1411	4.3
	1946	1.2
31 M	0228	4.0
	0801	1.3
	1459	4.1
	2028	1.5

HIGH WATERS – IMPORTANT NOTE. DOUBLE HIGH WATERS OCCUR AT SOUTHAMPTON. THE PREDICTIONS ARE FOR THE FIRST HIGH WATER.

ENGLAND, SOUTH COAST – SOUTHAMPTON
LAT 50°54'N   LONG 1°24'W

TIME ZONE GMT   TIMES AND HEIGHTS OF HIGH AND LOW WATERS   YEAR 1987

SEPTEMBER

Day	Time	M	Day	Time	M
1 TU	0323	3.9	16 W	0459	3.5
	0854	1.7		1016	2.3
	1558	3.9		1726	3.4
	2131	1.8		2304	2.2
2 W	0439	3.7	17 TH	0640	3.4
	1011	2.0		1157	2.4
	1716	3.8		1900	3.4
	2301	1.9			
3 TH	0618	3.7	18 F	0046	2.1
	1153	2.0		0808	3.6
	1846	3.8		1326	2.2
				2017	3.6
4 F	0041	1.7	19 SA	0159	1.8
	0749	3.9		0902	3.8
	1324	1.7		1425	1.9
	2003	4.0		2106	3.8
5 SA	0159	1.3	20 SU	0249	1.5
	0852	4.2		0941	4.0
	1432	1.3		1507	1.5
	2101	4.2		2143	4.0
6 SU	0259	0.9	21 M	0326	1.1
	0940	4.4		1011	4.2
	1525	0.9		1540	1.1
	2148	4.4		2213	4.2
7 M	0348	0.5	22 TU	0357	0.8
	1024	4.6		1039	4.4
○	1613	0.5		1611	0.9
	2233	4.6		2242	4.4
8 TU	0434	0.2	23 W	0427	0.5
	1105	4.8		1107	4.5
	1658	0.3	●	1643	0.7
	2314	4.7		2311	4.4
9 W	0518	0.1	24 TH	0459	0.5
	1145	4.8		1133	4.6
	1741	0.3		1715	0.6
	2356	4.7		2341	4.5
10 TH	0059	0.1	25 F	0531	0.5
	1225	4.8		1201	4.6
	1820	0.3		1747	0.6
11 F	0037	4.6	26 SA	0010	4.5
	0637	0.3		0602	0.6
	1305	4.6		1230	4.6
	1856	0.5		1819	0.6
12 SA	0119	4.5	27 SU	0042	4.4
	0710	0.6		0632	0.7
	1344	4.4		1303	4.5
	1928	0.7		1851	0.8
13 SU	0202	4.2	28 M	0120	4.3
	0742	0.9		0705	1.0
	1428	4.2		1342	4.3
	1959	1.1		1924	1.1
14 M	0249	3.9	29 TU	0207	4.1
	0815	1.4		0743	1.3
	1514	3.9		1430	4.1
	2039	1.5		2008	1.4
15 TU	0345	3.7	30 W	0306	3.9
	0903	1.9		0838	1.7
	1610	3.6		1534	3.9
	2136	2.0		2114	1.8

OCTOBER

Day	Time	M	Day	Time	M
1 TH	0430	3.7	16 F	0552	3.4
	1004	2.0		1107	2.4
	1700	3.7		1807	3.4
	2253	1.9		2358	2.2
2 F	0617	3.7	17 SA	0723	3.5
	1151	2.0		1241	2.2
	1840	3.8		1933	3.5
3 SA	0034	1.7	18 SU	0115	1.9
	0743	4.0		0821	3.7
	1317	1.7		1345	1.9
	1956	4.0		2028	3.7
4 SU	0147	1.3	19 M	0208	1.5
	0840	4.3		0900	4.0
	1419	1.2		1428	1.5
	2049	4.2		2107	4.0
5 M	0242	0.8	20 TU	0248	1.2
	0923	4.8		0934	4.2
	1508	0.8		1505	1.1
	2133	4.5		2140	4.2
6 TU	0329	0.5	21 W	0321	0.9
	1002	4.7		1003	4.3
	1551	0.5		1538	0.8
	2213	4.6		2212	4.3
7 W	0411	0.3	22 TH	0354	0.7
	1040	4.7		1033	4.5
○	1633	0.4	●	1612	0.7
	2252	4.7		2244	4.5
8 TH	0451	0.2	23 F	0429	0.6
	1118	4.8		1102	4.6
	1713	0.3		1648	0.6
	2332	4.7		2315	4.5
9 F	0531	0.3	24 SA	0505	0.6
	1154	4.7		1133	4.6
	1750	0.4		1725	0.5
				2349	4.5
10 SA	0010	4.6	25 SU	0541	0.6
	0606	0.5		1206	4.6
	1232	4.6		1800	0.6
	1824	0.5			
11 SU	0050	4.4	26 M	0026	4.4
	0638	0.6		0618	0.7
	1308	4.3		1242	4.5
	1854	0.8		1837	0.7
12 M	0130	4.2	27 TU	0108	4.3
	0707	1.1		0655	1.0
	1347	4.1		1326	4.3
	1923	1.2		1915	1.0
13 TU	0214	3.9	28 W	0159	4.1
	0739	1.5		0741	1.3
	1429	3.8		1418	4.1
	1959	1.6		2004	1.3
14 W	0306	3.7	29 TH	0305	4.0
	0822	2.0		0841	1.7
	1519	3.6		1524	3.9
	2050	1.9		2111	1.6
15 TH	0416	3.5	30 F	0430	3.9
	0929	2.3		1003	1.9
	1629	3.4		1651	3.8
	2212	2.2		2242	1.7
			31 SA	0603	3.9
				1137	1.8
				1823	3.8

NOVEMBER

Day	Time	M	Day	Time	M
1 SU	0012	1.5	16 M	0007	1.9
	0720	4.1		0718	3.7
	1254	1.5		1244	1.9
	1935	4.0		1930	3.6
2 M	0122	1.2	17 TU	0109	1.7
	0816	4.4		0807	3.9
	1354	1.2		1337	1.6
	2029	4.2		2021	3.9
3 TU	0215	0.9	18 W	0158	1.4
	0901	4.6		0847	4.1
	1443	0.8		1422	1.2
	2113	4.4		2102	4.1
4 W	0302	0.6	19 TH	0240	1.1
	0939	4.7		0924	4.3
	1525	0.6		1502	1.0
	2154	4.5		2141	4.2
5 TH	0345	0.5	20 F	0319	0.9
	1017	4.7		0958	4.4
○	1606	0.5		1542	0.8
	2233	4.6		2217	4.4
6 F	0425	0.5	21 SA	0400	0.8
	1053	4.7		1034	4.6
	1645	0.5	●	1623	0.6
	2311	4.5		2255	4.4
7 SA	0504	0.6	22 SU	0442	0.7
	1129	4.6		1109	4.6
	1722	0.5		1705	0.5
	2349	4.5		2335	4.5
8 SU	0539	0.8	23 M	0526	0.7
	1203	4.4		1148	4.5
	1755	0.7		1748	0.6
9 M	0028	4.3	24 TU	0018	4.4
	0611	1.0		0611	0.8
	1239	4.2		1231	4.5
	1825	0.9		1831	0.6
10 TU	0107	4.1	25 W	0106	4.4
	0641	1.2		0655	0.9
	1316	4.0		1318	4.3
	1856	1.2		1916	0.8
11 W	0151	3.9	26 TH	0201	4.3
	0715	1.5		0746	1.1
	1356	3.8		1413	4.2
	1931	1.5		2006	1.0
12 TH	0239	3.7	27 F	0304	4.2
	0756	1.8		0842	1.4
	1444	3.7		1517	4.0
	2018	1.8		2106	1.2
13 F	0339	3.6	28 SA	0416	4.1
	0853	2.1		0950	1.5
	1545	3.5		1632	3.9
	2123	2.0		2218	1.4
14 SA	0454	1.5	29 SU	0532	4.1
	1009	2.2		1104	1.6
	1703	3.4		1751	3.9
	2248	2.0		2334	1.4
15 SU	0613	3.5	30 M	0645	4.2
	1133	2.1		1215	1.5
	1825	3.5		1904	4.0

DECEMBER

Day	Time	M	Day	Time	M
1 TU	0044	1.3	16 W	0704	3.8
	0744	4.3		1236	1.7
	1318	1.3		1926	3.7
	2004	4.1			
2 W	0143	1.1	17 TH	0059	1.6
	0833	4.4		0757	4.0
	1412	1.0		1333	1.5
	2054	4.2		2021	3.9
3 TH	0234	1.0	18 F	0156	1.4
	0917	4.4		0844	4.2
	1459	0.9		1425	1.2
	2138	4.3		2110	4.1
4 F	0321	0.9	19 SA	0248	1.2
	0956	4.5		0927	4.3
	1542	0.8		1515	0.9
	2219	4.4		2156	4.2
5 SA	0404	0.9	20 SU	0337	1.0
	1034	4.4		1011	4.4
○	1623	0.7	●	1603	0.7
	2300	4.4		2241	4.4
6 SU	0443	0.9	21 M	0428	0.8
	1110	4.4		1054	4.5
	1700	0.7		1652	0.6
	2338	4.3		2327	4.4
7 M	0519	1.0	22 TU	0518	0.7
	1145	4.3		1138	4.5
	1736	0.8		1741	0.5
8 TU	0016	4.2	23 W	0015	4.5
	0552	1.1		0607	0.7
	1220	4.2		1225	4.5
	1807	0.9		1828	0.5
9 W	0054	4.1	24 TH	0104	4.5
	0625	1.2		0655	0.7
	1255	4.1		1312	4.4
	1839	1.1		1914	0.5
10 TH	0133	4.0	25 F	0154	4.5
	0657	1.4		0742	0.8
	1333	3.9		1405	4.3
	1913	1.2		2000	0.7
11 F	0215	3.9	26 SA	0248	4.4
	0735	1.6		0830	1.0
	1417	3.8		1500	4.2
	1952	1.4		2049	0.9
12 SA	0301	3.8	27 SU	0348	4.3
	0820	1.7		0923	1.2
	1505	3.7		1603	4.0
	2044	1.6		2143	1.1
13 SU	0357	3.7	28 M	0453	4.2
	0916	1.9		1021	1.4
	1605	3.6		1713	3.9
	2140	1.8		2249	1.3
14 M	0458	3.7	29 TU	0600	4.1
	1020	1.9		1128	1.5
	1711	3.6		1825	3.9
	2248	1.8		2358	1.5
15 TU	0603	3.7	30 W	0706	4.1
	1129	1.9		1236	1.5
	1822	3.6		1936	3.9
	2358	1.8			
			31 TH	0107	1.5
				0806	4.1
				1342	1.4
				2036	4.0

HIGH WATERS–IMPORTANT NOTE. DOUBLE HIGH WATERS OCCUR AT SOUTHAMPTON. THE PREDICTIONS ARE FOR THE FIRST HIGH WATER.

PORTSMOUTH

MEAN SPRING AND NEAP CURVES
Springs occur 2 days after New and Full Moon

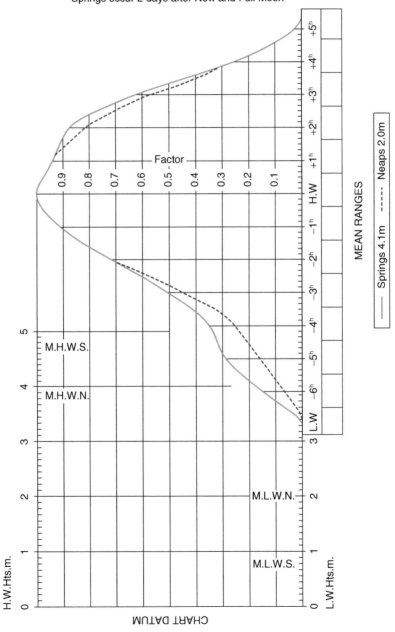

MEAN RANGES   —— Springs 4.1m   ----- Neaps 2.0m

TIME ZONE GMT   TIMES AND HEIGHTS OF HIGH AND LOW WATERS   YEAR 1987

SEPTEMBER

Day	TIME	M	Day	TIME	M	Day	TIME	M	Day	TIME	M
1 TU	0350	4.0	16 W	0523	3.6	2 W	0454	3.7	17 TH	0701	3.5
	0906	1.5		1049	2.3		1020	1.8		1229	2.3
	1621	4.0		1750	3.6		1734	3.8		1925	3.5
	2139	1.7		2334	2.2		2305	1.8			
3 TH	0628	3.7	18 F	0109	2.1	4 F	0044	1.7	19 SA	0215	1.8
	1158	1.9		0827	3.6		0804	3.8		0925	3.9
	1905	3.8		1350	2.1		1329	1.7		1444	1.7
				2043	3.7		2028	4.0		2135	3.9
5 SA	0203	1.4	20 SU	0300	1.5	6 SU	0303	1.0	21 M	0333	1.2
	0916	4.2		1006	4.1		1011	4.5		1037	4.4
	1436	1.4		1521	1.5		1529	1.0		1552	1.2
	2132	4.3		2215	4.1		2225	4.5		2245	4.3
7 M O	0353	0.7	22 TU	0403	1.0	8 TU	0438	0.4	23 W •	0433	0.9
	1100	4.7		1105	4.5		1145	4.9		1132	4.6
	1616	0.8		1621	1.1		1659	0.7		1650	1.0
	2313	4.7		2313	4.4		2356	4.8		2341	4.5
9 W	0521	0.5	24 TH	0505	0.8	10 TH	0037	4.8	25 F	0014	4.5
	1228	5.0		1203	4.6		0602	0.5		0538	0.7
	1742	0.7		1722	0.9		1309	5.0		1237	4.6
							1821	0.7		1755	0.8
11 F	0118	4.8	26 SA	0049	4.6	12 SA	0158	4.7	27 SU	0127	4.5
	0640	0.6		0612	0.7		0716	0.9		0644	0.8
	1348	4.9		1313	4.6		1425	4.7		1350	4.5
	1858	0.8		1828	0.8		1936	1.0		1902	0.8
13 SU	0238	4.5	28 M	0207	4.4	14 M	0321	4.2	29 TU	0249	4.2
	0753	1.2		0718	1.0		0835	1.6		0757	1.3
	1501	4.5		1428	4.4		1541	4.2		1510	4.2
	2015	1.3		1938	1.0		2102	1.7		2024	1.3
15 TU	0412	3.9	30 W	0339	4.0						
	0930	2.0		0851	1.6						
	1634	3.9		1605	3.9						
	2206	2.0		2126	1.7						

OCTOBER

Day	TIME	M	Day	TIME	M	Day	TIME	M	Day	TIME	M
1 TH	0452	3.7	16 F	0612	3.6	2 F	0632	3.7	17 SA	0018	2.2
	1012	1.9		1142	2.3		1157	2.0		0738	3.7
	1726	3.7		1836	3.5		1902	3.7		1307	2.1
	2259	1.8								1958	3.6
3 SA	0037	1.7	18 SU	0130	1.9	4 SU	0150	1.3	19 M	0217	1.6
	0800	4.0		0839	3.9		0902	4.3		0922	4.1
	1321	1.7		1404	1.8		1423	1.3		1443	1.5
	2020	4.0		2055	3.8		2117	4.3		2136	4.1
5 M	0245	1.0	20 TU	0253	1.3	6 TU	0331	0.7	21 W	0324	1.1
	0952	4.6		0956	4.4		1037	4.8		1027	4.6
	1512	1.0		1514	1.3		1555	0.8		1544	1.1
	2205	4.5		2207	4.3		2249	4.7		2238	4.5
7 W O	0415	0.5	22 TH	0357	0.9	8 TH	0456	0.5	23 F	0433	0.8
	1119	4.9		1058	4.7		1159	4.9		1130	4.8
	1636	0.7		1615	1.0		1715	0.7		1650	0.9
	2331	4.8		2310	4.6					2346	4.6
9 F	0012	4.8	24 SA	0508	0.8	10 SA	0053	4.7	25 SU	0025	4.6
	0535	0.8		1206	4.7		0610	0.8		0545	0.8
	1238	4.8		1726	0.8		1314	4.6		1245	4.7
	1752	0.8					1827	0.8		1804	0.8
11 SU	0133	4.6	26 M	0107	4.5	12 M	0211	4.5	27 TU	0153	4.4
	0643	1.1		0622	0.9		0717	1.4		0702	1.1
	1348	4.6		1325	4.6		1422	4.4		1408	4.4
	1901	1.0		1844	0.9		1939	1.3		1927	1.1
13 TU	0251	4.2	28 W	0242	4.2	14 W	0338	3.9	29 TH	0341	4.0
	0756	1.7		0748	1.4		0849	2.1		0849	1.7
	1500	4.2		1458	4.2		1549	3.9		1600	3.9
	2023	1.7		2018	1.3		2121	2.0		2125	1.6
15 TH	0444	3.7	30 F	0457	3.9				31 SA	0626	3.9
	1005	2.3		1011	1.9					1143	1.9
	1702	3.6		1721	3.8					1850	3.8
	2244	2.2		2252	1.7						

NOVEMBER

Day	TIME	M	Day	TIME	M	Day	TIME	M	Day	TIME	M
1 SU	0018	1.6	16 M	0020	1.9	2 M	0125	1.3	17 TU	0117	1.7
	0742	4.1		0733	3.9		0840	4.4		0823	4.1
	1301	1.6		1301	1.9		1359	1.3		1348	1.6
	2001	4.0		1955	3.7		2056	4.3		2043	4.0
3 TU	0217	1.0	18 W	0201	1.4	4 W	0303	0.8	19 TH	0241	1.2
	0926	4.6		0905	4.3		1009	4.8		0944	4.5
	1445	1.1		1427	1.4		1527	0.9		1504	1.2
	2141	4.5		2123	4.2		2224	4.7		2202	4.5
5 TH O	0347	0.8	20 F	0320	1.1	6 F	0428	0.8	21 SA •	0400	1.0
	1050	4.8		1021	4.7		1129	4.8		1100	4.8
	1608	0.8		1541	1.0		1647	0.8		1620	0.9
	2307	4.7		2240	4.6		2349	4.7		2323	4.6
7 SA	0506	0.9	22 SU	0441	0.9	8 SU	0031	4.6	23 M	0007	4.6
	1208	4.8		1141	4.8		0541	1.0		0523	0.9
	1723	0.8		1703	0.8		1244	4.6		1224	4.7
							1759	0.9		1746	0.8
9 M	0111	4.5	24 TU	0055	4.6	10 TU	0149	4.4	25 W	0145	4.5
	0614	1.2		0607	1.0		0648	1.5		0653	1.2
	1318	4.5		1309	4.6		1352	4.3		1357	4.4
	1833	1.1		1832	0.9		1909	1.3		1921	1.0
11 W	0229	4.2	26 TH	0240	4.4	12 TH	0312	4.0	27 F	0340	4.3
	0727	1.7		0745	1.4		0817	2.0		0846	1.6
	1430	4.1		1451	4.3		1516	3.9		1552	4.1
	1951	1.6		2015	1.2		2043	1.9		2118	1.3
13 F	0406	3.9	28 SA	0448	4.2	14 SA	0515	3.7	29 SU	0601	4.2
	0920	2.2		0956	1.7		1038	2.2		1113	1.7
	1617	3.7		1703	3.9		1732	3.5		1821	3.9
	2149	2.0		2230	1.5		2306	2.1		2345	1.5
15 SU	0629	3.7	30 M	0709	4.2						
	1158	2.1		1226	1.6						
	1851	3.6		1931	4.0						

DECEMBER

Day	TIME	M	Day	TIME	M	Day	TIME	M	Day	TIME	M
1 TU	0052	1.4	16 W	0005	1.8	2 W	0148	1.3	17 TH	0104	1.6
	0808	4.3		0715	4.0		0857	4.5		0810	4.2
	1326	1.4		1243	1.7		1416	1.2		1335	1.5
	2030	4.2		1943	3.9		2120	4.3		2038	4.1
3 TH	0236	1.2	18 F	0157	1.4	4 F	0321	1.1	19 SA	0246	1.2
	0941	4.6		0901	4.4		1023	4.6		0950	4.5
	1502	1.1		1425	1.3		1545	1.0		1513	1.1
	2204	4.5		2129	4.3		2248	4.5		2219	4.5
5 SA O	0402	1.0	20 SU •	0335	1.1	6 SU	0443	1.1	21 M	0422	1.0
	1104	4.6		1038	4.7		1143	4.6		1126	4.7
	1625	0.9		1600	0.9		1703	0.9		1648	0.8
	2332	4.5		2308	4.6					2358	4.6
7 M	0014	4.5	22 TU	0510	0.9	8 TU	0053	4.4	23 W	0047	4.6
	0519	1.2		1212	4.7		0554	1.3		0557	1.0
	1220	4.5		1736	0.7		1256	4.4		1259	4.6
	1738	1.0					1813	1.1		1823	0.7
9 W	0132	4.3	24 TH	0138	4.6	10 TH	0209	4.2	25 F	0230	4.6
	0629	1.4		0645	1.0		0707	1.6		0735	1.2
	1331	4.2		1347	4.5		1408	4.1		1438	4.4
	1848	1.2		1912	0.9		1926	1.4		2003	0.9
11 F	0248	4.0	26 SA	0325	4.6	12 SA	0331	3.9	27 SU	0423	4.5
	0749	1.7		0829	1.3		0837	1.9		0927	1.5
	1449	4.0		1533	4.3		1635	3.8		1633	4.1
	2008	1.6		2057	1.0		2058	1.7		2158	1.2
13 SU	0419	4.0	28 M	0523	4.3	14 M	0516	3.9	29 TU	0627	4.2
	0933	2.0		1032	1.5		1037	2.0		1142	1.6
	1630	3.7		1741	4.0		1733	3.6		1853	3.9
	2155	1.8		2304	1.4		2300	1.8			
15 TU	0616	3.9	30 W	0013	1.5				31 TH	0118	1.6
	1141	1.9		0729	4.2					0826	4.2
	1840	3.7		1250	1.5					1360	1.4
				2003	4.1					2102	4.1

ABERDEEN

MEAN SPRING AND NEAP CURVES
Springs occur 2 days after New and Full Moon

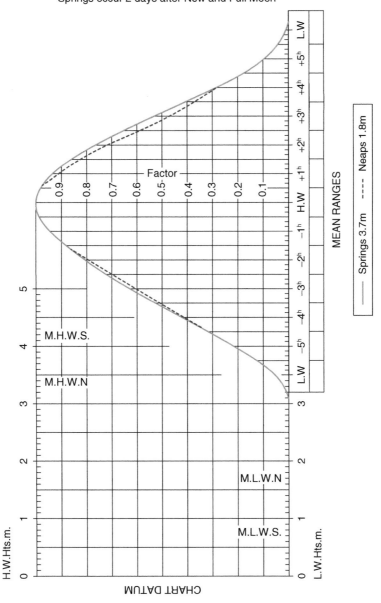

MEAN RANGES

——— Springs 3.7m   - - - - Neaps 1.8m

TIME ZONE GMT   TIMES AND HEIGHTS OF HIGH AND LOW WATERS   YEAR 1987

MAY

Day	TIME	M	Day	TIME	M
1 F	0257	3.9	16 SA	0246	4.1
	0857	0.8		0854	0.7
	1514	3.8		1519	4.1
	2110	1.1		2118	0.9
2 SA	0328	3.8	17 SU	0332	4.0
	0934	1.0		0945	0.8
	1555	3.6		1617	4.0
	2142	1.4		2209	1.2
3 SU	0403	3.6	18 M	0424	3.9
	1014	1.2		1042	0.9
	1641	3.4		1722	3.8
	2219	1.6		2306	1.5
4 M	0442	3.5	19 TU	0525	3.7
	1059	1.3		1150	1.0
	1734	3.2		1832	3.6
	2305	1.9			
5 TU	0532	3.3	20 W	0017	1.7
	1158	1.5		0632	3.6
	1839	3.1		1309	1.1
				1947	3.5
6 W	0008	2.0	21 TH	0137	1.7
	0635	3.2		0747	3.6
	1317	1.6		1427	1.0
	1955	3.1		2057	3.5
7 TH	0144	2.1	22 F	0251	1.6
	0752	3.2		0858	3.6
	1443	1.5		1534	0.9
	2110	3.2		2159	3.5
8 F	0312	1.9	23 SA	0353	1.4
	0907	3.2		1002	3.7
	1548	1.3		1630	0.8
	2209	3.3		2251	3.6
9 SA	0409	1.7	24 SU	0444	1.2
	1007	3.4		1057	3.8
	1635	1.1		1718	0.8
	2255	3.5		2336	3.7
10 SU	0452	1.4	25 M	0529	1.0
	1054	3.6		1143	3.9
	1716	0.9		1758	0.7
	2334	3.7			
11 M	0530	1.2	26 TU	0017	3.8
	1136	3.8		0611	0.9
	1754	0.7		1225	3.9
				1836	0.8
12 TU	0011	3.8	27 W ●	0053	3.9
	0607	1.0		0650	0.8
	1215	4.0		1306	3.9
	1831	0.6		1912	0.8
13 W ○	0048	4.0	28 TH	0128	3.9
	0646	0.8		0730	0.8
	1255	4.2		1344	3.8
	1910	0.5		1944	1.0
14 TH	0124	4.1	29 F	0202	3.9
	0726	0.7		0806	0.9
	1338	4.2		1423	3.8
	1949	0.6		2018	1.1
15 F	0204	4.1	30 SA	0234	3.9
	0809	0.7		0843	0.9
	1426	4.2		1501	3.7
	2032	0.7		2050	1.3
			31 SU	0307	3.8
				0921	1.0
				1543	3.6
				2125	1.4

JUNE

Day	TIME	M	Day	TIME	M
1 M	0341	3.7	16 TU	0416	4.1
	1000	1.1		1040	0.6
	1627	3.5		1713	3.9
	2203	1.6		2255	1.3
2 TU	0420	3.6	17 W	0512	4.0
	1044	1.2		1139	0.7
	1715	3.4		1812	3.8
	2247	1.7		2354	1.4
3 W	0505	3.5	18 TH	0611	3.9
	1133	1.3		1243	0.9
	1805	3.3		1913	3.6
	2340	1.8			
4 TH	0600	3.4	19 F	0057	1.5
	1232	1.4		0716	3.7
	1903	3.2		1351	1.0
				2016	3.4
5 F	0043	1.9	20 SA	0204	1.5
	0700	3.4		0823	3.6
	1338	1.4		1456	1.1
	2005	3.2		2118	3.4
6 SA	0154	1.8	21 SU	0310	1.4
	0806	3.4		0928	3.6
	1443	1.3		1556	1.1
	2105	3.3		2216	3.4
7 SU	0300	1.7	22 M	0410	1.3
	0908	3.4		1028	3.6
	1539	1.2		1648	1.1
	2200	3.4		2306	3.5
8 M	0357	1.5	23 TU	0505	1.2
	1004	3.6		1122	3.6
	1630	1.0		1734	1.1
	2249	3.6		2351	3.6
9 TU	0447	1.3	24 W	0554	1.1
	1057	3.8		1210	3.6
	1716	0.9		1815	1.1
	2334	3.8			
10 W	0536	1.1	25 TH	0032	3.7
	1146	3.9		0638	1.0
	1803	0.8		1253	3.6
				1853	1.1
11 TH ○	0018	3.9	26 F ●	0109	3.8
	0622	0.9		0719	1.0
	1236	4.1		1334	3.7
	1848	0.7		1928	1.2
12 F	0102	4.1	27 SA	0142	3.9
	0710	0.8		0755	0.9
	1328	4.2		1412	3.8
	1934	0.8		2002	1.2
13 SA	0147	4.2	28 SU	0215	3.9
	0759	0.6		0830	0.9
	1422	4.2		1450	3.8
	2022	0.8		2036	1.3
14 SU	0233	4.2	29 M	0247	3.9
	0851	0.6		0905	0.9
	1518	4.2		1528	3.8
	2111	1.0		2111	1.4
15 M	0322	4.2	30 TU	0321	3.9
	0943	0.6		0942	0.9
	1616	4.1		1606	3.7
	2202	1.2		2146	1.4

JULY

Day	TIME	M	Day	TIME	M
1 W	0357	3.8	16 TH	0447	4.2
	1021	1.0		1112	0.6
	1647	3.6		1739	3.8
	2226	1.5		2319	1.2
2 TH	0438	3.8	17 F	0540	4.0
	1102	1.1		1205	0.9
	1729	3.5		1831	3.6
	2306	1.5			
3 F	0522	3.7	18 SA	0011	1.3
	1149	1.2		0636	3.8
	1814	3.4		1303	1.1
	2354	1.6		1927	3.4
4 SA	0612	3.6	19 SU	0112	1.5
	1241	1.3		0740	3.6
	1906	3.3		1408	1.3
				2030	3.3
5 SU	0050	1.7	20 M	0223	1.6
	0710	3.5		0850	3.4
	1338	1.3		1517	1.5
	2005	3.3		2136	3.3
6 M	0155	1.7	21 TU	0341	1.5
	0813	3.5		1002	3.3
	1442	1.3		1620	1.5
	2107	3.4		2237	3.4
7 TU	0304	1.6	22 W	0448	1.4
	0919	3.5		1106	3.4
	1545	1.2		1715	1.5
	2207	3.5		2329	3.5
8 W	0410	1.4	23 TH	0544	1.3
	1026	3.7		1158	3.5
	1645	1.1		1800	1.4
	2304	3.7			
9 TH	0512	1.2	24 F	0012	3.7
	1127	3.8		0628	1.1
	1740	1.0		1243	3.6
	2354	3.9		1839	1.4
10 F	0608	0.9	25 SA ●	0049	3.8
	1227	4.0		0706	1.0
	1834	0.9		1321	3.7
				1914	1.3
11 SA ○	0043	4.1	26 SU	0123	3.9
	0702	0.7		0741	0.9
	1323	4.2		1358	3.9
	1924	0.9		1947	1.2
12 SU	0131	4.3	27 M	0154	4.0
	0754	0.5		0813	0.8
	1416	4.3		1432	3.9
	2012	0.9		2019	1.2
13 M	0220	4.4	28 TU	0225	4.0
	0843	0.3		0846	0.7
	1508	4.3		1505	3.9
	2058	0.9		2050	1.2
14 TU	0308	4.4	29 W	0257	4.1
	0932	0.3		0918	0.7
	1559	4.2		1538	3.9
	2145	1.0		2122	1.2
15 W	0357	4.4	30 TH	0331	4.0
	1021	0.4		0952	0.8
	1649	4.0		1613	3.8
	2231	1.1		2156	1.2
			31 F	0406	4.0
				1028	0.9
				1648	3.7
				2231	1.3

AUGUST

Day	TIME	M	Day	TIME	M
1 SA	0444	3.9	16 SU	0553	3.8
	1106	1.0		1208	1.3
	1729	3.6		1835	3.3
	2312	1.4			
2 SU	0529	3.8	17 M	0021	1.5
	1151	1.2		0653	3.4
	1817	3.4		1307	1.6
				1938	3.2
3 M	0001	1.5	18 TU	0135	1.7
	0622	3.6		0809	3.2
	1245	1.3		1427	1.8
	1914	3.4		2054	3.2
4 TU	0104	1.7	19 W	0317	1.7
	0731	3.5		0938	3.1
	1352	1.5		1556	1.9
	2023	3.4		2207	3.3
5 W	0225	1.7	20 TH	0438	1.6
	0851	3.5		1054	3.3
	1510	1.5		1659	1.8
	2135	3.5		2305	3.5
6 TH	0349	1.5	21 F	0532	1.4
	1012	3.6		1146	3.5
	1626	1.4		1746	1.6
	2241	3.7			
7 F	0502	1.2	22 SA	0612	1.1
	1122	3.8		1227	3.7
	1730	1.2		1822	1.4
	2337	4.0			
8 SA	0601	0.9	23 SU	0027	3.8
	1222	4.1		0646	0.9
	1824	1.0		1302	3.8
				1855	1.3
9 SU ○	0029	4.2	24 M ●	0059	4.0
	0653	0.5		0719	0.8
	1314	4.3		1334	4.0
	1912	0.9		1924	1.2
10 M	0116	4.4	25 TU	0130	4.1
	0741	0.3		0748	0.7
	1404	4.4		1405	4.0
	1955	0.8		1954	1.1
11 TU	0202	4.6	26 W	0159	4.2
	0826	0.1		0819	0.6
	1449	4.4		1434	4.0
	2037	0.7		2023	1.0
12 W	0246	4.6	27 TH	0207	4.2
	0911	0.2		0850	0.6
	1532	4.3		1505	4.0
	2118	0.8		2053	1.0
13 TH	0331	4.5	28 F	0301	4.2
	0953	0.3		0921	0.7
	1616	4.1		1536	3.9
	2157	0.9		2125	1.1
14 F	0416	4.4	29 SA	0334	4.2
	1037	0.6		0953	0.8
	1658	3.8		1610	3.8
	2240	1.0		2159	1.2
15 SA	0502	4.1	30 SU	0410	4.0
	1120	1.0		1030	1.0
	1744	3.6		1648	3.7
	2325	1.3		2238	1.3
			31 M	0454	3.8
				1112	1.2
				1736	3.6
				2327	1.5

LERWICK

MEAN SPRING AND NEAP CURVES
Springs occurs 1 day after New and Full Moon

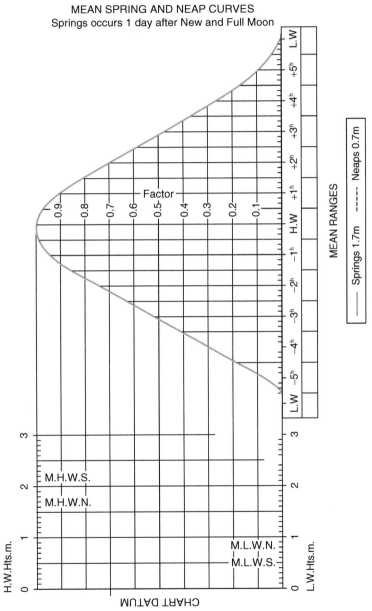

SHETLAND ISLANDS – LERWICK
LAT 60°09'N   LONG 1°08'W

TIME ZONE GMT   TIMES AND HEIGHTS OF HIGH AND LOW WATERS   YEAR 1987

(Times in GMT; heights (M) in metres. ○ = full moon, ● = new moon.)

SEPTEMBER

Day	Time	M	Time	M	Time	M	Time	M
1 TU	0327	1.8	0937	0.9	1557	1.7	2203	1.0
2 W	0439	1.7	1054	1.1	1711	1.7	2345	1.0
3 TH	0623	1.7	1237	1.1	1846	1.8		
4 F	0125	0.9	0755	1.8	1356	1.0	2001	1.9
5 SA	0231	0.7	0856	1.9	1452	0.9	2058	2.1
6 SU	0324	0.5	0945	2.1	1539	0.7	2146	2.2
7 M ○	0411	0.3	1029	2.1	1622	0.6	2231	2.4
8 TU	0454	0.2	1110	2.2	1703	0.5	2314	2.4
9 W	0536	0.2	1150	2.2	1742	0.5	2356	2.4
10 TH	0616	0.3	1228	2.1	1822	0.5		
11 F	0038	2.4	0655	0.4	1306	2.1	1901	0.5
12 SA	0119	2.2	0733	0.6	1344	2.0	1941	0.7
13 SU	0201	2.0	0810	0.8	1422	1.9	2025	0.8
14 M	0245	1.9	0849	1.0	1505	1.8	2118	0.9
15 TU	0338	1.7	0937	1.1	1559	1.7	2246	1.1
16 W	0500	1.6	1129	1.2	1730	1.6		
17 TH	0049	1.0	0710	1.6	1320	1.2	1914	1.7
18 F	0157	1.0	0818	1.6	1416	1.1	2013	1.8
19 SA	0243	0.8	0859	1.7	1455	1.0	2055	1.9
20 SU	0319	0.6	0932	1.8	1527	0.9	2131	2.0
21 M	0350	0.6	1003	1.9	1556	0.8	2204	2.1
22 TU	0419	0.6	1032	2.0	1625	0.7	2236	2.2
23 W ●	0447	0.5	1102	2.1	1654	0.7	2309	2.2
24 TH	0517	0.5	1132	2.1	1724	0.7	2342	2.2
25 F	0547	0.5	1204	2.1	1756	0.7		
26 SA	0016	2.2	0620	0.6	1238	2.1	1829	0.7
27 SU	0053	2.2	0655	0.7	1313	2.1	1906	0.8
28 M	0133	2.1	0733	0.8	1353	2.0	1949	0.8
29 TU	0220	2.0	0819	0.9	1438	1.9	2043	0.9
30 W	0319	1.8	0918	1.1	1535	1.8	2200	1.0

OCTOBER

Day	Time	M	Time	M	Time	M	Time	M
1 TH	0440	1.7	1046	1.2	1652	1.8	2348	0.9
2 F	0628	1.7	1232	1.1	1827	1.8		
3 SA	0118	0.8	0749	1.8	1344	1.0	1943	1.9
4 SU	0219	0.6	0843	2.0	1436	0.9	2039	2.1
5 M	0308	0.4	0928	2.1	1520	0.7	2127	2.2
6 TU	0351	0.3	1008	2.1	1601	0.6	2211	2.4
7 W ○	0432	0.3	1045	2.2	1640	0.5	2253	2.4
8 TH	0510	0.3	1122	2.2	1719	0.5	2334	2.4
9 F	0547	0.4	1158	2.2	1757	0.5		
10 SA	0015	2.3	0623	0.6	1234	2.1	1835	0.6
11 SU	0054	2.2	0656	0.7	1310	2.1	1915	0.7
12 M	0135	2.0	0729	0.9	1348	2.0	1958	0.8
13 TU	0218	1.8	0803	1.1	1429	1.9	2050	1.0
14 W	0310	1.7	0845	1.2	1519	1.8	2208	1.1
15 TH	0424	1.6	1003	1.3	1631	1.7		
16 F	0011	1.0	0635	1.6	1242	1.3	1819	1.7
17 SA	0121	1.0	0744	1.6	1342	1.2	1930	1.8
18 SU	0205	0.9	0825	1.7	1421	1.1	2017	1.9
19 M	0240	0.7	0857	1.9	1453	1.0	2055	2.0
20 TU	0311	0.7	0927	2.0	1523	0.8	2130	2.1
21 W	0341	0.6	0958	2.1	1553	0.8	2205	2.2
22 TH ●	0412	0.5	1029	2.1	1624	0.7	2240	2.2
23 F	0444	0.5	1102	2.2	1658	0.6	2317	2.3
24 SA	0518	0.6	1136	2.2	1734	0.6	2355	2.2
25 SU	0555	0.6	1213	2.2	1812	0.7		
26 M	0037	2.2	0634	0.7	1252	2.2	1855	0.7
27 TU	0123	2.1	0718	0.9	1335	2.1	1944	0.8
28 W	0216	2.0	0809	1.0	1424	2.0	2045	0.8
29 TH	0319	1.9	0913	1.1	1523	1.9	2204	0.8
30 F	0441	1.8	1038	1.2	1637	1.8	2341	0.8
31 SA	0616	1.8	1213	1.1	1803	1.9		

NOVEMBER

Day	Time	M	Time	M	Time	M	Time	M
1 SU	0100	0.7	0729	1.8	1321	1.0	1918	2.0
2 M	0159	0.6	0821	1.9	1413	0.9	2016	2.1
3 TU	0246	0.5	0904	2.0	1458	0.7	2106	2.2
4 W	0329	0.4	0943	2.1	1539	0.6	2151	2.2
5 TH ○	0408	0.3	1020	2.1	1619	0.6	2234	2.3
6 F	0445	0.5	1056	2.2	1659	0.5	2315	2.2
7 SA	0520	0.6	1132	2.2	1738	0.6	2355	2.2
8 SU	0554	0.7	1155	2.3	1803	0.6		
9 M	0035	2.1	0626	0.9	1245	2.1	1858	0.7
10 TU	0115	0.9	0659	1.0	1322	1.0	1941	0.8
11 W	0158	1.8	0734	1.1	1403	1.9	2030	0.9
12 TH	0247	1.7	0817	1.2	1449	1.8	2131	1.0
13 F	0347	1.6	0917	1.2	1546	1.8	2254	1.0
14 SA	0510	1.6	1053	1.3	1657	1.7		
15 SU	0016	0.9	0639	1.6	1236	1.2	1819	1.7
16 M	0109	0.9	0732	1.7	1328	1.1	1922	1.8
17 TU	0150	0.8	0811	1.8	1408	1.0	2010	1.9
18 W	0226	0.7	0847	1.9	1444	0.9	2053	2.0
19 TH	0302	0.7	0922	2.0	1520	0.8	2134	2.1
20 F	0338	0.6	0958	2.1	1558	0.7	2215	2.2
21 SA ●	0416	0.6	1035	2.2	1637	0.7	2257	2.2
22 SU	0456	0.7	1114	2.3	1719	0.6	2314	2.2
23 M	0538	0.7	1155	2.3	1803	0.6		
24 TU	0028	2.2	0622	0.8	1238	2.2	1852	0.6
25 W	0118	2.1	0710	0.9	1324	2.2	1945	0.6
26 TH	0212	2.0	0803	0.9	1414	2.1	2045	0.6
27 F	0313	1.9	0903	1.0	1510	2.0	2156	0.7
28 SA	0424	1.8	1015	1.1	1615	1.9	2316	0.7
29 SU	0543	1.7	1136	1.0	1731	1.9		
30 M	0031	0.6	0655	1.8	1247	1.0	1847	1.9

DECEMBER

Day	Time	M	Time	M	Time	M	Time	M
1 TU	0132	0.6	0751	1.8	1346	0.9	1952	2.0
2 W	0223	0.6	0837	1.9	1436	0.8	2046	2.0
3 TH	0307	0.6	0919	2.0	1521	0.7	2135	2.1
4 F	0346	0.7	0958	2.1	1604	0.6	2219	2.1
5 SA ○	0424	0.7	1035	2.1	1646	0.6	2300	2.1
6 SU	0459	0.8	1113	2.2	1726	0.6	2341	2.0
7 M	0533	0.8	1149	2.2	1806	0.6		
8 TU	0020	2.0	0607	0.9	1226	2.1	1845	0.7
9 W	0059	1.9	0641	0.9	1304	2.1	1925	0.7
10 TH	0140	1.8	0717	1.0	1342	2.0	2007	0.8
11 F	0222	1.7	0756	1.0	1423	1.9	2052	0.8
12 SA	0308	1.7	0841	1.1	1507	1.8	2143	0.9
13 SU	0401	1.6	0935	1.1	1557	1.8	2240	0.9
14 M	0500	1.6	1042	1.1	1657	1.7	2343	0.9
15 TU	0608	1.6	1200	1.1	1807	1.7		
16 W	0043	0.9	0710	1.7	1309	1.1	1917	1.8
17 TH	0136	0.8	0801	1.8	1404	1.0	2016	1.9
18 F	0224	0.8	0848	2.0	1452	0.8	2108	2.0
19 SA	0310	0.7	0931	2.1	1538	0.7	2157	2.1
20 SU ●	0356	0.7	1015	2.2	1624	0.6	2245	2.2
21 M	0441	0.7	1058	2.3	1710	0.5	2333	2.2
22 TU	0527	0.7	1142	2.3	1758	0.4		
23 W	0021	2.2	0614	0.7	1227	2.3	1847	0.4
24 TH	0110	2.1	0701	0.7	1313	2.3	1938	0.4
25 F	0201	2.0	0750	0.8	1400	2.2	2033	0.4
26 SA	0254	1.9	0842	0.9	1452	2.1	2133	0.5
27 SU	0352	1.8	0941	0.9	1548	2.0	2240	0.6
28 M	0457	1.7	1049	1.0	1655	1.9	2353	0.7
29 TU	0608	1.7	1207	0.9	1813	1.8		
30 W	0101	0.7	0715	1.7	1318	0.9	1929	1.8
31 TH	0159	0.8	0811	1.8	1419	0.8	2032	1.8

ULLAPOOL
MEAN SPRING AND NEAP CURVES
Spring occurs 1 day after New and Full Moon.

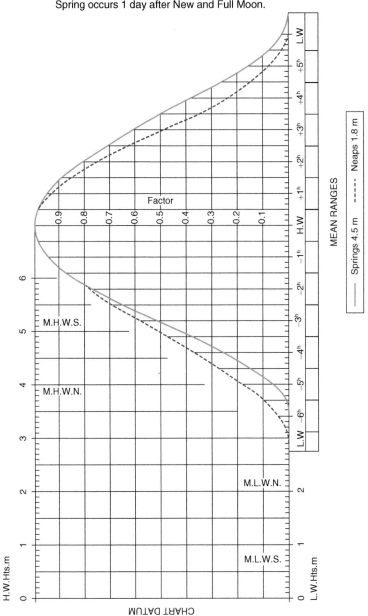

MEAN RANGES

——— Springs 4.5 m   - - - - - Neaps 1.8 m

TIME ZONE GMT   TIMES AND HEIGHTS OF HIGH AND LOW WATERS   YEAR 1987

MAY

Day	TIME	M		Day	TIME	M
1 F	0245	1.1		16 SA	0239	0.7
	0834	4.6			0835	4.8
	1453	1.3			1500	0.9
	2042	4.6			2056	4.9
2 SA	0322	1.3		17 SU	0326	0.8
	0912	4.3			0931	4.5
	1526	1.5			1546	1.2
	2116	4.3			2154	4.6
3 SU	0401	1.6		18 M	0418	1.0
	0957	4.0			1041	4.2
	1602	1.8			1638	1.5
	2200	4.1			2305	4.3
4 M	0445	1.8		19 TU	0518	1.3
	1058	3.7			1204	3.9
	1643	2.1			1744	1.8
	2306	3.8				
5 TU	0542	2.1		20 W	0024	4.2
	1219	3.5			0634	1.4
	1739	2.3			1326	3.9
					1909	1.9
6 W	0035	3.7		21 TH	0137	4.2
	0710	2.2			0801	1.4
	1341	3.5			1437	3.9
	1912	2.5			2033	1.8
7 TH	0156	3.7		22 F	0243	4.2
	0847	2.1			0914	1.3
	1450	3.7			1538	4.1
	2053	2.4			2138	1.6
8 F	0300	3.8		23 SA	0341	4.4
	0947	1.9			1010	1.2
	1545	3.9			1628	4.3
	2153	2.1			2229	1.4
9 SA	0351	4.1		24 SU	0432	4.5
	1032	1.6			1056	1.1
	1629	4.2			1711	4.5
	2237	1.8			2315	1.3
10 SU	0434	4.3		25 M	0517	4.6
	1110	1.3			1137	1.0
	1706	4.5			1748	4.6
	2317	1.5			2357	1.2
11 M	0512	4.6		26 TU	0557	4.7
	1147	1.1			1215	1.0
	1741	4.8			1821	4.7
	2355	1.3				
12 TU	0549	4.9		27 W	0036	1.1
	1223	0.9			0634	4.7
	1815	5.0 •			1251	1.0
					1853	4.8
13 W O	0034	1.0		28 TH	0115	1.1
	0627	5.0			0709	4.7
	1300	0.7			1325	1.1
	1851	5.2			1923	4.7
14 TH	0114	0.8		29 F	0153	1.1
	0706	5.1			0744	4.6
	1338	0.7			1359	1.2
	1928	5.2			1955	4.7
15 F	0155	0.7		30 SA.	0230	1.2
	0748	5.0			0821	4.4
	1418	0.8			1433	1.4
	2009	5.1			2028	4.6
				31 SU	0308	1.3
					0900	4.3
					1508	1.5
					2106	4.4

JUNE

Day	TIME	M		Day	TIME	M
1 M	0348	1.5		16 TU	0414	0.7
	0944	4.1			1036	4.3
	1545	1.7			1632	1.2
	2148	4.2			2253	4.6
2 TU	0431	1.6		17 W	0510	0.9
	1035	3.9			1144	4.2
	1626	1.9			1730	1.5
	2239	4.1			2357	4.5
3 W	0519	1.8		18 TH	0610	1.1
	1135	3.7			1252	4.0
	1715	2.1			1835	1.6
	2339	3.9				
4 TH	0618	1.9		19 F	0102	4.3
	1241	3.7			0717	1.3
	1818	2.2			1358	4.0
					1945	1.7
5 F	0043	3.9		20 SA	0205	4.2
	0729	1.9			0825	1.5
	1346	3.8			1501	4.0
	1935	2.2			2053	1.8
6 SA	0147	3.9		21 SU	0307	4.2
	0837	1.9			0927	1.5
	1447	3.9			1557	4.1
	2047	2.1			2154	1.7
7 SU	0249	4.0		22 M	0405	4.2
	0934	1.7			1022	1.5
	1540	4.1			1647	4.2
	2145	1.9			2249	1.7
8 M	0345	4.2		23 TU	0457	4.3
	1023	1.5			1110	1.5
	1627	4.4			1730	4.3
	2236	1.7			2338	1.6
9 TU	0436	4.4		24 W	0542	4.3
	1109	1.3			1153	1.5
	1711	4.6			1807	4.3
	2324	1.4				
10 W	0524	4.7		25 TH	0023	1.4
	1154	1.1			0622	4.4
	1752	4.9			1232	1.4
					1840	4.6
11 TH	0011	1.1		26 F	0104	1.3
	0610	4.8			0659	4.5
	1237	0.9 •			1309	1.4 O
	1834	5.0			1913	4.7
12 F	0058	0.9		27 SA	0143	1.2
	0656	4.9			0734	4.5
	1322	0.8			1344	1.4
	1917	5.1			1945	4.7
13 SA	0145	0.7		28 SU	0220	1.2
	0745	4.9			0809	4.4
	1406	0.8			1419	1.4
	2003	5.1			2017	4.7
14 SU	0233	0.6		29 M	0257	1.2
	0836	4.8			0845	4.4
	1452	0.9			1453	1.4
	2054	5.0			2051	4.6
15 M	0322	0.6		30 TU	0333	1.3
	0933	4.6			0923	4.3
	1540	1.0			1529	1.5
	2151	4.8			2127	4.5

JULY

Day	TIME	M		Day	TIME	M
1 W	0409	1.4		16 TH	0445	0.7
	1003	4.2			1105	4.4
	1606	1.6			1702	1.2
	2206	4.4			2319	4.7
2 TH	0448	1.5		17 F	0534	1.0
	1048	4.1			1204	4.2
	1648	1.8			1754	1.5
	2250	4.3				
3 F	0531	1.7		18 SA	0017	4.4
	1140	4.0			0627	1.4
	1736	1.9			1309	4.0
					1853	1.8
4 SA	0621	1.8		19 SU	0122	4.1
	1240	3.9			0727	1.7
	1833	2.1			1419	3.9
					2003	2.0
5 SU	0038	4.1		20 M	0232	3.9
	0723	1.8			0838	1.9
	1346	4.0			1527	3.9
	1941	2.1			2121	2.1
6 M	0147	4.0		21 TU	0343	3.9
	0831	1.8			0950	2.0
	1453	4.1			1627	4.0
	2052	2.0			2232	2.0
7 TU	0302	4.1		22 W	0444	4.0
	0938	1.7			1051	1.9
	1554	4.2			1716	4.2
	2200	1.9			2331	1.8
8 W	0410	4.2		23 TH	0533	4.1
	1039	1.5			1140	1.8
	1648	4.5			1756	4.4
	2302	1.6				
9 TH	0510	4.5		24 F	0017	1.6
	1134	1.3			0613	4.3
	1737	4.7			1221	1.6
	2358	1.2			1830	4.5
10 F	0603	4.7		25 SA	0057	1.4
	1224	1.1			0648	4.4
	1823	5.0 •			1257	1.5
					1900	4.7
11 SA	0050	0.9		26 SU	0133	1.2
	0652	4.9			0721	4.5
	1312	0.9			1330	1.4
	1909	5.1			1929	4.8
12 SU	0139	0.6		27 M	0206	1.1
	0741	5.0			0752	4.6
	1358	0.7			1403	1.3
	1955	5.2			1958	4.9
13 M	0227	0.4		28 TU	0238	1.1
	0829	4.9			0822	4.6
	1443	0.7			1434	1.3
	2043	5.3			2027	4.9
14 TU	0313	0.3		29 W	0309	1.1
	0919	4.8			0853	4.6
	1528	0.8			1507	1.3
	2133	5.2			2057	4.8
15 W	0359	0.7		30 TH	0341	1.2
	1011	4.6			0927	4.5
	1614	0.9			1541	1.4
	2225	5.0			2130	4.7
				31 F	0414	1.3
					1004	4.4
					1618	1.5
					2207	4.6

AUGUST

Day	TIME	M		Day	TIME	M
1 SA	0450	1.4		16 SU	0536	1.6
	1047	4.3			1207	3.9
	1659	1.7			1803	1.9
	2250	4.4				
2 SU	0531	1.6		17 M	0031	3.9
	1140	4.1			0626	2.0
	1748	1.9			1332	3.7
	2343	4.2			1912	2.2
3 M	0623	1.8		18 TU	0159	3.7
	1250	4.0			0740	2.3
	1848	2.1			1459	3.7
					2055	2.3
4 TU	0057	4.0		19 W	0325	3.6
	0733	2.0			0925	2.4
	1414	4.0			1608	3.8
	2007	2.1			2229	2.2
5 W	0236	3.9		20 TH	0432	3.8
	0902	2.0			1043	2.2
	1530	4.1			1701	4.1
	2137	2.0			2326	1.9
6 TH	0401	4.0		21 F	0521	4.0
	1022	1.8			1130	1.9
	1633	4.3			1740	4.3
	2254	1.6				
7 F	0506	4.3		22 SA	0007	1.6
	1125	1.4			0559	4.3
	1726	4.7			1207	1.7
	2354	1.2			1812	4.6
8 SA	0558	4.6		23 SU	0042	1.3
	1216	1.1			0630	4.5
	1813	5.0			1240	1.5
					1840	4.8
9 SU	0044	0.7		24 M	0113	1.1
	0645	4.9			0659	4.7
	1302	0.8 •			1310	1.3
	1857	5.3			1906	4.9
10 M	0129	0.3		25 TU	0142	1.0
	0728	5.1			0726	4.8
	1344	0.6			1340	1.2
	1940	5.5			1931	5.0
11 TU	0212	0.1		26 W	0211	0.9
	0810	5.2			0753	4.9
	1426	0.5			1410	1.1
	2022	5.5			1958	5.1
12 W	0253	0.1		27 TH	0240	0.9
	0852	5.1			0821	4.9
	1506	0.5			1441	1.1
	2105	5.4			2026	5.1
13 TH	0333	0.3		28 F	0309	1.0
	0934	4.9			0851	4.9
	1547	0.7			1514	1.2
	2149	5.1			2057	5.0
14 F	0413	0.6		29 SA	0340	1.1
	1018	4.6			0925	4.7
	1628	1.1			1549	1.3
	2234	4.8			2131	4.8
15 SA	0453	1.1		30 SU	0415	1.3
	1105	4.2			1004	4.5
	1712	1.5			1629	1.5
	2325	4.3			2212	4.5
				31 M	0454	1.6
					1053	4.3
					1714	1.8
					2304	4.1

TIME ZONE GMT   TIMES AND HEIGHTS OF HIGH AND LOW WATERS   YEAR 1987

SEPTEMBER

Date	Day	Time	M	Time	M	Time	M	Time	M
1	TU	0542	1.8	1207	4.0	1812	2.0		
2	W	0032	3.8	0652	2.1	1350	3.9	1941	2.2
3	TH	0235	3.7	0845	2.2	1516	4.0	2136	2.0
4	F	0401	4.0	1018	1.9	1621	4.3	2253	1.5
5	SA	0501	4.3	1117	1.4	1713	4.7	2346	1.0
6	SU	0549	4.7	1203	1.0	1758	5.1		
7	M o	0030	0.5	0630	5.0	1245	0.7	1838	5.4
8	TU	0110	0.2	0708	5.2	1324	0.4	1918	5.6
9	W	0149	0.1	0744	5.3	1402	0.4	1956	5.6
10	TH	0226	0.1	0820	5.2	1439	0.5	2034	5.5
11	F	0302	0.4	0855	5.0	1517	0.7	2111	5.1
12	SA	0337	0.8	0855	4.7	1517	1.1	2111	4.7
13	SU	0413	1.2	1007	4.3	1635	1.5	2235	4.2
14	M	0449	1.7	1054	3.9	1721	2.0	2340	3.8
15	TU	0533	2.2	1234	3.6	1828	2.3		
16	W	0127	3.5	0640	2.5	1424	3.6	2036	2.4
17	TH	0302	3.5	0903	2.5	1540	3.8	2217	2.2
18	F	0411	3.7	1026	2.3	1633	4.0	2306	1.8
19	SA	0459	4.0	1109	2.0	1712	4.3	2342	1.5
20	SU	0534	4.3	1142	1.7	1743	4.6		
21	M	0014	1.3	0603	4.5	1213	1.5	1810	4.8
22	TU	0043	1.1	0630	4.8	1242	1.3	1836	5.0
23	W	0111	0.9	0656	4.9	1312	1.1	1901	5.1
24	TH	0139	0.8	0722	5.1	1342	1.0	1928	5.2
25	F	0208	0.8	0750	5.1	1414	1.0	1957	5.2
26	SA	0239	0.9	0820	5.1	1448	1.0	2030	5.0
27	SU	0311	1.0	0854	4.9	1525	1.2	2106	4.8
28	M	0346	1.3	0934	4.6	1606	1.4	2149	4.4
29	TU	0427	1.6	1025	4.3	1653	1.7	2250	4.0
30	W	0517	1.9	1151	3.9	1756	2.0		

OCTOBER

Date	Day	Time	M	Time	M	Time	M	Time	M
1	TH	0048	3.7	0635	2.2	1340	3.9	1941	2.1
2	F	0239	3.7	0845	2.2	1502	4.0	2136	1.8
3	SA	0354	4.0	1008	1.8	1604	4.4	2240	1.3
4	SU	0448	4.4	1059	1.4	1654	4.8	2327	0.8
5	M	0531	4.8	1142	1.0	1737	5.1		
6	TU	0007	0.3	0609	5.1	1221	0.7	1816	5.4
7	W o	0045	0.3	0644	5.2	1259	0.5	1853	5.5
8	TH	0121	0.2	0717	5.3	1336	0.5	1929	5.5
9	F	0156	0.4	0749	5.2	1413	0.6	2004	5.3
10	SA	0231	0.6	0820	5.0	1450	0.9	2040	4.9
11	SU	0304	1.0	0852	4.7	1527	1.2	2117	4.5
12	M	0338	1.4	0925	4.4	1607	1.6	2201	4.1
13	TU	0413	1.8	1008	4.0	1653	2.0	2307	3.7
14	W	0455	2.2	1133	3.7	1757	2.3		
15	TH	0051	3.5	0557	2.5	1333	3.6	1957	2.4
16	F	0224	3.5	0810	2.6	1452	3.7	2135	2.1
17	SA	0332	3.7	0943	2.4	1549	3.9	2226	1.8
18	SU	0422	4.0	1029	2.1	1631	4.2	2304	1.6
19	M	0459	4.3	1105	1.8	1705	4.5	2336	1.3
20	TU	0530	4.5	1138	1.6	1735	4.7		
21	W	0007	1.1	0558	4.8	1210	1.3	1803	4.9
22	TH o	0037	1.0	0626	5.0	1242	1.1	1832	5.1
23	F	0107	0.9	0654	5.2	1315	1.0	1903	5.2
24	SA	0139	0.8	0725	5.2	1351	1.0	1936	5.1
25	SU	0213	0.9	0758	5.1	1428	1.0	2013	5.0
26	M	0249	1.0	0836	4.9	1509	1.1	2055	4.7
27	TU	0328	1.3	0921	4.6	1554	1.3	2149	4.3
28	W	0413	1.6	1022	4.3	1647	1.6	2311	3.9
29	TH	0510	1.9	1152	4.1	1758	1.8		
30	F	0059	3.7	0636	2.1	1324	4.0	1941	1.8
31	SA	0227	3.8	0827	2.1	1438	4.2	2113	1.5

NOVEMBER

Date	Day	Time	M	Time	M	Time	M	Time	M
1	SU	0334	4.1	0940	1.7	1539	4.5	2212	1.2
2	M	0426	4.4	1031	1.4	1630	4.7	2259	0.9
3	TU	0509	4.7	1115	1.1	1714	5.0	2340	0.7
4	W	0546	4.9	1156	0.9	1754	5.2		
5	TH o	0018	0.6	0621	5.1	1235	0.8	1831	5.2
6	F	0054	0.6	0653	5.1	1314	0.8	1907	5.2
7	SA	0129	0.8	0725	5.1	1351	0.9	1943	5.0
8	SU	0204	1.0	0756	4.9	1429	1.1	2019	4.7
9	M	0238	1.2	0829	4.7	1508	1.3	2058	4.4
10	TU	0312	1.5	0906	4.4	1549	1.6	2145	4.1
11	W	0349	1.8	0951	4.2	1635	1.8	2245	3.8
12	TH	0432	2.1	1055	3.9	1733	2.1		
13	F	0004	3.6	0527	2.4	1221	3.8	1855	2.2
14	SA	0125	3.6	0652	2.5	1340	3.8	2026	2.1
15	SU	0234	3.7	0830	2.4	1443	3.9	2128	1.9
16	M	0330	3.9	0932	2.2	1535	4.1	2213	1.7
17	TU	0414	4.2	1018	2.0	1618	4.3	2252	1.5
18	W	0451	4.5	1058	1.8	1656	4.6	2328	1.3
19	TH	0525	4.7	1136	1.5	1732	4.8		
20	F	0003	1.1	0558	5.0	1214	1.3	1808	5.0
21	SA o	0039	1.0	0632	5.1	1253	1.1	1846	5.1
22	SU	0116	0.9	0708	5.2	1334	1.0	1926	5.0
23	M	0155	0.9	0746	5.2	1417	0.9	2009	4.9
24	TU	0236	1.1	0830	5.0	1502	1.0	2100	4.6
25	W	0320	1.2	0922	4.8	1552	1.1	2202	4.3
26	TH	0410	1.5	1026	4.6	1648	1.3	2319	4.1
27	F	0509	1.7	1141	4.4	1755	1.4		
28	SA	0042	4.0	0623	1.9	1256	4.3	1914	1.5
29	SU	0157	4.0	0747	1.9	1405	4.4	2032	1.4
30	M	0302	4.1	0859	1.8	1507	4.5	2135	1.3

DECEMBER

Date	Day	Time	M	Time	M	Time	M	Time	M
1	TU	0358	4.3	0957	1.6	1603	4.6	2227	1.2
2	W	0445	4.5	1048	1.4	1652	4.7	2312	1.1
3	TH	0527	4.7	1134	1.3	1736	4.8		
4	F	0604	4.8	1217	1.2	1817	4.8		
5	SA	0032	1.1	0638	4.9	1259	1.2	1855	4.8
6	SU	0109	1.2	0712	4.9	1339	1.2	1932	4.7
7	M	0145	1.2	0745	4.9	1418	1.2	2009	4.6
8	TU	0221	1.4	0819	4.8	1458	1.3	2048	4.4
9	W	0256	1.5	0856	4.6	1537	1.5	2130	4.2
10	TH	0333	1.7	0936	4.4	1619	1.6	2217	4.0
11	F	0413	1.9	1022	4.3	1704	1.8	2312	3.9
12	SA	0458	2.1	1115	4.1	1757	2.0		
13	SU	0014	3.8	0553	2.3	1214	4.0	1900	2.1
14	M	0120	3.8	0702	2.4	1318	4.0	2008	2.1
15	TU	0224	3.9	0816	2.4	1423	4.0	2110	2.0
16	W	0321	4.1	0920	2.2	1524	4.2	2202	1.8
17	TH	0411	4.3	1015	2.0	1618	4.4	2250	1.6
18	F	0455	4.6	1105	1.8	1707	4.6	2334	1.4
19	SA	0537	4.8	1152	1.5	1753	4.8		
20	SU	0018	1.2	0617	5.0	1239	1.2	1837	4.9
21	M	0101	1.1	0658	5.2	1325	1.0	1923	5.0
22	TU	0145	1.0	0742	5.2	1412	0.8	2011	4.9
23	W	0229	1.0	0828	5.2	1459	0.7	2102	4.8
24	TH	0315	1.0	0919	5.1	1548	0.8	2158	4.6
25	F	0403	1.2	1015	5.0	1639	0.9	2300	4.4
26	SA	0456	1.4	1116	4.8	1734	1.1		
27	SU	0006	4.2	0553	1.6	1219	4.6	1834	1.3
28	M	0115	4.1	0659	1.8	1324	4.4	1941	1.5
29	TU	0223	4.1	0810	1.9	1431	4.3	2050	1.7
30	W	0327	4.1	0920	1.9	1536	4.3	2153	1.7
31	TH	0424	4.3	1024	1.9	1635	4.3	2249	1.7

MILFORD HAVEN
MEAN SPRING AND NEAP CURVES
Springs occur 2 days after New and Full Moon.

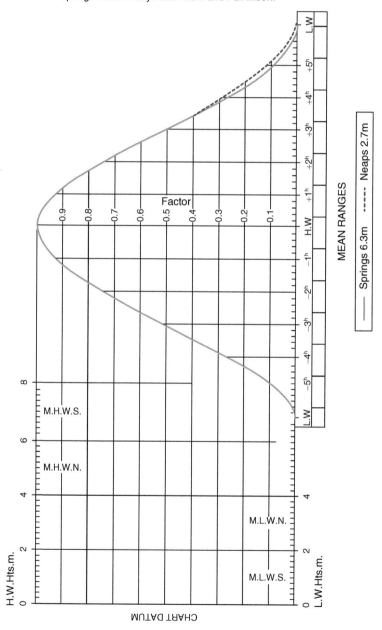

MEAN RANGES

—— Springs 6.3m   ----- Neaps 2.7m

Factor

0.9 0.8 0.7 0.6 0.5 0.4 0.3 0.2 0.1

H.W.

M.H.W.S.

M.H.W.N.

M.L.W.N.

M.L.W.S.

H.W.Hts.m.

L.W.Hts.m.

CHART DATUM

TIME ZONE GMT   TIMES AND HEIGHTS OF HIGH AND LOW WATERS   YEAR 1987

SEPTEMBER

Date	Time	m	Date	Time	m
1 TU	0433	2.1	16 W	0542	2.9
	1045	5.6		1204	4.9
	1701	2.3		1855	3.1
	2315	5.5			
2 W	0534	2.5	17 TH	0109	4.6
	1158	5.3		0745	3.1
	1824	2.5		1411	4.9
				2054	2.8
3 TH	0043	5.2	18 F	0300	4.9
	0719	2.6		0918	2.7
	1338	5.3		1531	5.3
	2019	2.4		2156	2.4
4 F	0227	5.3	19 SA	0359	5.4
	0904	2.3		1010	2.2
	1511	5.7		1617	5.8
	2142	1.9		2240	1.9
5 SA	0352	5.8	20 SU	0440	5.9
	1014	1.7		1049	1.8
	1620	6.4		1655	6.2
	2245	1.2		2315	1.5
6 SU	0451	6.4	21 M	0515	6.3
	1109	1.1		1125	1.4
	1713	7.0		1727	6.6
	2337	0.7		2347	1.2
7 M	0540	6.9	22 TU	0546	6.6
O	1156	0.7		1156	1.2
	1800	7.4		1758	6.8
8 TU	0022	0.4	23 W	0017	1.0
	0624	7.2		0617	6.8
	1238	0.4		1225	1.0
	1842	7.6		1828	7.0
9 W	0103	0.2	24 TH	0046	0.9
	0703	7.3		0646	6.9
	1319	0.3		1255	0.9
	1923	7.7		1859	7.0
10 TH	0141	0.3	25 F	0116	0.8
	0741	7.3		0716	6.9
	1355	0.4		1326	0.9
	2001	7.5		1928	7.0
11 F	0218	0.5	26 SA	0145	0.9
	0818	7.1		0747	6.7
	1432	0.7		1358	1.0
	2037	7.1		2001	6.9
12 SA	0251	0.9	27 SU	0216	1.1
	0853	6.7		0819	6.7
	1507	1.1		1430	1.2
	2112	6.6		2034	6.7
13 SU	0325	1.4	28 M	0249	1.6
	0927	6.2		0853	6.4
	1541	1.7		1505	1.5
	2149	6.1		2111	6.3
14 M	0359	2.0	29 TU	0325	1.8
	1003	5.8		0934	6.0
	1619	2.2		1549	1.9
	2230	5.5		2157	5.8
15 TU	0438	2.5	30 W	0412	2.2
	1049	5.3		1027	5.6
	1712	2.7		1649	2.4
	2325	4.9		2304	5.3

OCTOBER

Date	Time	m	Date	Time	m
1 TH	0523	2.6	16 F	0022	4.6
	1151	5.3		0652	3.1
	1828	2.6		1321	4.9
				2012	2.9
2 F	0043	5.1	17 SA	0218	4.8
	0726	2.7		0837	2.8
	1337	5.3		1447	5.2
	2022	2.3		2118	2.5
3 SA	0230	5.4	18 SU	0321	5.3
	0900	2.2		0934	2.4
	1505	5.9		1538	5.7
	2135	1.7		2203	2.0
4 SU	0343	6.0	19 M	0403	5.8
	1002	1.6		1014	1.9
	1606	6.5		1617	6.1
	2231	1.1		2240	1.6
5 M	0435	6.5	20 TU	0438	6.2
	1052	1.1		1049	1.6
	1655	7.1		1652	6.5
	2318	0.7		2312	1.3
6 TU	0519	7.0	21 W	0511	6.5
	1134	0.7		1122	1.3
	1737	7.4		1725	6.8
	2358	0.5		2344	1.1
7 W	0600	7.2	22 TH	0543	6.8
	1215	0.5		1154	1.0
●	1818	7.5	●	1756	7.0
8 TH	0036	0.4	23 F	0015	0.9
	0636	7.3		0614	7.0
	1252	0.5		1228	0.9
	1856	7.5		1829	7.1
9 F	0113	0.5	24 SA	0048	0.9
	0713	7.2		0648	7.1
	1328	0.6		1302	0.9
	1933	7.3		1904	7.1
10 SA	0147	0.8	25 SU	0121	0.9
	0748	7.0		0723	7.0
	1404	0.9		1337	1.0
	2008	6.9		1941	6.9
11 SU	0220	1.1	26 M	0157	1.1
	0822	6.7		0759	6.8
	1437	1.3		1415	1.2
	2042	6.5		2019	6.6
12 M	0253	1.6	27 TU	0233	1.4
	0856	6.5		0840	6.5
	1511	1.8		1457	1.5
	2117	5.9		2104	6.2
13 TU	0324	2.1	28 W	0317	1.8
	0932	5.8		0928	6.1
	1548	2.3		1549	1.9
	2156	5.4		2159	5.8
14 W	0402	2.6	29 TH	0412	2.2
	1016	5.3		1031	5.7
	1638	2.8		1658	2.3
	2249	4.9		2311	5.4
15 TH	0501	3.0	30 F	0530	2.6
	1125	4.9		1154	5.5
	1810	3.1		1836	2.4
			31 SA	0042	5.3
				0719	2.5
				1324	5.6
				2005	2.1

NOVEMBER

Date	Time	m	Date	Time	m
1 SU	0212	5.5	16 M	0219	5.2
	0839	2.1		0833	2.6
	1442	6.1		1440	5.5
	2112	1.6		2108	2.3
2 M	0318	6.0	17 TU	0312	5.5
	0938	1.6		0924	2.2
	1541	6.5		1527	5.9
	2206	1.2		2152	1.9
3 TU	0409	6.5	18 W	0353	6.0
	1027	1.2		1006	1.8
	1630	6.9		1607	6.2
	2251	1.0		2231	1.6
4 W	0454	6.8	19 TH	0431	6.3
	1111	0.9		1045	1.5
	1713	7.1		1647	6.6
	2332	0.8		2309	1.3
5 TH	0533	7.0	20 F	0508	6.7
	1151	0.8		1125	1.2
O	1754	7.2		1725	6.8
				2347	1.1
6 F	0010	0.8	21 SA	0547	6.9
	0612	7.0		1204	1.0
	1229	0.8	●	1805	7.0
	1832	7.1			
7 SA	0048	0.9	22 SU	0025	1.0
	0649	7.0		0627	7.1
	1306	1.0		1243	1.0
	1909	6.9		1846	7.0
8 SU	0123	1.1	23 M	0104	1.0
	0724	6.8		0707	7.1
	1342	1.2		1327	1.0
	1945	6.6		1931	6.9
9 M	0157	1.4	24 TU	0147	1.1
	0759	6.6		0752	6.9
	1418	1.5		1412	1.1
	2020	6.3		2018	6.7
10 TU	0230	1.4	25 W	0230	1.4
	0834	6.7		0840	6.7
	1453	1.4		1501	1.4
	2057	6.3		2108	6.3
11 W	0303	2.1	26 TH	0321	1.7
	0912	5.9		0934	6.4
	1531	2.3		1556	1.6
	2136	5.5		2204	6.0
12 TH	0342	2.4	27 F	0417	2.0
	0956	5.5		1034	6.1
	1617	2.6		1702	1.9
	2226	5.1		2308	5.7
13 F	0433	2.8	28 SA	0527	2.2
	1052	5.2		1142	5.7
	1723	2.8		1818	2.0
	2333	4.9			
14 SA	0550	3.0	29 SU	0019	5.6
	1208	5.1		0649	2.2
	1852	2.8		1255	5.9
				1931	1.9
15 SU	0103	4.9	30 M	0134	5.7
	0721	2.9		0802	2.0
	1335	5.2		1405	6.1
	2012	2.6		2037	1.8

DECEMBER

Date	Time	m	Date	Time	m
1 TU	0242	5.9	16 W	0159	5.3
	0904	1.8		0818	2.5
	1533	6.3		1420	5.6
	2134	1.6		2054	2.2
2 W	0338	6.2	17 TH	0257	5.6
	0959	1.6		0915	2.2
	1600	6.5		1518	5.9
	2224	1.4		2148	1.9
3 TH	0427	6.4	18 F	0349	6.0
	1047	1.4		1009	1.8
	1649	6.6		1610	6.2
	2308	1.3		2237	1.6
4 F	0512	6.6	19 SA	0438	6.4
	1132	1.3		1058	1.5
	1734	6.6		1701	6.5
	2350	1.3		2323	1.3
5 SA	0554	6.7	20 SU	0526	6.7
	1212	1.3		1146	1.2
O	1815	6.6	•	1750	6.8
6 SU	0029	1.3	21 M	0010	1.1
	0634	6.7		0614	7.0
	1252	1.3		1235	1.0
	1855	6.6		1839	6.9
7 M	0106	1.4	22 TU	0056	1.0
	0712	6.7		0702	7.1
	1330	1.4		1323	0.8
	1933	6.4		1927	6.9
8 TU	0141	1.5	23 W	0142	1.0
	0747	6.5		0751	7.2
	1406	1.6		1412	0.8
	2008	6.2		2016	6.9
9 W	0216	1.7	24 TH	0230	1.0
	0822	6.4		0840	7.1
	1440	1.8		1501	0.9
	2044	6.0		2105	6.7
10 TH	0249	1.9	25 F	0318	1.2
	0858	6.1		0929	6.9
	1517	2.0		1552	1.1
	2121	5.8		2156	6.4
11 F	0325	2.1	26 SA	0410	1.4
	0936	5.9		1021	6.7
	1555	2.2		1645	1.4
	2202	5.5		2248	6.1
12 SA	0407	2.3	27 SU	0504	1.7
	1020	5.5		1116	6.4
	1641	2.4		1743	1.7
	2248	5.3		2344	5.9
13 SU	0457	2.5	28 M	0605	1.9
	1112	5.4		1215	6.1
	1739	2.5		1845	1.9
	2346	5.1			
14 M	0600	2.7	29 TU	0048	5.7
	1211	5.3		0713	2.1
	1845	2.6		1320	5.9
				1951	2.0
15 TU	0052	5.1	30 W	0157	5.6
	0710	2.7		0825	2.1
	1317	5.4		1429	5.8
	1952	2.5		2058	2.1
			31 TH	0304	5.7
				0931	2.1
				1534	5.9
				2159	2.0

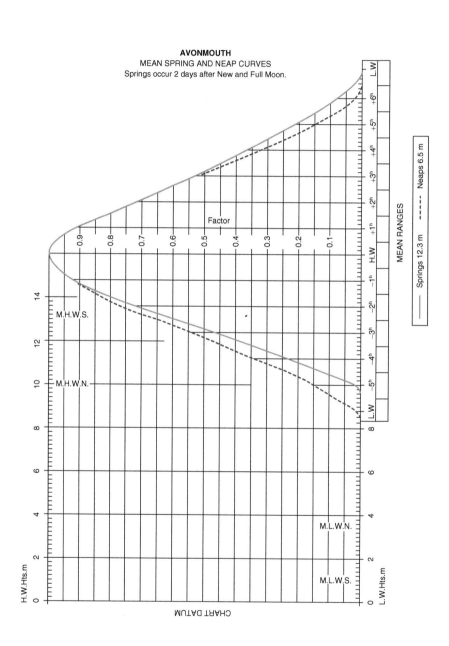

AVONMOUTH
MEAN SPRING AND NEAP CURVES
Springs occur 2 days after New and Full Moon.

Factor

0.9 0.8 0.7 0.6 0.5 0.4 0.3 0.2 0.1

M.H.W.S.

M.H.W.N.

M.L.W.N.

M.L.W.S.

H.W.Hts.m

L.W.Hts.m

CHART DATUM

MEAN RANGES

Springs 12.3 m   Neaps 6.5 m

ENGLAND, WEST COAST-PORT OF BRISTOL (AVONMOUTH)
LAT 51°30'N   LONG 2°43'W

TIME ZONE GMT   TIMES AND HEIGHTS OF HIGH AND LOW WATERS   YEAR 1987

MAY

Day	TIME	M	TIME	M	TIME	M	TIME	M
1 F	0324	1.2	0903	12.7	1535	1.2	2112	12.4
2 SA	0348	1.5	0932	11.9	1559	1.6	2141	11.7
3 SU	0412	2.0	1002	11.0	1623	2.2	2207	10.8
4 M	0438	2.6	1033	10.2	1651	2.8	2238	10.1
5 TU	0511	3.2	1109	9.4	1726	3.4	2322	9.4
6 W	0554	3.8	1203	8.9	1814	4.0		
7 TH	0027	9.0	0655	4.2	1327	8.7	1926	4.2
8 F	0209	9.2	0823	4.1	1501	9.3	2103	3.9
9 SA	0327	10.0	0949	3.4	1603	10.3	2220	3.1
10 SU	0423	11.0	1052	2.7	1654	11.3	2325	2.5
11 M	0512	11.9	1157	2.2	1742	12.1		
12 TU	0028	1.9	0558	12.5	1257	1.7	1827	12.7
13 W O	0123	1.4	0645	13.0	1348	1.3	1910	13.0
14 TH	0211	1.1	0728	13.2	1432	1.1	1954	13.2
15 F	0253	0.9	0813	13.1	1511	1.1	2036	13.0
16 SA	0331	1.0	0858	12.8	1548	1.4	2121	12.6
17 SU	0409	1.3	0946	12.2	1623	1.8	2210	12.0
18 M	0447	1.8	1037	11.5	1702	2.4	2302	11.3
19 TU	0530	2.3	1133	10.9	1750	2.9		
20 W	0001	10.8	0627	2.8	1234	10.5	1853	3.3
21 TH	0106	10.7	0741	2.8	1340	10.4	2015	3.1
22 F	0216	10.8	0854	2.5	1453	10.8	2127	2.6
23 SA	0329	11.3	0959	2.1	1604	11.3	2233	2.2
24 SU	0434	11.8	1104	1.9	1702	1.8	2340	1.9
25 M	0526	12.2	1205	1.7	1750	12.2		
26 TU	0035	1.8	0612	12.4	1256	1.6	1832	12.5
27 W•	0119	1.7	0653	12.5	1337	1.6	1910	12.6
28 TH	0157	1.6	0731	12.5	1412	1.5	1947	12.6
29 F	0230	1.5	0808	12.4	1444	1.4	2020	12.4
30 SA	0303	1.6	0843	12.0	1515	1.6	2051	12.0
31 SU	0331	1.9	0914	11.5	1542	2.0	2121	11.5

JUNE

Day	TIME	M	TIME	M	TIME	M	TIME	M
1 M	0359	2.3	0945	10.9	1610	2.4	2152	10.9
2 TU	0428	2.7	1017	10.4	1640	2.8	2226	10.5
3 W	0501	3.0	1054	10.0	1712	3.1	2306	10.2
4 TH	0540	3.3	1137	9.7	1754	3.4	2358	10.0
5 F	0628	3.5	1234	9.6	1849	3.6		
6 SA	0103	10.0	0727	3.5	1349	9.8	1955	3.5
7 SU	0222	10.4	0836	3.3	1505	10.4	2115	3.3
8 M	0331	11.0	0953	2.9	1607	11.1	2233	2.8
9 TU	0431	11.7	1105	2.5	1704	11.9	2344	2.3
10 W	0526	12.2	1217	2.1	1757	12.4		
11 TH	0052	1.8	0621	12.6	1320 O	1.7	1849	12.8
12 F	0148	1.4	0713	12.8	1413	1.4	1940	13.0
13 SA	0239	1.2	0805	12.8	1501	1.3	2037	13.0
14 SU	0325	1.1	0856	12.8	1545	1.5	2118	12.8
15 M	0409	1.3	0945	12.5	1627	1.7	2207	12.5
16 TU	0452	1.5	1034	12.1	1709	2.1	2257	12.1
17 W	0536	1.9	1122	11.7	1753	2.4	2347	11.7
18 TH	0622	2.2	1214	11.3	1842	2.7		
19 F	0041	11.4	0716	2.4	1309	11.0	1940	2.8
20 SA	0140	11.2	0813	2.5	1409	10.9	2042	2.8
21 SU	0246	11.2	0912	2.5	1517	11.0	2143	2.8
22 M	0352	11.3	1012	2.5	1621	11.2	2247	2.7
23 TU	0452	11.5	1113	2.5	1716	11.5	2350	2.5
24 W	0544	11.7	1212	2.3	1805	11.8		
25 TH	0043	2.3	0631	11.8	1303	2.1	1849	12.0
26 F	0130	2.1	0714	11.9	1348	1.9	1928	12.0
27 SA	0211	2.0	0754	11.9	1427	1.9	2005	12.0
28 SU	0249	2.0	0829	11.7	1503	2.0	2037	11.8
29 M	0324	2.2	0901	11.5	1535	2.2	2108	11.6
30 TU	0355	2.4	0932	11.3	1606	2.4	2141	11.5

JULY

Day	TIME	M	TIME	M	TIME	M	TIME	M
1 W	0424	2.6	1003	11.1	1634	2.6	2213	11.4
2 TH	0454	2.7	1037	10.9	1704	2.7	2249	11.2
3 F	0526	2.8	1113	10.7	1737	2.9	2330	11.0
4 SA	0601	2.9	1154	10.5	1817	3.1		
5 SU	0018	10.8	0645	3.1	1248	10.3	1907	3.3
6 M	0121	10.6	0740	3.3	1406	10.3	2015	3.5
7 TU	0243	10.7	0900	3.3	1527	10.7	2153	3.4
8 W	0355	11.2	1031	3.0	1633	11.4	2315	2.8
9 TH	0501	11.7	1149	2.5	1736	12.0		
10 F	0028	2.2	0605	12.1	1302	2.0	1836	12.5
11 SA	0134	1.7	0706	12.5	1402	1.6	1931	12.9
12 SU	0232	1.2	0759	12.9	1456	1.3	2020	13.2
13 M	0322	1.0	0849	13.1	1543	1.2	2108	13.3
14 TU	0407	0.9	0934	13.0	1626	1.3	2153	13.3
15 W	0448	1.0	1017	12.8	1704	1.6	2238	12.9
16 TH	0526	1.4	1101	12.4	1740	2.0	2322	12.4
17 F	0600	1.8	1143	11.8	1814	2.4		
18 SA	0007	11.8	0636	2.3	1228	11.2	1852	2.9
19 SU	0056	11.2	0719	2.8	1319	10.8	1941	3.3
20 M	0154	10.7	0815	3.2	1420	10.5	2050	3.6
21 TU	0303	10.5	0924	3.4	1535	10.5	2203	3.5
22 W	0419	10.6	1033	3.2	1648	10.8	2316	3.1
23 TH	0523	11.0	1140	2.8	1746	11.3		
24 F	0018	2.7	0615	11.4	1241	2.4	1834	11.7
25 SA	0112	2.3	0700	11.7	1330	2.1	1914	11.9
26 SU	0157	2.1	0740	11.8	1413	2.0	1949	12.0
27 M	0236	2.1	0812	11.8	1451	2.0	2022	12.0
28 TU	0312	2.1	0843	11.9	1525	2.0	2051	12.1
29 W	0345	2.1	0912	11.9	1556	2.1	2122	12.2
30 TH	0414	2.2	0943	11.9	1624	2.2	2153	12.2
31 F	0441	2.3	1013	11.8	1649	2.3	2226	12.0

AUGUST

Day	TIME	M	TIME	M	TIME	M	TIME	M
1 SA	0505	2.4	1044	11.4	1715	2.6	2259	11.6
2 SU	0532	2.7	1116	11.0	1746	2.9	2337	11.0
3 M	0605	3.0	1158	10.5	1827	3.4		
4 TU	0029	10.5	0652	3.4	1302	10.1	1924	3.8
5 W	0155	10.2	0804	3.9	1450	10.1	2124	4.0
6 TH	0327	10.4	1009	3.6	1610	10.8	2257	3.2
7 F	0445	11.0	1133	2.9	1725	11.6		
8 SA	0017	2.4	0600	11.9	1252	2.1	1858	12.5
9 SU	0127	1.6	0659	12.6	1355 O	1.4	1921	13.2
10 M	0223	0.9	0748	13.2	1447	1.0	2008	13.7
11 TU	0312	0.5	0833	13.5	1532	0.7	2050	13.9
12 W	0355	0.4	0914	13.6	1612	0.8	2131	13.8
13 TH	0431	0.6	0953	13.4	1644	1.2	2212	13.4
14 F	0501	1.1	1030	12.8	1712	1.7	2249	12.8
15 SA	0527	1.8	1106	12.8	1736	1.7	2327	12.8
16 SU	0553	2.4	1142	11.3	1803	2.9		
17 M	0007	11.0	0622	3.0	1222	10.5	1838	3.5
18 TU	0059	10.2	0706	3.7	1323	9.9	1935	4.1
19 W	0215	9.6	0823	4.2	1454	9.6	2129	4.2
20 TH	0355	9.7	1006	3.9	1630	10.1	2255	3.5
21 F	0508	10.4	1122	3.2	1730	10.9		
22 SA	0000	2.8	0558	11.1	1221	2.5	1817	11.6
23 SU	0052	2.2	0641	1101	1310	2.5	1855	11.6
24 M	0137	2.0	0717	11.9	1354 •	1.9	1927	12.2
25 TU	0218	1.9	0748	12.1	1432	1.8	1957	12.4
26 W	0253	1.8	0818	12.3	1507	1.7	2027	12.7
27 TH	0355	1.8	0847	12.5	1538	1.7	2057	12.8
28 F	0355	1.8	0917	12.5	1606	1.8	2128	12.7
29 SA	0420	1.9	0946	12.3	1630	2.0	2159	12.3
30 SU	0441	2.2	1014	11.8	1651	2.4	2230	11.7
31 M	0504	2.6	1044	11.1	1718	2.9	2305	11.0

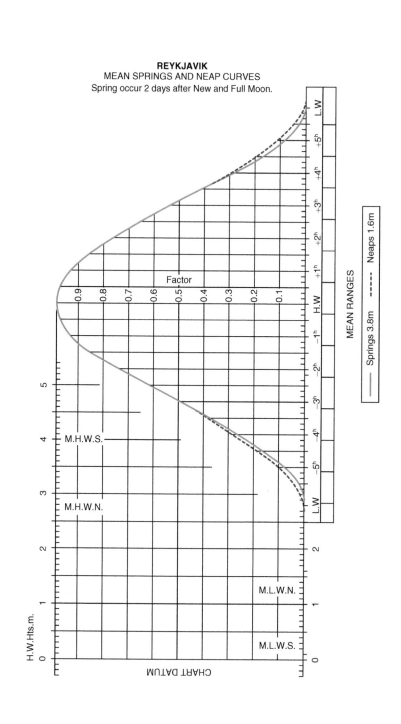

REYKJAVIK
MEAN SPRINGS AND NEAP CURVES
Spring occur 2 days after New and Full Moon.

MEAN RANGES

Springs 3.8m ----- Neaps 1.6m

TIME ZONE GMT   TIMES AND HEIGHTS OF HIGH AND LOW WATERS

JANUARY

Day	Time	M	Time	M	Time	M	Time	M
1 TH	0048	0.3	0704	4.4	1328	0.3	1930	3.9
2 F	0135	0.3	0755	4.4	1418	0.3	2020	3.8
3 SA	0223	0.4	0846	4.3	1505	0.4	2112	3.8
4 SU	0312	0.5	0936	4.1	1555	0.5	2203	3.6
5 M	0403	0.7	1027	3.9	1645	0.7	2257	3.5
6 TU	0458	1.0	1120	3.6	1737	1.0	2353	3.3
7 W	0558	1.3	1217	3.4	1832	1.2		
8 TH	0053	3.2	0706	1.5	1319	3.1	1934	1.4
9 F	0159	3.2	0820	1.5	1426	3.0	2040	1.4
10 SA	0307	3.2	0932	1.5	1535	3.0	2145	1.4
11 SU	0406	3.4	1031	1.4	1631	3.1	2237	1.3
12 M	0455	3.5	1118	1.2	1718	3.2	2320	1.1
13 TU	0536	3.7	1156	1.1	1757	3.3		
14 W	0000	1.0	0611	3.8	1231	0.9	1832	3.4
15 TH ○	0035	0.8	0646	3.9	1304	0.8	1906	3.5
16 F	0107	0.8	0719	4.0	1337	0.7	1937	3.6
17 SA	0140	0.7	0751	4.0	1409	0.7	2009	3.6
18 SU	0212	0.7	0822	4.0	1442	0.7	2042	3.6
19 M	0244	0.8	0856	3.9	1514	0.7	2115	3.6
20 TU	0319	0.9	0931	3.8	1549	0.8	2152	3.5
21 W	0359	1.0	1010	3.7	1628	0.9	2235	3.4
22 TH	0445	1.2	1057	3.5	1715	1.1	2330	3.3
23 F	0544	1.3	1154	3.3	1811	1.2		
24 SA	0038	3.3	0657	1.5	1304	3.2	1921	1.3
25 SU	0155	3.3	0820	1.4	1423	3.1	2042	1.3
26 M	0314	3.4	0942	1.4	1542	3.2	2159	1.1
27 TU	0423	3.8	1051	1.0	1649	3.4	2302	0.8
28 W	0520	4.0	1146	0.7	1746	3.6	2354	0.5
29 TH ●	0611	4.3	1234	0.4	1835	3.9		
30 F	0042	0.8	0657	4.4	1319	0.2	1921	4.0
31 SA	0127	0.2	0742	4.5	1402	0.1	2006	4.1

FEBUARY

Day	Time	M	Time	M	Time	M	Time	M
1 SU	0209	0.2	0826	4.5	1443	0.2	2049	4.0
2 M	0251	0.3	0910	4.3	1524	0.4	2131	3.9
3 TU	0334	0.6	0952	4.0	1603	0.6	2214	3.7
4 W	0419	0.9	1035	3.7	1644	0.9	2259	3.5
5 TH	0506	1.2	1122	3.3	1729	1.3	2353	3.3
6 F	0604	1.5	1218	3.0	1825	1.5		
7 SA	0057	3.1	0720	1.8	1331	2.8	1941	1.7
8 SU	0218	3.0	0900	1.6	1500	2.8	2115	1.7
9 M	0342	3.1	1020	1.6	1619	2.9	2226	1.5
10 TU	0441	3.3	1109	1.4	1709	3.1	2312	1.3
11 W	0523	3.6	1146	1.1	1747	3.3	2349	1.0
12 TH	0558	3.8	1218	0.9	1819	3.5		
13 F ○	0021	0.8	0629	3.9	1248	0.7	1849	3.6
14 SA	0050	0.7	0659	4.1	1316	0.6	1917	3.8
15 SU	0120	0.5	0728	4.2	1345	0.5	1945	3.9
16 M	0151	0.5	0757	4.2	1413	0.4	2015	3.9
17 TU	0222	0.5	0827	4.1	1443	0.5	2044	3.9
18 W	0254	0.6	0858	4.0	1515	0.6	2118	3.8
19 TH	0331	0.8	0935	3.8	1550	0.8	2157	3.7
20 F	0413	1.0	1017	3.6	1631	1.0	2247	3.5
21 SA	0506	1.3	1113	3.3	1725	1.3	2358	3.3
22 SU	0624	1.5	1234	3.0	1845	1.5		
23 M	0128	3.2	0802	1.6	1408	2.9	2026	1.5
24 TU	0303	3.3	0942	1.4	1542	3.0	2157	1.2
25 W	0420	3.6	1049	1.0	1651	3.3	2259	0.9
26 TH	0516	4.0	1139	0.6	1742	3.7	2347	0.5
27 F	0601	4.2	1221	0.3	1824	4.0		
28 SA ●	0031	0.3	0643	4.4	1300	0.1	1904	4.2

MARCH

Day	Time	M	Time	M	Time	M	Time	M
1 SU	0110	0.1	0723	4.5	1337	0.1	1942	4.3
2 M	0148	0.1	0801	4.4	1412	0.1	2019	4.2
3 TU	0226	0.3	0837	4.2	1447	0.3	2056	4.1
4 W	0303	0.5	0914	3.9	1521	0.6	2132	3.9
5 TH	0341	0.8	0950	3.6	1555	0.9	2210	3.6
6 F	0420	1.2	1031	3.3	1633	1.3	2257	3.3
7 SA	0508	1.5	1123	2.9	1720	1.6		
8 SU	0000	3.0	0619	1.8	1238	2.7	1839	1.8
9 M	0127	2.9	0813	1.9	1420	2.6	2042	1.8
10 TU	0308	3.0	1002	1.7	1559	2.8	2207	1.6
11 W	0417	3.2	1049	1.4	1649	3.0	2252	1.3
12 TH	0459	3.5	1122	1.1	1725	3.3	2326	1.0
13 F	0533	3.7	1151	0.8	1754	3.5	2357	0.8
14 SA	0603	3.9	1219	0.6	1822	3.8		
15 SU ○	0027	0.6	0631	4.1	1246	0.4	1849	3.9
16 M	0056	0.4	0659	4.2	1314	0.3	1917	4.1
17 TU	0126	0.3	0728	4.2	1342	0.3	1945	4.1
18 W	0158	0.3	0759	4.2	1413	0.3	2016	4.1
19 TH	0232	0.4	0833	4.2	1446	0.4	2051	4.0
20 F	0308	0.6	0910	3.8	1521	0.7	2132	3.8
21 SA	0353	0.9	0956	3.5	1603	1.0	2226	3.5
22 SU	0452	1.2	1059	3.1	1702	1.3	2346	3.2
23 M	0617	1.5	1228	2.8	1835	1.5		
24 TU	0121	3.1	0802	1.5	1409	2.8	2027	1.5
25 W	0258	3.3	0939	1.2	1542	3.0	2155	1.2
26 TH	0412	3.6	1037	0.9	1642	3.4	2249	0.8
27 F	0501	3.9	1120	0.5	1726	3.7	2332	0.5
28 SA	0543	4.1	1158	0.3	1804	4.0		
29 SU ●	0011	0.2	0621	4.3	1234	0.1	1839	4.2
30 M	0048	0.1	0656	4.3	1307	0.1	1913	4.3
31 TU	0124	0.2	0731	4.2	1340	0.2	1947	4.2

APRIL

Day	Time	M	Time	M	Time	M	Time	M
1 W	0159	0.3	0806	4.0	1412	0.4	2020	4.1
2 TH	0233	0.5	0840	3.7	1444	0.6	2056	3.9
3 F	0308	0.8	0915	3.5	1517	0.9	2132	3.6
4 SA	0346	1.1	0955	3.2	1552	1.2	2217	3.3
5 SU	0433	1.4	1045	2.9	1640	1.5	2320	3.0
6 M	0537	1.6	1157	2.6	1754	1.7		
7 TU	0041	2.9	0716	1.8	1331	2.6	1947	1.8
8 W	0215	2.9	0911	1.6	1514	2.7	2125	1.6
9 TH	0332	3.1	1007	1.4	1612	3.0	2216	1.3
10 F	0419	3.6	1042	0.8	1648	3.5	2252	0.7
11 SA	0454	3.6	1113	0.6	1719	3.5	2325	0.7
12 SU	0526	3.8	1142	0.6	1747	3.8	2356	0.5
13 M	0556	4.0	1211	0.4	1815	4.0		
14 TU ○	0027	0.3	0628	4.1	1241	0.3	1846	4.2
15 W	0100	0.2	0700	4.1	1313	0.2	1919	4.2
16 TH	0135	0.2	0735	4.0	1347	0.3	1955	4.2
17 F	0215	0.3	0815	3.9	1425	0.4	2036	4.0
18 SA	0258	0.5	0900	3.6	1505	0.7	2125	3.8
19 SU	0350	0.8	0956	3.3	1557	1.0	2230	3.5
20 M	0458	1.1	1109	3.0	1708	1.3	2350	3.2
21 TU	0621	1.3	1232	2.8	1841	1.4		
22 W	0117	3.2	0755	1.3	1405	2.9	2020	1.3
23 TH	0242	3.3	0915	1.0	1524	3.1	2135	1.0
24 F	0348	3.5	1009	0.8	1619	3.4	2227	0.7
25 SA	0435	3.7	1051	0.5	1701	3.7	2309	0.5
26 SU	0518	3.9	1129	0.4	1737	3.9	2349	0.4
27 M	0554	3.9	1204	0.3	1812	4.1		
28 TU ●	0025	0.3	0631	3.9	1236	0.3	1846	4.0
29 W	0100	0.3	0706	3.8	1310	0.4	1920	4.1
30 TH	0135	0.4	0740	3.7	1342	0.5	1954	4.0

YEKATERININSKAYA
MEAN SPRING AND NEAP CURVES
Springs occur 2 days after New and Full Moon

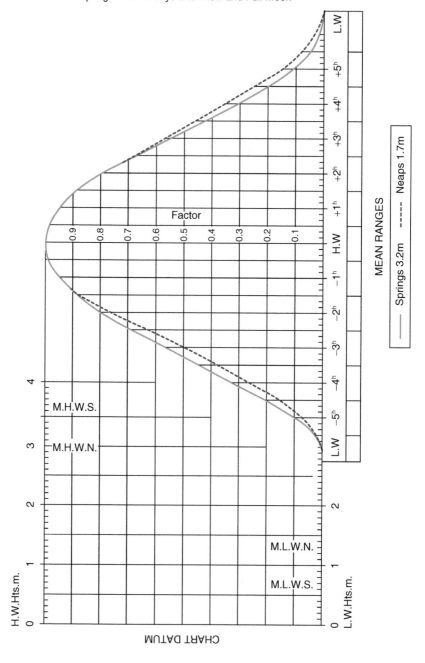

MEAN RANGES

—— Springs 3.2m   - - - - - Neaps 1.7m

TIME ZONE -0300   TIMES AND HEIGHTS OF HIGH AND LOW WATERS   YEAR 1987

JANUARY

Day	TIME	M	Day	TIME	M
1 TH	0154	0.5	16 F	0230	1.0
	0751	3.6		0830	3.2
	1412	0.6		1440	1.1
	2017	3.9		2043	3.5
2 F	0245	0.4	17 SA	0303	0.9
	0847	3.7		0905	3.3
	1506	0.6		1514	1.1
	2106	3.9		2113	3.5
3 SA	0335	0.3	18 SU	0336	0.9
	0942	3.6		0937	3.3
	1557	0.7		1548	1.1
	2156	3.9		2142	3.5
4 SU	0426	0.3	19 M	0408	0.9
	1037	3.6		1009	3.3
	1648	0.8		1622	1.1
	2246	3.8		2213	3.5
5 M	0516	0.5	20 TU	0443	0.9
	1132	3.4		1045	3.2
	1739	1.0		1659	1.2
	2337	3.6		2259	3.4
6 TU	0607	0.6	21 W	0520	0.9
	1228	3.3		1126	3.2
	1832	1.2		1740	1.2
				2330	3.4
7 W	0030	3.4	22 TH	0601	0.9
	0701	0.8		1214	3.2
	1326	3.1		1826	1.3
	1928	1.4			
8 TH	0126	1.2	23 F	0018	3.3
	0758	3.0		0648	1.1
	1427	1.5		1309	3.2
	2031	3.2		1922	1.4
9 F	0225	3.1	24 SA	0114	3.1
	0859	1.2		0743	1.1
	1530	3.0		1412	3.2
	2139	1.5		2029	1.5
10 SA	0329	3.0	25 SU	0217	3.1
	1002	1.3		0849	1.1
	1631	3.1		1520	3.2
	2245	1.5		2143	1.4
11 SU	0435	3.0	26 M	0329	3.0
	1101	1.3		1002	1.1
	1726	3.1		1630	3.3
	2344	1.5		2255	1.2
12 M	0536	2.9	27 TU	0446	3.1
	1154	1.3		1115	1.0
	1814	3.3		1734	3.5
13 TU	0034	1.3	28 W	0001	1.0
	0628	3.0		0557	3.3
	1242	1.2		1220	0.9
	1856	3.4		1830	3.7
14 W	0117	1.1	29 TH •	0057	0.7
	0713	3.1		0658	3.5
	1324	1.1		1318	0.7
	1934	3.5		1920	3.9
15 TH O	0155	1.0	30 F	0148	0.4
	0754	3.2		0751	3.7
	1403	1.1		1409	0.6
	2010	3.5		2006	4.0
			31 SA	0235	0.3
				0840	3.8
				1456	0.6
				2051	4.1

FEBUARY

Day	TIME	M	Day	TIME	M
1 SU	0320	0.2	16 M	0310	0.7
	0928	3.8		0912	3.5
	1541	0.6		1525	0.9
	2135	4.0		2114	3.7
2 M	0404	0.2	17 TU	0340	0.6
	1014	3.7		0941	3.5
	1625	0.7		1557	0.9
	2220	3.9		2144	3.6
3 TU	0448	0.4	18 W	0413	0.6
	1101	3.7		1015	3.5
	1708	0.7		1631	1.0
	2305	3.9		2219	3.6
4 W	0532	0.6	19 TH	0448	0.7
	1148	3.4		1054	3.4
	1753	1.1		1710	1.0
	2351	3.5		2300	3.5
5 TH	0617	0.8	20 F	0527	0.8
	1239	3.2		1140	3.3
	1842	1.3		1755	1.2
				2347	3.3
6 F	0042	3.2	21 SA	0613	0.9
	0705	1.2		1233	3.2
	1335	3.0		1849	1.3
	1941	1.5			
7 SA	0140	2.9	22 SU	0044	3.1
	0805	1.4		0710	1.1
	1440	2.9		1338	3.1
	2054	1.6		2000	1.4
8 SU	0251	2.8	23 M	0154	3.0
	0918	1.5		0825	1.3
	1554	3.0		1456	3.1
	2216	1.6		2124	1.4
9 M	0413	2.7	24 TU	0323	2.9
	1036	1.5		0954	1.3
	1700	3.1		1620	3.2
	2327	1.5		2249	1.2
10 TU	0524	2.8	25 W	0456	3.1
	1140	1.4		118	1.2
	1754	3.2		1727	3.4
				2356	0.9
11 W	0020	1.3	26 TH	0603	3.3
	0619	2.9		1221	0.9
	1231	1.3		1820	3.7
	1838	3.3			
12 TH	0102	1.2	27 F	0049	0.6
	0702	3.1		0656	3.6
	1313	1.2		1311	0.7
	1916	3.5		1906	3.9
13 F O	0138	1.0	28 SA •	0134	0.3
	0740	3.2		0742	3.7
	1349	1.0		1355	0.6
	1950	3.6		1948	4.0
14 SA	0210	0.9			
	0813	3.3			
	1423	1.0			
	2020	3.6			
15 SU	0241	0.8			
	0843	3.4			
	1454	0.9			
	2047	3.6			

MARCH

Day	TIME	M	Day	TIME	M
1 SU	0217	0.2	16 M	0209	0.5
	0824	3.8		0813	3.5
	1437	0.5		1457	0.8
	2029	4.0		2042	3.7
2 M	0257	0.1	17 TU	0238	0.1
	0905	3.8		0841	3.8
	1517	0.5		1457	0.5
	2110	4.0		2042	4.0
3 TU	0337	0.2	18 W	0309	0.4
	0946	3.7		0911	3.6
	1556	0.6		1530	0.7
	2150	3.8		2115	3.6
4 W	0415	0.4	19 TH	0342	0.4
	1026	3.6		0946	3.8
	1635	0.8		1605	0.7
	2231	3.6		2153	3.6
5 TH	0454	0.4	20 F	0419	0.5
	1026	3.4		1027	3.5
	1635	1.0		1645	0.8
	2231	3.3		2236	3.4
6 F	0532	0.9	21 SA	0500	0.7
	1153	3.2		1114	3.4
	1800	1.2		1732	1.0
				2328	3.2
7 SA	0001	3.1	22 SU	0549	0.9
	0615	1.2		1210	3.2
	1245	3.0		1830	1.2
	1853	1.5			
8 SU	0058	2.8	23 M	0030	3.0
	0709	1.5		0653	1.2
	1352	2.9		1318	3.1
	2007	1.8		1946	1.3
9 M	0217	2.6	24 TU	0152	2.9
	0831	1.7		0821	1.4
	1513	2.8		1445	3.1
	2140	1.7		2116	1.2
10 TU	0350	2.6	25 W	0339	2.9
	1008	1.7		0958	1.3
	1627	2.9		1609	3.2
	2300	1.5		2240	1.0
11 W	0503	2.7	26 TH	0459	3.1
	1120	1.5		1114	1.2
	1725	3.1		1711	3.4
	2354	1.3		2342	0.8
12 TH	0557	2.9	27 F	0556	3.3
	1210	1.3		1209	0.9
	1810	3.3		1801	3.8
13 F	0035	1.1	28 SA	0031	0.5
	0639	3.1		0642	3.6
	1250	1.1		1255	0.7
	1848	3.4		1845	3.7
14 SA	0109	0.9	29 SU •	0113	0.3
	0714	3.3		0723	3.7
	1325	1.0		1335	0.8
	1920	3.5		1925	3.8
15 SU O	0140	0.8	30 M	0153	0.2
	0745	3.4		0801	3.8
	1356	0.9		1414	0.5
	1947	3.6		2004	3.9
			31 TU	0230	0.2
				0838	3.8
				1452	0.5
				2042	3.8

APRIL

Day	TIME	M	Day	TIME	M
1 W	0307	0.3	16 TH	0238	0.3
	0915	3.6		0843	3.7
	1528	0.6		1504	0.5
	2121	3.6		2050	3.6
2 TH	0342	0.5	17 F	0315	0.3
	0953	3.5		0922	3.7
	1605	0.7		1543	0.6
	2200	3.4		2133	3.5
3 F	0417	0.7	18 SA	0356	0.5
	1031	3.3		1006	3.6
	1643	0.9		1627	0.7
	2241	3.2		2223	3.3
4 SA	0453	1.0	19 SU	0443	1.0
	1113	3.2		1105	3.6
	1724	1.1		1724	0.7
	2327	2.9		2318	3.3
5 SU	0532	1.2	20 M	0540	1.0
	1202	3.0		1157	3.2
	1814	1.3		1824	1.0
6 M	0023	2.7	21 TU	0033	3.0
	0624	1.5		0654	1.2
	1306	2.8		1309	3.1
	1922	1.5		1941	1.0
7 TU	0145	2.6	22 W	0205	2.9
	0745	1.7		0820	1.3
	1428	2.8		1431	3.1
	2051	1.6		2104	3.0
8 W	0316	2.6	23 TH	0335	3.0
	0927	1.7		0946	0.3
	1542	2.8		1545	3.1
	2215	1.5		2219	0.9
9 TH	0426	2.7	24 F	0443	3.1
	1043	1.5		1054	1.1
	1642	3.0		1644	3.3
	2313	1.3		2317	0.7
10 F	0521	2.9	25 SA	0536	3.3
	1135	1.3		1146	0.9
	1729	3.1		1734	3.4
	2335	1.1			
11 SA	0603	3.1	26 SU	0005	0.5
	1216	1.1		0620	3.4
	1807	3.2		1231	0.8
				1818	3.5
12 SU	0030	0.9	27 M	0047	0.4
	0639	3.3		0659	3.5
	1251	1.0		1311	0.7
	1839	3.4		1859	3.6
13 M	0102	0.7	28 TU •	0126	0.3
	0709	3.4		0736	3.6
	1323	0.8		1350	0.6
	1908	3.5		1938	3.5
14 TU O	0132	0.5	29 W	0202	0.4
	0738	3.6		0812	3.6
	1355	0.7		1427	0.6
	1938	3.6		2017	3.5
15 W	0204	0.3	30 TH	0238	0.5
	0808	3.7		0848	3.5
	1428	0.8		1504	0.7
	2012	3.6		2056	3.3

ESBJERG
MEAN SPRING AND
NEAP CURVES
Spring occurs 3 days after
New and Full Moon

DENMARK – ESBJERG
LAT 55°28'N   LONG 8°27'E

TIME ZONE -0100   TIMES AND HEIGHTS OF HIGH AND LOW WATERS   YEAR 1987

MAY

Date	Day	TIME	M	TIME	M	TIME	M	TIME	M
1	F	0430	1.3	1031	0.0	1639	1.4	2251	0.0
2	SA	0459	1.3	1104	0.0	1710	1.4	2328	0.0
3	SU	0530	1.2	1137	0.1	1744	1.4		
4	M	0006	0.1	0606	1.2	1213	0.1	1823	1.4
5	TU	0048	0.1	0649	1.1	1254	0.2	1911	1.3
6	W	0139	0.2	0742	1.1	1344	0.3	2008	1.3
7	TH	0247	0.3	0852	1.0	1453	0.4	2123	1.3
8	F	0415	0.3	1031	1.1	1621	0.4	2253	1.4
9	SA	0530	0.2	1155	1.2	1735	0.3		
10	SU	0008	1.5	0624	0.1	1253	1.3	1830	0.2
11	M	0105	1.6	0709	0.0	1341	1.4	1918	0.1
12	TU	0154	1.6	0750	-0.1	1424	1.4	2001	0.0
13 ○	W	0239	1.6	0831	-0.1	1505	1.5	2043	-0.1
14	TH	0322	1.6	0910	-0.1	1544	1.5	2126	-0.2
15	F	0405	1.6	0950	-0.1	1622	1.4	2208	-0.2
16	SA	0446	3.6	1031	2.3	1700	3.6	2252	2.2
17	SU	0528	1.4	1113	-0.1	1739	1.5	2338	-0.2
18	M	0612	1.4	1157	-0.1	1822	1.5		
19	TU	0028	-0.2	0659	1.3	1244	0.0	1912	1.5
20	W	0123	-0.1	0754	1.2	1339	0.1	2010	1.5
21	TH	0227	-0.1	0900	1.2	1443	0.1	2120	1.5
22	F	0341	0.1	1014	1.2	1557	0.1	2237	1.5
23	SA	0455	0.0	1126	1.2	1710	0.1	2347	1.6
24	SU	0558	0.0	1226	1.3	1813	0.0		
25	M	0048	1.6	0651	-0.1	1318	1.4	1906	-0.1
26	TU	0139	1.5	0754	-0.1	1402	1.4	1954	-0.1
27 ●	W	0224	1.5	0846	-0.1	1441	1.4	2037	-0.1
28	TH	0303	1.4	0857	0.0	1517	1.3	2118	-0.1
29	F	0338	1.3	0934	0.0	1549	1.3	2156	0.0
30	SA	0410	1.2	1008	0.0	1621	1.3	2234	0.0
31	SU	0441	1.1	1042	0.0	1653	1.3	2311	0.0

JUNE

Date	Day	TIME	M	TIME	M	TIME	M	TIME	M
1	M	0513	1.1	1115	0.1	1728	1.4	2348	0.0
2	TU	0549	1.1	1151	0.1	1806	1.4		
3	W	0029	0.1	0630	1.1	1231	0.1	1849	1.4
4	TH	0115	0.1	0716	1.3	1317	0.2	1939	1.4
5	F	0208	0.2	0811	1.1	1412	0.3	2307	1.4
6	SA	0313	0.2	0918	1.1	1519	0.3	2148	1.5
7	SU	0424	0.2	1039	1.2	1634	0.3	2306	1.5
8	M	0530	0.2	1153	1.3	1741	0.2		
9	TU	0015	1.5	0625	0.1	1252	1.3	1839	0.0
10	W	0113	1.5	0714	0.0	1343	1.4	1931	-0.1
11 ○	TH	0206	1.5	0801	-0.1	1431	1.4	2019	-0.2
12	F	0256	1.5	0846	-0.1	1516	1.4	2106	-0.2
13	SA	0344	1.4	0930	-0.2	1601	1.4	2153	-0.3
14	SU	0431	1.3	1014	-0.2	1644	1.4	2240	-0.3
15	M	0517	1.3	1058	-0.2	1728	1.5	2328	-0.3
16	TU	0602	1.3	1143	-0.1	1813	1.5		
17	W	0017	-0.3	0649	1.3	1230	-0.1	1902	1.6
18	TH	0110	-0.2	0739	1.3	1321	-0.1	1958	1.6
19	F	0206	-0.1	0836	1.2	1418	0.0	2101	1.6
20	SA	0310	0.0	0941	1.2	1524	0.1	2211	1.5
21	SU	0419	0.1	1049	1.3	1636	0.1	2320	1.5
22	M	0525	0.1	1152	1.3	1744	0.1		
23	TU	0021	1.5	0622	0.1	1247	1.3	1843	0.0
24	W	0115	1.4	0712	0.0	1335	1.3	1934	0.0
25	TH	0202	1.3	0757	0.0	1419	1.3	2021	0.0
26 ●	F	0244	1.2	0837	0.0	1458	1.3	2103	0.0
27	SA	0322	1.1	0915	0.0	1534	1.3	2143	0.0
28	SU	0357	1.1	0951	0.0	1609	1.3	2220	0.0
29	M	0431	1.4	1025	0.0	1642	1.3	2256	0.0
30	TU	0504	1.1	1059	0.0	1717	1.4	2332	0.0

JULY

Date	Day	TIME	M	TIME	M	TIME	M	TIME	M
1	W	0538	1.1	1134	0.1	1753	1.5		
2	TH	0009	0.0	0614	1.2	1211	0.1	1831	1.5
3	F	0049	0.1	0652	1.2	1252	0.1	1914	1.5
4	SA	0134	0.1	0737	1.2	1339	0.2	2003	1.5
5	SU	0225	0.2	0829	1.3	1434	0.2	2100	1.5
6	M	0325	0.2	0933	1.2	1540	0.2	2210	1.5
7	TU	0434	0.2	1048	1.3	1654	0.1	2325	1.5
8	W	0542	0.1	1201	1.3	1803	0.0		
9	TH	0035	1.5	0641	0.0	1303	1.3	1904	-0.1
10	F	0136	1.4	0735	-0.1	1359	1.4	1959	-0.2
11 ○	SA	0233	1.4	0824	-0.1	1451	1.4	2050	-0.3
12	SU	0326	1.4	0911	-0.2	1541	1.4	2139	-0.3
13	M	0416	1.3	0957	-0.2	1628	1.5	2227	-0.3
14	TU	0503	1.3	1042	-0.2	1714	1.6	2314	-0.3
15	W	0546	1.3	1126	-0.2	1800	1.6		
16	TH	0001	-0.3	0629	1.3	1212	-0.2	1847	1.6
17	F	0048	-0.2	0713	1.3	1259	-0.1	1937	1.6
18	SA	0138	-0.1	0802	1.3	1350	-0.1	2033	1.5
19	SU	0232	0.1	0859	1.3	1449	0.0	2138	1.5
20	M	0335	0.2	1005	1.3	1559	0.1	2247	1.4
21	TU	0446	0.2	1113	1.2	1714	0.1	2352	1.3
22	W	0551	0.2	1215	1.2	1821	0.1		
23	TH	0050	1.2	0648	0.1	1309	1.3	1917	0.0
24	F	0142	1.2	0737	0.1	1358	1.3	2006	0.0
25 ●	SA	0228	1.2	0820	0.0	1441	1.3	2049	-0.1
26	SU	0310	1.1	0859	0.0	1521	1.3	2128	-0.1
27	M	0348	1.1	0935	0.0	1558	1.3	2205	-0.1
28	TU	0423	1.1	1008	0.0	1632	1.4	2239	0.0
29	W	0455	1.2	1041	0.0	1706	1.5	2312	0.0
30	TH	0525	1.2	1114	0.0	1739	1.5	2346	0.0
31	F	0556	1.3	1149	0.0	1813	1.6		

AUGUST

Date	Day	TIME	M	TIME	M	TIME	M	TIME	M
1	SA	0022	0.1	0629	1.3	1227	0.0	1850	1.6
2	SU	0101	0.1	0707	1.3	1310	0.0	1933	1.6
3	M	0146	0.1	0752	1.3	1400	0.1	2025	1.5
4	TU	0240	0.2	0858	1.3	1500	0.1	2130	1.4
5	W	0346	0.2	0958	1.3	1615	0.1	2247	1.4
6	TH	0503	0.2	1116	1.3	1735	0.0		
7	F	0005	1.3	0613	0.1	1229	1.3	1844	-0.1
8	SA	0114	1.3	0713	0.0	1332	1.4	1942	-0.3
9 ○	SU	0214	1.3	0805	-0.1	1429	1.4	2034	-0.3
10	M	0308	1.3	0853	-0.2	1521	1.5	2123	-0.4
11	TU	0357	1.3	0938	-0.2	1610	1.6	2209	-0.4
12	W	0442	1.3	1022	-0.3	1656	1.6	2254	-0.3
13	TH	0522	1.3	1106	-0.3	1740	1.6	2337	-0.2
14	F	0601	1.4	1149	-0.2	1824	1.6		
15	SA	0020	-0.1	0638	1.4	1233	-0.2	1908	1.5
16	SU	0103	0.0	0720	1.3	1320	-0.1	1957	1.4
17	M	0150	0.1	0809	1.3	1413	0.0	2055	1.3
18	TU	0245	0.2	0911	1.2	1519	0.1	2207	1.2
19	W	0357	0.3	1027	1.2	1642	0.2	2321	1.1
20	TH	0518	0.3	1140	1.2	1800	0.1		
21	F	0027	1.1	0624	0.2	1242	1.2	1859	0.0
22	SA	0124	1.1	0716	0.1	1336	1.3	1948	0.0
23	SU	0213	1.2	0800	0.1	1422	1.3	2030	-0.1
24 ●	M	0256	1.2	0839	0.0	1503	1.4	2107	-0.1
25	TU	0334	1.2	0914	0.0	1542	1.4	2142	-0.1
26	W	0409	1.3	0947	0.0	1617	1.5	2215	-0.1
27	TH	0440	1.3	1019	0.0	1650	1.5	2246	0.0
28	F	0508	1.3	1052	0.0	1722	1.6	2319	0.0
29	SA	0536	1.4	1127	0.0	1754	1.6	2353	0.0
30	SU	0606	1.4	1204	0.0	1829	1.6		
31	M	0031	0.1	0641	1.4	1245	0.0	1910	1.5

POINT DE GRAVE
MEAN SPRING AND NEAP CURVES
Spring occurs 1 day after New and Full Moon.

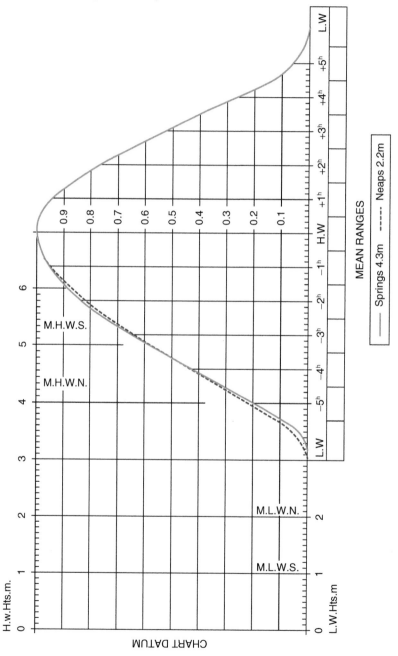

MEAN RANGES

—— Springs 4.3m ----- Neaps 2.2m

FRANCE, WEST COAST – POINTE DE GRAVE
LAT 45°34'N   LONG 1°04'W

TIME ZONE –0100   TIMES AND HEIGHTS OF HIGH AND LOW WATERS   YEAR 1987

MAY

Day	Time	m	Time	m	Time	m	Time	m
1 F	0042	1.1	0647	4.9	1252	1.3	1859	4.9
2 SA	0116	1.3	0721	4.7	1324	1.5	1933	4.7
3 SU	0151	1.5	0759	4.4	1400	1.8	2013	4.4
4 M	0232	1.8	0846	4.2	1445	2.0	2105	4.2
5 TU	0325	2.0	0950	3.9	1545	2.3	2213	4.1
6 W	0439	2.2	1114	3.9	1706	2.3	2338	4.0
7 TH	0558	2.2	1234	4.0	1820	2.2		
8 F	0055	4.2	0703	2.0	1334	4.2	1919	2.0
9 SA	0153	4.4	0756	1.7	1421	4.5	2010	1.7
10 SU	0241	4.7	0842	1.5	1503	4.7	2056	1.4
11 M	0323	4.9	0925	1.2	1542	5.0	2141	1.2
12 TU	0403	5.1	1006	1.0	1621	5.1	2224	1.0
13 W ○	0443	5.2	1047	0.9	1700	5.3	2306	0.8
14 TH	0524	5.3	1128	0.9	1742	5.3	2349	0.7
15 F	0607	5.2	1210	0.9	1825	5.3		
16 SA	0032	0.8	0652	5.1	1252	1.0	1913	5.2
17 SU	0118	0.9	0743	4.9	1339	1.3	2006	5.0
18 M	0208	1.1	0841	4.6	1432	1.5	2106	4.8
19 TU	0306	1.3	0948	4.4	1535	1.7	2216	4.5
20 W	0414	1.5	1106	4.3	1647	1.8	2331	4.6
21 TH	0527	1.6	1222	4.3	1759	1.7		
22 F	0044	4.7	0637	1.6	1328	4.5	1907	1.6
23 SA	0149	4.8	0740	1.5	1422	4.6	2005	1.4
24 SU	0243	4.8	0834	1.4	1509	4.8	2058	1.3
25 M	0330	4.9	0921	1.3	1549	4.9	2145	1.2
26 TU	0410	4.9	1004	1.2	1626	4.9	2229	1.1
27 W ●	0447	4.9	1043	1.2	1700	5.0	2309	1.1
28 TH	0521	4.9	1120	1.2	1733	5.0	2346	1.2
29 F	0554	4.8	1154	1.3	1806	4.9		
30 SA	0021	1.2	0628	4.7	1228	1.4	1839	4.8
31 SU	0056	1.4	0702	4.5	1302	1.5	1914	4.7

JUNE

Day	Time	m	Time	m	Time	m	Time	m
1 M	0132	1.5	0740	4.4	1340	1.7	1953	4.6
2 TU	0212	1.7	0823	4.3	1422	1.9	2039	4.4
3 W	0258	1.8	0915	4.1	1514	2.0	2134	4.3
4 TH	0353	1.9	1019	4.1	1614	2.1	2239	4.2
5 F	0456	2.0	1129	4.1	1718	2.0	2348	4.3
6 SA	0559	1.9	1235	4.2	1819	1.9		
7 SU	0053	4.4	0659	1.8	1331	4.4	1917	1.7
8 M	0151	4.5	0754	1.6	1421	4.6	2012	1.5
9 TU	0244	4.7	0845	1.4	1508	4.8	2105	1.3
10 W	0334	4.9	0934	1.2	1554	5.0	2156	1.1
11 TH ○	0422	5.1	1022	1.0	1640	5.2	2245	0.9
12 F	0510	5.2	1109	1.0	1728	5.3	2334	0.8
13 SA	0600	5.1	1156	1.0	1817	5.3		
14 SU	0023	0.7	0650	5.1	1243	1.0	1909	5.3
15 M	0113	0.8	0742	5.0	1333	1.1	2003	5.2
16 TU	0204	0.9	0836	4.8	1425	1.3	2059	5.1
17 W	0258	1.1	0934	4.6	1521	1.4	2159	4.9
18 TH	0354	1.3	1036	4.5	1621	1.5	2302	4.7
19 F	0454	1.5	1144	4.4	1724	1.6		
20 SA	0009	4.6	0558	1.6	1251	4.4	1829	1.7
21 SU	0115	4.5	0703	1.7	1351	4.4	1933	1.6
22 M	0215	4.5	0803	1.7	1443	4.5	2032	1.6
23 TU	0306	4.6	0856	1.6	1528	4.6	2124	1.5
24 W	0350	4.6	0941	1.5	1606	4.7	2210	1.4
25 TH	0428	4.6	1022	1.4	1641	4.8	2251	1.3
26 F ●	0504	4.7	1059	1.4	1715	4.9	2328	1.3
27 SA	0539	4.7	1135	1.4	1749	4.9		
28 SU	0004	1.3	0613	4.7	1210	1.4	1823	4.8
29 M	0039	1.4	0646	4.6	1244	1.5	1856	4.8
30 TU	0114	1.4	0720	4.6	1321	1.5	1931	4.7

JULY

Day	Time	m	Time	m	Time	m	Time	m
1 W	0151	1.5	0756	4.5	1359	1.6	2010	4.6
2 TH	0230	1.6	0837	4.4	1442	1.7	2055	4.6
3 F	0313	1.7	0927	4.3	1529	1.8	2146	4.5
4 SA	0402	1.8	1026	4.2	1623	1.8	2246	4.4
5 SU	0458	1.8	1134	4.2	1723	1.9	2354	4.4
6 M	0602	1.8	1242	4.3	1828	1.8		
7 TU	0105	4.4	0707	1.7	1344	4.5	1932	1.6
8 W	0211	4.6	0810	1.6	1441	4.7	2035	1.4
9 TH	0312	4.7	0909	1.4	1534	5.0	2134	1.2
10 F	0408	4.9	1003	1.2	1626	5.2	2230	0.9
11 SA ○	0501	5.1	1055	1.0	1717	5.4	2323	0.8
12 SU	0552	5.2	1144	0.9	1808	5.5		
13 M	0014	0.7	0641	5.2	1233	0.9	1859	5.5
14 TU	0102	0.7	0728	5.1	1321	0.9	1949	5.4
15 W	0150	0.8	0816	5.0	1408	1.0	2038	5.3
16 TH	0237	1.0	0903	4.8	1457	1.2	2129	5.0
17 F	0324	1.2	0954	4.4	1548	1.4	2222	4.7
18 SA	0415	1.5	1053	4.3	1644	1.6	2324	4.5
19 SU	0514	1.8	1206	4.2	1749	1.9		
20 M	0036	4.3	0622	2.0	1320	4.2	1902	1.9
21 TU	0148	4.2	0733	2.0	1423	4.3	2010	1.9
22 W	0249	4.3	0834	1.9	1512	4.4	2106	1.7
23 TH	0336	4.4	0923	1.7	1552	4.6	2152	1.6
24 F	0415	4.5	1004	1.6	1627	4.7	2234	1.4
25 SA ●	0450	4.6	1042	1.5	1700	4.9	2311	1.4
26 SU	0523	4.7	1117	1.4	1733	4.9	2346	1.3
27 M	0554	4.8	1152	1.3	1804	4.9		
28 TU	0020	1.1	0625	4.8	1225	1.3	1835	4.9
29 W	0053	1.3	0654	4.8	1259	1.3	1906	4.9
30 TH	0125	1.3	0726	4.7	1333	1.4	1939	4.9
31 F	0159	1.4	0801	4.7	1410	1.4	2017	4.8

AUGUST

Day	Time	m	Time	m	Time	m	Time	m
1 SA	0236	1.5	0842	4.5	1450	1.5	2100	4.7
2 SU	0317	1.6	0932	4.4	1536	1.7	2154	4.5
3 M	0406	1.8	1039	4.3	1634	1.8	2305	4.3
4 TU	0510	1.9	1158	4.2	1745	1.9		
5 W	0030	4.3	0628	2.0	1315	4.4	1903	1.8
6 TH	0151	4.4	0745	1.8	1422	4.6	2017	1.5
7 F	0301	4.6	0852	1.6	1521	5.0	2122	1.2
8 SA	0400	4.9	0950	1.3	1615	5.3	2219	0.9
9 SU ○	0451	5.1	1042	1.0	1705	5.5	2310	0.7
10 M	0537	5.3	1130	0.8	1753	5.6	2358	0.6
11 TU	0621	5.3	1216	0.7	1839	5.7		
12 W	0043	0.6	0703	5.3	1301	0.7	1924	5.6
13 TH	0126	0.7	0744	5.1	1344	0.9	2007	5.3
14 F	0208	1.0	0824	4.9	1426	1.1	2049	5.0
15 SA	0249	1.3	0905	4.6	1511	1.4	2134	4.6
16 SU	0334	1.6	0954	4.3	1603	1.8	2229	4.3
17 M	0427	2.0	1105	4.1	1709	2.1	2350	4.0
18 TU	0539	2.3	1247	4.0	1833	2.2		
19 W	0126	3.9	0704	2.3	1405	4.1	1950	2.1
20 TH	0236	4.1	0812	2.1	1458	4.3	2048	1.9
21 F	0323	4.3	0903	1.9	1537	4.6	2132	1.7
22 SA	0358	4.5	0944	1.7	1610	4.8	2211	1.5
23 SU	0430	4.7	1021	1.5	1640	4.9	2248	1.3
24 M ●	0500	4.8	1055	1.3	1711	5.0	2321	1.2
25 TU	0529	4.9	1129	1.2	1740	5.1	2354	1.2
26 W	0557	4.9	1202	1.2	1809	5.1		
27 TH	0026	1.1	0626	5.0	1234	1.2	1838	5.1
28 F	0057	1.2	0656	4.9	1307	1.2	1909	5.1
29 SA	0129	1.2	0729	4.9	1341	1.3	1944	4.9
30 SU	0203	1.3	0806	4.7	1418	1.4	2024	4.7
31 M	0241	1.5	0852	4.5	1502	1.6	2116	4.5

No.	PLACE	Lat. N.		Long. W.		TIME DIFFERENCES				HEIGHT DIFFERENCES (IN METERS)				M.L.
						High Water		Low Water		MHWS	MHWN	MLWN	MLWS	Z_0 m.
						(Zone G.M.T)								
14	PLYMOUTH (DEVONPORT) (see page 2)					0000 and 1200	0600 and 1800	0000 and 1200	0600 and 1800	5.5	4.4	2.2	0.8	
	England													
1	Isles of Scilly													
	St. Mary's	49	55	6	19	−0030	−0110	−0100	−0020	+0.2	−0.1	−0.2	−0.1	3.13
2	Penzance (Newlyn)	50	06	5	33	−0040	−0105	−0045	−0020	+0.1	0.0	−0.2	0.0	3.08
2a	Porthleven	50	05	5	19	−0045	−0105	−0035	−0025	0.0	−0.1	−0.2	0.0	3.08
3	Lizard Point	49	57	5	12	−0045	−0055	−0040	−0030	−0.2	−0.2	−0.3	−0.2	2.99
4	Coverack	50	01	5	05	−0030	−0040	−0020	−0010	−0.2	−0.2	−0.3	−0.2	2.99
4a	Helford River (Entrance)	50	05	5	05	−0030	−0035	−0015	−0010	−0.2	−0.2	−0.3	−0.2	⊙
	River Fal													
5	Falmouth	50	09	5	03	−0030	−0030	−0010	−0010	−0.2	−0.2	−0.3	−0.2	3.00
5a	Truro	50	16	5	03	−0020	−0025	§	§	−2.0	−2.0	§	§	⊙
7	Mevagissey	50	16	4	47	−0010	−0015	−0005	+0005	−0.1	−0.1	−0.2	−0.1	3.14
7a	Par	50	21	4	42	−0005	−0015	0000	−0010	−0.4	−0.4	−0.4	−0.2	3.12
	River Fowey													
8	Fowey	50	20	4	38	−0010	−0015	−0010	−0005	−0.1	−0.1	−0.2	−0.2	3.14
8a	Lostwithiel	50	24	4	40	+0005	−0010	§	§	−4.1	−4.1	§	§	⊙
11	Looe	50	21	4	27	−0010	−0010	−0005	−0005	−0.1	−0.2	−0.2	−0.2	⊙
12	Whitsand Bay	50	20	4	15	0000	0000	0000	0000	0.0	+0.1	−0.1	+0.2	⊙
	River Tamar													
14	PLYMOUTH (DEVONPORT)	50	22	4	11			STANDARD PORT		See Table V				3.35
14a	Jupiter Point	50	23	4	14	+0010	+0005	0000	−0005	0.0	0.0	+0.1	0.0	3.35
14b	Saltash	50	24	4	12	0000	+0010	0000	−0005	+0.1	+0.1	+0.1	+0.1	3.36
14c	Cargreen	50	26	4	12	0000	+0010	+0020	+0020	0.0	0.0	−0.1	0.0	3.39
14d	Cotehele Quay	50	29	4	13	0000	+0020	+0045	+0045	−0.9	−0.9	−0.8	−0.4	2.40
	River Tavy													
14e	Lopwell	50	28	4	09	⊙	⊙	§	§	−2.6	−2.7	§	§	⊙
	River Lynher													
14f	St. Germans	50	23	4	18	0000	0000	+0020	+0020	−0.3	−0.1	0.0	+0.2	3.34
15	Turnchapel	50	22	4	07	0000	0000	+0010	−0015	0.0	+0.1	+0.2	+0.1	3.32
15a	Bovisand Pier	50	20	4	08	0000	−0020	0000	−0010	−0.2	−0.1	0.0	+0.1	3.30
	River Yealm													
17	Entrance	50	18	4	04	+0006	+0006	+0002	+0002	−0.1	−0.1	−0.1	−0.1	3.18
14	PLYMOUTH (DEVONPORT) (see page 2)					0100 and 1300	0600 and 1800	0100 and 1300	0600 and 1800	5.5	4.4	2.2	0.8	
	Salcombe River													
20	Salcombe	50	13	3	47	0000	+0010	+0005	−0005	−0.2	−0.3	−0.1	−0.1	3.10
21	Start Point	50	13	3	39	+0005	+0030	−0005	+0005	−0.2	−0.4	−0.1	−0.1	3.20 ★
	River Dart													
23	Darmouth	50	21	3	34	+0025	+0040	+0015	0000	−0.8	−0.6	−0.1	−0.2	2.80 ★
23a	Greenway Quay	50	23	3	35	+0030	+0045	+0025	+0005	−0.6	−0.6	−0.2	−0.2	2.84
23b	Totness	50	26	3	41	+0025	+0040	+0015	0030	−2.0	−2.1	§	§	1.36
25	Torquay	50	28	3	31	+0025	+0045	+0010	0000	−0.6	−0.7	−0.2	−0.1	2.89 ★
26	Teignmouth (Approaches)	50	33	3	30	+0025	+0040	0000	0000	−0.7	−0.8	−0.3	−0.2	2.74 ★
26a	Exmouth (Approaches)	50	36	3	23	+0030	+0050	+0015	+0005	−0.9	−1.0	−0.5	−0.3	⊙
	River Exe													
27	Exmouth Dock	50	37	3	25	+0040	+0100	+0050	+0020	−1.5	−1.6	−0.9	−0.6	⊙
27a	Starcross	50	38	3	27	+0040	+0110	⊙	⊙	−1.4	−1.5	⊙	⊙	⊙
27b	Topsham	50	54	3	28	+0045	+0105	⊙	⊙	−1.5	−1.6	⊙	⊙	⊙
28	Lyme Regis	50	43	2	56	+0040	+0100	+0005	−0005	−1.2	−1.3	−0.5	−0.2	2.44 ★
29	Bridport (West Bay)	50	42	2	45	+0025	+0040	0000	0000	−1.4	−1.4	−0.0	−0.2	2.43 ★
30	Chesil Beach	50	37	2	33	+0040	+0055	−0005	+0010	−1.6	−1.5	−0.5	0.0	2.28 ★
31	Chesil Cove	50	34	2	28	+0035	+0050	−0010	+0005	−1.5	−1.6	−0.5	−0.2	2.27 ★

No data.
§ Dries out except for river water.
★ See notes on page 344.
C For intermediate heights, use harmonic constants (see Part III) AND n.p.159.
∫ For intermediate heights, see pages xxii to xxiv.
X M.L. inferred.

No.	PLACE	Lat. N.	Long. W.	High Water		Low Water		MHWS	MHWN	MLWN	MLWS	Z₀ m.	
				(Zone G.M.T)								M.L.	
33	PORTLAND	(see page 6)		0100 and 1300	0700 and 1900	0100 and 1300	0700 and 1900	2.1	1.4	0.7	0.2	1.00	
34	LULWORTH Cove	50 37	2 15	−0015	−0005	−0005	+0005	+0.2	+0.1	+0.2	+0.1	⊙	★
65	PORTSMOUTH	(see page 14)		0000 and 1200	0600 and 1800	0500 and 1700	1100 and 2300	4.7	3.8	1.8	0.6		
35	Swanage	50 37	1 57	−0250	+0105	−0105	−0105	−2.7	−2.2	−0.7	−0.3	1.19	*j
	Poole Harbour												
36	Entrance	50 40	1 56	−0240	+0105	−0100	−0030	−2.7	−2.2	−0.7	−0.3	1.49	*j
36a	Town Quay	50 43	1 59	−0210	+0140	−0015	−0005	−2.6	−2.2	−0.7	−0.2	1.49	*j
36b	Pottery Pier	50 42	1 59	−0150	+0200	−0010	0000	−2.7	−2.1	−0.6	0.0	1.49	*j
36c	Wareham (River Frome)	50 41	2 06	−0140	+0205	+0110	+0035	−2.5	−2.1	−0.7	+0.1	⊙	§*
36d	Cleavel Point	50 40	2 00	−0220	+0130	−0025	−0015	−2.6	−2.3	−0.7	−0.3	⊙	*
37	Bournemouth	50 43	1 52	−0240	+0055	−0050	−0030	−2.7	−2.2	−0.8	−0.3	1.49	*j
38	Christchurch(Entrance)	50 43	1 45	−0230	+0030	−0035	−0035	−2.9	−2.4	−1.2	−0.2	1.17	*j
38a	Christchurch(Tuckton)	50 44	1 47	−0205	+0110	+0110	+0105	−3.0	−2.5	−1.0	+0.1	1.14	§*
39	Hurst Point	50 42	1 33	−0115	−0005	−0030	−0025	−2.0	−1.5	−0.5	−0.1	1.97	*j
40	Lymington	50 46	1 32	−0110	+0005	−0020	−0020	−1.7	−1.2	−0.5	−0.1	2.04	*j
42	Bucklers Hard	50 48	1 25	−0040	−0010	+0010	−0010	−1.0	−0.8	−0.2	−0.3	2.40	*j
43	Stanson Point	50 47	1 21	−0050	−0010	−0005	−0010	−0.9	−0.6	−0.2	0.0	2.39	*j
	Isle of wight												
45	Yarmouth	50 42	1 30	−0105	+0005	−0025	−0030	−1.6	−1.3	−0.4	0.0	2.03	*j
46	Totland Bay	50 41	1 33	−0130	−0045	−0040	−0040	−2.0	−1.5	−0.5	−0.1	1.88	*j
48	Freshwater	50 40	1 31	−0210	+0025	−0040	−0020	−2.1	−1.5	−0.4	0.0	1.62	*j
51	Ventnor	50 36	1 12	−0025	−0030	−0025	−0030	−0.8	−0.6	−0.2	+0.2	2.33	*j
53	Sandown	50 39	1 09	0000	+0005	+0010	+0025	−0.6	−0.5	−0.2	0.0	2.41	*j
54	Bembridge Harbour	50 42	1 06	−0010	+0005	+0020	0000	−1.6	−1.5	−1.4	−0.6	⊙	*j
58	Ryde	50 44	1 07	−0010	+0010	−0005	−0010	−0.2	−0.1	0.0	+0.1	2.76	*j
	Median River												
60	Cowes	50 46	1 18	−0015	+0015	0000	−0020	−0.5	−0.3	−0.1	0.0	2.67	*j
60a	Folly Inn	50 44	1 17	−0015	+0015	0000	−0020	−0.6	−0.4	−0.1	+0.2	⊙	*j
60b	Newport	50 42	1 17	⊙	⊙	⊙	⊙	−0.6	−0.4	+0.1	+1.3	⊙	*j
62	SOUTHAMPTON			0400 and 1600	1100 and 2300	0000 and 1200	0600 and 1800	4.5	3.7	1.8	0.5		
		(see page 10)											
61	Calshot Castle	50 49	1 18	+0015	+0030	+0015	+0005	0.0	0.0	+0.2	+0.3	2.96	*j
62	SOUTHAMPTON	50 54	1 24	STANDARD PORT					See Table V			2.87	
62a	Redbridge	50 54	1 28	−0020	+0005	0000	−0005	−0.1	−0.1	−0.1	−0.1	2.82	*
	River Hamble												
63	Warsash	50 51	1 18	+0020	+0010	+0010	0000	0.0	+0.1	+0.3	+0.3	2.95	*
63a	Bursledon	50 53	1 18	+0020	+0020	+0010	+0010	+0.1	+0.1	+0.2	+0.2	3.05	*
65	PORTSMOUTH	(see page 14)		0500 and 1700	1000 and 2200	0000 and 1200	0600 and 1800	4.7	3.8	1.8	0.6		
64	Lee−on−the−Solent	50 48	1 12	−0005	+0005	−0015	−0010	−0.2	−0.1	+0.1	+0.2	⊙	*j
65	PORTSMOUTH	50 48	1 07	STANDARD PORT					See Table V			2.84	
	Chichester Harbour												
68	Entrance	50 47	0 56	−0010	+0005	+0015	+0020	+0.2	+0.2	0.0	+0.1	2.83	j
68a	Northney	50 50	0 58	+0010	+0015	+0015	+0025	+0.2	0.0	−0.2	−0.3	0.73	j
68b	Bosham	50 50	0 52	0000	+0010	⊙	⊙	+0.2	+0.1	⊙	⊙	⊙	
68c	Itchenor	50 48	0 52	−0005	+0005	+0005	+0025	+0.1	0.0	−0.2	−0.2	2.90	j
68d	Dell Quary	50 49	0 49	+0005	+0015	⊙	⊙	+0.2	+0.1	⊙	⊙	⊙	
69	Selsey Bill	50 43	0 47	−0005	−0005	−0035	+0035	+0.6	+0.6	0.0	0.0	2.94	j
70	Nab Tower	50 40	0 57	−0015	0000	−0015	+0015	−0.2	0.0	+0.2	0.0	2.58	j

SEASONAL CHANGES IN MEAN LEVEL

No.	Jan 1	Feb 1	Mar 1	Apr 1	May 1	June 1	July 1	Aug 1	Sep 1	Oct 1	Nov 1	Dec 1	Jan 1
1 − 60b	Negligible												
61 − 63a	+0.1	0.0	−0.1	−0.1	−0.1	0.0	0.0	0.0	0.0	0.0	+0.1	+0.1	+0.1
64 − 70	Negligible												

No.	PLACE	Lat. N.		Long. W.		High Water (Zone G.M.T.)		Low Water		MHWS	MHWN	MLWN	MLWS
244	ABERDEEN	(see page 70)				0000 and 1200	0600 and 1800	0100 and 1300	0700 and 1900	4.3	3.4	1.6	0.6
	River Tay												
235	Bar	56	27	2	38	+0100	+0057	+0100	+0110	+0.9	+0.8	+0.3	+0.1
236	Dundee	56	27	2	58	+0132	+0129	+0125	+0150	+1.0	+0.9	+0.3	+0.1
236a	Newburgh	56	21	3	14	+0212	+0203	+0249	+0337	−0.2	−0.4	−1.1	−0.5
236b	Perth	56	24	3	27	+0217	+0225	+0512	+0528	−0.9	−1.4	−1.2	−0.3
241	Arbroath	56	33	2	35	+0056	+0037	+0034	+0055	+0.7	+0.7	+0.2	+0.1
242	Montrose	56	42	2	27	+0100	+0100	+0030	+0040	+0.5	+0.5	+0.3	+0.1
243	Stonehaven	56	58	2	12	+0013	+0008	+0013	+0009	+0.2	+0.2	+0.1	0.0
244	ABERDEEN	57	09	2	05	STANDARD PORT				See Table V			
245	Peterhead	57	30	1	46	−0035	−0045	−0035	−0040	−0.5	−0.3	−0.1	−0.1
246	Fraserburgh	57	41	2	00	−0045	−0115	−0110	−0045	−0.4	−0.3	−0.1	0.0
244	ABERDEEN	(see page 70)				0200 and 1400	0900 and 2100	0400 and 1600	0900 and 2100	4.3	3.4	1.6	0.6
247	Banff	57	40	2	31	−0100	−0150	−0150	−0050	−0.8	−0.6	−0.5	−0.2
247a	Whitehills	57	41	2	35	−0122	−0137	−0117	−0127	−0.4	−0.3	+0.1	+0.1
248	Buckie	57	40	2	58	−0130	−0145	−0125	−0140	−0.2	−0.2	0.0	+0.1
249	Lossiemouth	57	43	3	18	−0125	−0200	−0130	−0130	−0.2	−0.2	0.0	0.0
250	Burghead	57	42	3	29	−0120	−0150	−0135	−0120	−0.2	−0.2	0.0	0.0
253	Nairn	57	36	3	52	−0120	−0150	−0135	−0130	0.0	−0.1	0.0	+0.1
254	McDermott Base	57	36	3	59	−0110	−0140	−0120	−0115	−0.1	−0.1	+0.1	+0.3
244	ABERDEEN	(see page 70)				0300 and 1500	1000 and 2200	0000 and 1200	0700 and 1900	4.3	3.4	1.6	0.6
	Inverness Firth												
255	Fortrose	57	35	4	08	−0125	−0125	−0125	−0125	0.0	0.0	⊙	⊙
256	Inverness	57	30	4	15	−0050	−0150	0200	−0105	+0.5	+0.3	+0.2	+0.1
	Cromarty Firth												
258	Cromarty	57	41	4	02	−0120	−0155	−0155	−0120	0.0	0.0	+0.1	+0.2
259	Invergordon	57	41	4	10	−0105	−0200	−0200	−0110	+0.1	+0.1	+0.1	+0.1
260	Dingwall	57	36	4	25	−0045	−0145	⊙	⊙	+0.1	+0.2	⊙	⊙
244	ABERDEEN	(see page 70)				0300 and 1500	0800 and 2000	0200 and 1400	0800 and 2000	4.3	3.4	1.6	0.6
	Dornoch Firth												
261	Portmahomack	57	50	3	50	−0120	−0210	−0140	−0110	−0.2	−0.1	+0.1	+0.1
262	Meikle Ferry	57	51	4	08	−0100	−0140	−0120	−0055	+0.1	0.0	−0.1	0.0
264	Golspie	57	58	3	59	−0130	−0215	−0155	−0130	−0.3	−0.3	−0.1	0.0
267	Wick	58	26	3	05	−0155	−0220	−0210	−0220	−0.9	−0.7	−0.2	−0.1
268	Duncansby Head	58	39	3	03	−0320	−0320	−0320	−0320	−1.2	−1.0	⊙	⊙
244	ABERDEEN	(see page 70)				0300 and 1500	1100 and 2300	0200 and 1400	0900 and 2100	4.3	3.4	1.6	0.6
	Orkney Islands												
270	Muckle Skerry	58	41	2	55	−0230	−0230	−0230	−0230	−1.7	−1.4	−0.6	−0.2
271	Burrayness	58	51	2	52	−0200	−0200	−0155	−0155	−1.0	−0.9	−0.3	0.0
272	Deer Sound	58	58	2	50	−0245	−0245	−0245	−0245	−1.1	−0.9	−0.3	0.0
273	Kirkwall	58	59	2	58	−0305	−0245	−0305	−0250	−1.4	−1.2	−0.5	−0.2
275	Kettlet Pier	59	14	2	36	−0230	−0230	−0225	−0225	−1.1	−0.9	−0.3	0.0
277	Pierowall	59	19	2	58	−0355	−0355	−0355	−0355	−0.6	−0.6	−0.2	0.0
279	Eynhallow Sound	59	08	3	05	−0400	−0400	−0355	−0355	−0.6	−0.6	−0.2	−0.1
280	Stormness	58	58	3	18	−0430	−0355	−0415	−0420	−0.7	−0.8	−0.1	−0.1
282	Widewall Bay	58	49	3	01	−0400	−0400	−0400	−0400	−0.7	−0.7	−0.3	−0.2

⊙ No data

* See notes on page 344

C For intermediate heights, use harmonic constants (see Part III) and N.P.159

X M.L. inferred

No.	PLACE	Lat. N.	Long W.	High Water (Zone G.M.T)		Low Water		MHWS	MHWN	MLWN	MLWS	M.L. Z_om.	
287	LERWICK	(See page 74)		0000 and 1200	0600 and 1800	0100 and 1300	0800 and 2000	2.2	1.6	0.9	0.5		
285	Fair Isle	59 33	1 38	−0020	−0025	−0020	−0035	0.0	+0.1	0.0	−0.1	1.40	
	Shetland Islands												
285a	Sumburgh	59 53	1 16	+0002	+0002	+0005	+0005	−0.5	−0.3	−0.2	−0.2	1.0	x
287	LERWICK	60 09	1 08	STANDARD PORT					See Table V			1.29	
288	Dury Voe	60 21	1 10	−0015	−0015	−0010	−0010	+0.1	+0.2	0.0	0.0	1.24	
289	Out Skerries	60 25	0 45	−0025	−0025	−0010	−0010	+0.1	+0.1	0.0	0.0	1.28	
289a	Toft Pier	60 28	1 12	−0105	−0100	−0125	−0115	+0.1	+0.1	−0.2	−0.2	1.23	
290	Burra Voe(Yell Sound)	60 30	1 03	−0025	−0025	−0025	−0025	+0.2	+0.2	0.0	00	1.31	
290a	Mid Yell	60 36	1 03	−0030	−0020	−0035	−0025	+0.2	+0.2	+0.1	0.0	1.52	
290b	Balta Sound	60 45	0 50	−0055	−0055	−0045	−0045	+0.1	+0.2	0.0	−0.1	1.32	
291	Burra Firth	60 48	0 52	−0110	−0110	−0115	−0115	+0.3	+0.3	0.0	0.0	⊙	
292	Bluemull Sound	60 42	1 00	−0135	−0135	−0155	−0155	+0.4	+0.3	+0.1	+0.0	⊙	
293	Sullom Voe	60 27	1 18	−0135	−0125	−0135	−0120	+0.1	+0.3	0.0	−0.2	1.32	
294	Hillswick	60 29	1 29	−0220	−0220	−0200	−0200	0.0	0.0	−0.1	0.0	⊙	
295	Scalloway	60 08	1 16	−0150	−0150	−0150	−0150	−0.6	−0.3	−0.3	0.0	⊙	
296	Quendale Bay	59 54	1 20	−0025	−0025	−0030	−0030	−0.4	−0.1	−0.2	0.0	1.14	
296a	Foula	60 07	2 03	−0140	−0130	−0140	−0120	−0.2	−0.1	−0.1	−0.1	1.23	
244	ABERDEEN	(see page 70)		0300 and 1500	1000 and 2200	0100 and 1300	0800 and 2000	4.3	3.4	1.6	0.6		
297	Stroma	58 40	3 08	−0320	−0320	−0320	−0320	−1.2	−1.1	−0.3	−0.1	⊙	
298	Scrabster	58 37	3 33	−0455	−0510	−0500	−0445	+0.7	+0.3	+0.5	+0.2	2.94	
334	ULLAPOOL	(see page 78)		0100 and 1300	0700 and 1900	0300 and 1500	0900 and 2100	5.2	3.9	2.1	0.7		
299	Sule Skerry	59 05	4 24	+0100	+0120	+0110	+0100	−1.2	−0.9	−0.4	−0.1	2.3	x
	Lach Eriboll												
300	Portnancon	58 30	4 42	+0055	+0105	+0055	+0100	0.0	+0.1	+0.1	+0.2	3.02	
301	Kyle of Durness	58 36	4 47	+0030	+0030	+0050	+0050	−0.6	−0.4	−0.3	−0.1	⊙	
304	Rona	59 08	5 49	+0010	+0030	+0010	+0030	−1.8	−1.4	−0.8	−0.3	1.73	
	Hebrides												
308	Stornoway	58 12	6 23	−0010	−0010	−0010	−0010	−0.4	−0.2	−0.1	0.0	2.86	
309	Each Shell	58 00	6 25	−0023	−0010	−0010	−0027	−0.4	−0.3	−0.2	0.0	2.75	
310	E. Lach Tarbert	57 54	6 48	−0035	−0020	−0020	−00300	−0.2	−0.2	0.0	+0.1	2.93	
311	Lach Maddy	57 36	7 06	−0054	−0024	−0026	−0040	−0.4	−0.3	−0.2	0.0	⊙	
311a	Lach Carnan	57 22	7 16	−0100	−0020	−0030	−0050	−0.7	−0.7	−0.2	−0.1	2.5	x
312	Lach Skiport	57 20	7 16	−0110	−0035	−0034	−0034	−0.6	−0.6	−0.4	−0.2	⊙	
313	Lach Boisdale	57 09	7 16	−0105	−0030	−0035	−0045	−0.9	−0.9	−0.5	−0.2	⊙	
314	Barra(North Bay)	57 00	7 24	−0113	−0041	−0044	−0058	−1.0	−0.7	−0.3	−0.1	⊙	
314a	Castle Bay	56 57	7 29	−0125	−0050	−0055	−0110	−0.9	−0.8	−0.4	−0.1	2.35	
316	Barra Head	56 47	7 38	−0125	−0050	−0055	−0105	−1.2	−0.9	−0.3	+0.1	⊙	
317	Shillay	57 31	7 41	−0113	−0053	−0057	−0117	−1.0	−0.9	−0.8	−0.3	⊙	
318	Balivanich	57 29	7 23	−0113	−0027	−0041	−0055	−1.1	−0.8	−0.6	−0.2	⊙	
318a	Scolpaig	57 39	7 29	−0046	−0046	−0049	−0049	−1.3	−0.9	−0.5	0.0	2.3	x
319	Leverburgh	57 46	7 01	−0051	−0030	−0025	−0035	−0.6	−0.4	−0.2	−0.1	2.59	
320	W.Loch Tarbert	57 55	6 55	−0103	−0043	−0024	−0044	−1.0	−0.7	−0.8	−0.3	2.13	
321	Little Bernera	58 16	6 52	−0031	−0021	−0027	−0037	−0.9	−0.8	−0.5	−0.2	⊙	
321a	Carloway	58 17	6 47	−0050	+0010	−0045	−0025	−1.0	−0.7	−0.5	−0.1	2.33	

SEASONAL CHANGES IN MEAN LEVEL

No.	Jan. 1	Feb. 1	Mar. 1	Apr. 1	May 1	June 1	July 1	Aug 1	Sep 1	Oct 1	Nov. 1	Dec. 1	Jan.1
235−246	+0.1	0.0	0.0	−0.1	−0.1	−0.1	0.0	0.0	0.0	0.0	+0.1	+0.1	+0.1
247−260	+0.1	0.0	−0.1	−0.1	−0.1	−0.1	0.0	0.0	0.0	0.0	+0.1	+0.1	+0.1
261−298	+0.1	0.0	0.0	−0.1	−0.1	−0.1	−0.1	0.0	0.0	0.0	+0.1	+0.1	+0.1
299−334	+0.1	0.0	0.0	0.0	−0.1	−0.1	−0.1	0.0	0.0	0.0	0.0	+0.1	+0.1

No.	PLACE	Lat.N.		Long.W.		TIME DIFFERENCES				HEIGHT DIFFERENCES (IN METRES)				M.L.	
						High Water		Low Water		MHWS	MHWN	MLWN	MLWS	Z_0 m.	
						(Zone G.M.T.)									
478	**HOLYHEAD**	(see page 94)				0000 and 1200	0600 and 1800	0500 and 1700	1100 and 2300	**5.7**	**4.5**	**2.0**	**0.7**		
475	Caernarvon	53	09	4	16	−0030	−0030	+0015	−0005	−0.4	−0.4	−0.1	−0.1	3.04	
475a	Fort Belan	53	07	4	20	−0040	−0015	−0025	−0005	−1.0	−0.9	−0.2	−0.1	2.83	
476	Trwyn Dinmor	53	19	4	03	+0025	+0015	+0050	+0035	+1.9	+1.5	+0.5	+0.2	4.23	
476a	Moelfre	53	20	4	14	+0025	+0020	+0050	+0035	+1.9	+1.4	+0.5	+0.2	4.17	
477	Amlwch	53	25	4	20	+0020	+0010	+0035	+0025	+1.6	+1.3	+0.5	+0.2	4.08	
477a	Cemaes Bay	53	25	4	27	+0020	+0025	+0040	+0035	+1.0	+0.7	+0.3	+0.1	3.67	
478	**HOLYHEAD**	53	19	4	37	**STANDARD PORT**				See Table V				3.21	
479	Trearddur Bay	53	16	4	37	−0045	−0025	−0015	−0015	−0.4	−0.4	0.0	+0.1	3.08	
479a	Porth Trecastell	53	12	4	30	−0045	−0025	−0005	−0015	−0.6	−0.6	0.0	0.0	2.97	
480	Llanddwyn Island	53	08	4	25	−0115	−0055	−0030	−0020	−0.7	−0.5	−0.1	0.0	2.95	
480a	Trevor	53	00	4	25	−0115	−0100	−0030	−0020	−0.8	−0.9	−0.2	−0.1	2.55	
481	Porth Dinllaen	52	57	4	34	−0120	−0105	−0035	−0025	−1.0	−1.0	−0.2	−0.2	2.51	
481a	Porth Ysgaden	52	54	4	39	−0125	−0110	−0040	−0035	−1.1	−1.0	−0.1	−0.1	2.45	
482	Bardsey Island	52	46	4	47	−0220	−0240	−0145	−0140	−1.2	−1.2	−0.5	−0.1	⊙	
496	**MILFORD HAVEN**	(see page 98)				0100 and 1300	0800 and 2000	0100 and 1300	0700 and 1900	**7.0**	**5.2**	**2.5**	**0.7**		
	Cardigan Bay														
482a	Aberdaron	52	48	4	43	+0210	+0200	+0240	+0310	−2.4	−1.9	−0.6	−0.2	2.55	cx
482b	St. Tudwal's Roads	52	49	4	29	+0155	+0145	+0240	+0310	−2.2	−1.9	−0.7	−0.2	2.52	c
483	Pwllheli	52	53	4	24	+0210	+0150	+0245	+0320	−2.0	−1.8	−0.6	−0.2	2.59	c
483a	Criccieth	52	55	4	14	+0210	+0155	+0255	+0320	−2.0	−1.8	−0.7	−0.3	2.56	
484	Porthmadog	52	55	4	08	+0235	+0210	⊙	⊙	−1.9	−1.8	⊙	⊙	2.64	c
485	Barmouth	52	43	4	03	+0215	+0205	+0310	+0320	−2.0	−1.7	−0.7	0.0	2.61	c
486	Aberdovey	52	32	4	03	+0215	+0200	+0230	+0305	−2.0	−1.7	−0.5	0.0	2.70	c
487	Aberystwyth	52	24	4	05	+0145	+0130	+0210	+0245	−2.0	−1.7	−0.7	0.0	⊙	
488	New Quay	52	13	4	21	+0150	+0125	+0155	+0230	−2.1	−1.8	−0.6	−0.1	2.6	cx
488a	Aberporth	52	08	4	33	+0135	+0120	+0150	+0220	−2.1	−1.8	−0.6	−0.1	2.39	
489	Port Cardigan	52	07	4	42	+0140	+0120	+0220	+0130	−2.3	−1.8	−0.5	0.0	⊙	
489a	Cardigan (Town)	52	05	4	40	+0220	+0150	⊙	⊙	−2.2	−1.6	⊙	⊙	2.52	
490	Fishguard	52	00	4	58	+0115	+0100	+0110	+0135	−2.2	−1.8	−0.5	+0.1	2.65	
491	Porthgain	51	57	5	11	+0055	+0045	+0045	+0100	−2.5	−1.8	−0.6	0.0	2.80	
492	Ramsey Sound	51	53	5	19	+0030	+0030	+0030	+0030	−1.9	−1.3	−0.3	0.0	3.18	
492a	Solva	51	52	5	12	+0015	+0010	+0030	+0015	−1.5	−1.0	−0.2	0.0	3.32	
492b	Little Haven	51	46	5	06	+0010	+0010	+0025	+0015	−1.1	−0.8	−0.2	0.0	3.54	
493	Martin's Haven	51	44	5	15	+0010	+0010	+0015	+0015	−0.8	−0.5	+0.1	+0.1	⊙	
494	Skomer Island	51	44	5	17	−0005	−0005	+0005	+0005	−0.4	−0.1	0.0	0.0	⊙	
495	Dale Roads	51	42	5	09	−0005	−0005	−0008	−0008	0.0	0.0	0.0	−0.1	3.83	
496	MILFORD HAVEN	51	42	5	01	**STANDARD PORT**				See Table V					
	Cleddau River														
497	Neyland	51	42	4	57	+0002	+0010	0000	0000	0.0	0.0	0.0	0.0	⊙	
498	Black Tar	51	45	4	54	+0010	+0020	+0005	0000	+0.1	+0.1	0.0	−0.1	⊙	
499	Haverfordwest	51	48	4	58	+0010	+0025	⊙	⊙	−4.8	−4.9	§	§	⊙	
501	Stackpole Quay	51	37	4	54	−0005	+0025	−0010	−0010	+0.9	+0.7	+0.2	+0.3	4.28	
502	Tenby	51	40	4	42	−0015	−0010	−0015	−0020	+1.4	+1.1	+0.5	+0.2	4.49	
	Towy River														
504	Ferryside	51	46	4	22	0000	−0010	+0220	0000	−0.3	−0.7	−1.7	−0.6	⊙	
504a	Carmarthen	51	51	4	18	+0010	0000	⊙	⊙	−4.4	−4.8	§	§	⊙	
	Burry Inlet														
505	Burry Port	51	41	4	15	+0003	+0003	+0007	+0007	+1.6	+1.4	+0.5	+0.4	4.75	
505a	Llanelli	51	40	4	10	−0003	−0003	+0150	+0020	+0.8	+0.6	⊙	⊙	⊙	
508	Mumbles	51	34	3	58	+0005	+0010	−0020	−0015	+2.3	+1.7	+0.6	+0.2	5.1	x
509	SWANSEA	51	37	3	55	**STANDARD PORT**				See Table V				5.24	
510	Port Talbot	51	35	3	49	−0005	+0005	−0015	−0030	+2.6	+2.1	+0.8	+0.3	5.33	
512	Porthcawl	51	28	3	42	0000	0000	0000	−0015	+2.9	+2.3	+0.8	+0.3	5.31	

⊙ No data

* See notes on page 344.

§ Dries out except for river water.

a Data approxiamte

c For intermediate heights, use harmonic constants (see Part III) and N.P.

X M.L. inferred.

No.	PLACE	Lat. N		Long. W		High Water		Low Water		MHWS	MHWN	MLWN	MLWS	Z₀ m.	
						(Zone G.M.T.)									
523	PORT OF BRISTOL	(see page 106)				0600 and	1100 and	0300 and	0800 and	13.2	10.0	3.5	0.9		
	(AVONMOUTH)					1800	2300	1500	2000						
513	Barry	51	23	3	16	−0030	−0015	−0125	−0030	−1.8	−1.3	+0.2	0.0	6.11	
513a	Flatholm	51	23	3	07	−0015	−0015	−0045	−0045	−1.4	−1.2	+0.2	+0.1	6.2	x
513b	Steepholm	51	20	3	06	−0020	−0020	−0050	−0050	−1.6	−1.4	+0.1	−0.1	6.1	x
514	Cardiff	51	27	3	09	−0015	−0015	−0100	−0030	−1.0	−0.6	+0.1	0.0	6.45	
515	Newport	51	33	2	59	−0020	−0010	0000	−0020	−1.1	−1.0	−0.6	−0.7	6.04	
	River Wye														
516	Chepstow	51	39	2	40	+0020	+0020	⊙	⊙	⊙	⊙	⊙	⊙	⊙	
523	PORT OF BRISTOL	(see page 106)				0000 and	0600 and	0000 and	0700 and	13.2	10.0	3.5	0.9		
	(AVONMOUTH)					1200	1800	1200	1900						
	England														
	River Severn														
517	Sudbrook	51	35	2	43	+0010	+0010	+0025	+0015	+0.2	+0.1	−0.1	+0.1	⊙	
518	Beachley(Aust.)	51	36	2	38	+0010	+0015	+0040	+0025	−0.2	−0.2	−0.5	−0.3	6.43	
519	Inward Rocks	51	39	2	37	+0020	+0020	+0105	+0045	−1.0	−1.1	−1.4	−0.6	5.66	* c
520	Narlwood Rocks	51	39	2	36	+0025	+0025	+0120	+0100	−1.9	−2.0	−2.3	−0.8	⊙	*
521	White House	51	40	2	33	+0025	+0025	+0145	+0120	−3.0	−3.1	−3.6	−1.0	3.93	* c
522	Berkeley	51	42	2	30	+0030	+0045	+0245	+0220	−3.8	−3.9	−3.4	−0.5	3.43	* c
522a	Sharpness Dock	51	43	2	29	+0035	+0050	+0305	+0245	−3.9	−4.2	−3.3	−0.4	⊙	*
522b	Wellhouse Rock	51	44	2	29	+0040	+0055	+0320	+0305	−4.1	−4.4	−3.1	−0.2	3.26	* c
522c	Epney	51	42	2	24	+0130	⊙	⊙	⊙	−9.4	⊙	⊙	⊙	⊙	*
522d	Minsterworth	51	50	2	23	+0140	⊙	⊙	⊙	−10.1	⊙	⊙	⊙	⊙	*
522e	Llanthony	51	51	2	21	+0215	⊙	⊙	⊙	−10.7	⊙	⊙	⊙	⊙	*
523	PORT OF BRISTOL	(see page 106)				0200 and	0800 and	0300 and	0800 and	13.2	10.0	3.5	0.9	6.92	
	(AVONMOUTH)					1400	2000	1500	2000						
	River Avon														
523a	Shirehampton	51	29	2	41	0000	0000	+0035	+0010	−0.7	−0.7	−0.8	0.0	⊙	
523b	Sea Mills	51	29	2	39	+0005	+0005	+0105	+0030	−1.4	−1.5	−1.7	−0.1	⊙	
524	Bristol(Cumberland Basin)	51	27	2	37	+0010	+0010	§	§	−2.9	−3.0	§	§	⊙	
524a	Portishead	51	30	2	45	−0002	0000	⊙	⊙	−0.1	−0.1	⊙	⊙	⊙	
525	Clevedon	51	27	2	52	−0010	−0020	−0025	−0015	−0.4	−0.2	+0.2	0.0	6.8	x
526	English and Welsh Grounds	51	28	2	59	−0008	−0008	−0030	−0030	−0.5	−0.8	−0.3	0.0	6.5	ax
527	Weston−super−Mare	51	21	2	59	−0020	−0030	−0130	−0030	−1.2	−1.0	−0.8	−0.2	6.1	x
	River Parrell														
528	Burnham	51	14	3	00	−0020	−0025	−0030	0000	−2.3	−1.9	−1.4	−1.1	⊙	
529	Bridgwater	51	08	3	00	−0015	−0030	+0305	+0455	−8.6	−8.1	§	§	⊙	*
530	Hinkley Point	51	13	3	08	−0020	−0025	−0100	−0040	−1.7	−1.6	+0.1	−0.1	5.0	x
531	Watchet	51	11	3	20	−0035	−0050	−0145	−0040	−1.9	−1.5	+0.1	+0.1	5.88	
532	Minehead	51	13	3	28	−0035	−0045	−0100	−0100	−2.6	−1.9	−0.1	0.0	5.71	
533	Porlock Bay	51	13	3	38	−0045	−0055	−0205	−0050	−3.0	−2.2	−0.1	−0.1	5.62	
534	Lynmouth	51	14	3	49	−0055	−0115	⊙	⊙	−3.6	−2.7	⊙	⊙	⊙	
496	MILFORD HAVEN	(see page 98)				0100 and	0700 and	0100 and	0700 and	7.0	5.2	2.5	0.7		
						1300	1900	1300	1900						
535	Ilfracombe	51	13	4	07	−0030	−0015	−0035	−0055	+2.2	+1.7	+0.5	0.0	4.98	
	Rivers Taw and Torridge														
536	Appledore	51	03	4	12	−0020	−0025	+0015	−0045	+0.5	0.0	−0.9	−0.5	3.64	* c
537	Yelland Marsh	51	04	4	10	−0010	−0015	+0100	−0015	−0.4	−0.9	−1.7	−1.1	2.52	* c
538	Fremington	51	05	4	07	−0010	−0015	+0030	−0030	−0.5	−1.2	−1.6	+0.1	⊙	*
539	Barnstaple	51	05	4	04	0000	−0015	−0155	−0245	−2.9	−3.8	−2.2	−0.4	⊙	*
540	Bideford	51	05	4	12	−0020	−0025	0000	0000	−1.1	−1.6	−2.5	−0.7	⊙	*
541	Clovelly	51	00	4	24	−0030	−0030	−0020	−0040	+1.3	+1.1	+0.2	+0.2	⊙	
542	Lundy Island	51	10	4	40	−0030	−0030	−0020	−0040	+1.0	+0.7	+0.2	+0.1	4.2	x
543	Bude	50	50	4	33	−0040	−0040	−0035	−0045	+0.7	+0.6	⊙	⊙	⊙	
544	Boscastle	50	41	4	42	−0045	−0010	−0110	−0100	+0.3	+0.4	+0.2	+0.2	4.02	

SEASONAL CHANGES IN MEAN LEVEL

No.	Jan. 1	Feb. 1	Mar. 1	Apr. 1	May. 1	June 1	July 1	Aug. 1	Sep. 1	Oct. 1	Nov. 1	Dec. 1	Jan. 1
475–482	+0.1	0.0	0.0	−0.1	−0.1	−0.1	0.0	0.0	0.0	0.0	+0.1	+0.1	+0.1
482a–512	0.0	0.0	0.0	0.0	0.0	0.0	−0.1	0.0	0.0	0.0	+0.1	+0.1	0.0
513–534	0.0	0.0	0.0	0.0	−0.1	−0.1	0.0	0.0	.0	0.0	0.0	0.0	0.0
535–544	+0.1	0.0	0.0	−0.1	−0.1	−0.1	0.0	0.0	0.0	0.0	+0.1	+0.1	+0.1

No.	PLACE	Lat. N.	Long. W.	TIME DIFFERENCES (Zone G.M.T)				HEIGHT DIFFERENCES (IN METERS)				M.L.
				High Water	Water	Low	Water	MHWS	MHWN	MLWN	MLWS	Z_0 m.
751	COBH (RINGASKIDDY)	(see page 130)		0500 and 1700	1100 and 2300	0500 and 1700	1100 and 2300	4.2	3.3	1.4	0.5	
741	Crookhaven	51 28	9 43	−0057	−0033	−0048	−0112	−0.8	−0.6	−0.4	−0.1	⊙
742	Skull	51 31	9 32	−0040	−0015	−0015	−0110	−0.9	−0.6	−0.2	−0.1	1.85
743	Baltimore	51 29	9 23	−005	−0005	−0010	−0050	−0.5	−0.3	+0.1	+0.1	2.12
744	Castleownshend	51 32	9 10	−0012	−0025	−0010	−0040	−0.6	−0.4	−0.2	0.0	2.2 x
745	Clonakilty Bay	51 35	8 50	−0033	−0011	−0019	−0041	−0.3	−0.2	⊙	⊙	⊙
746	Courtmacsherry	51 38	8 42	−0029	−0007	+0005	−0017	−0.4	−0.3	−0.2	−0.1	⊙
747	Kinsale	51 42	8 31	−0019	−0005	−0009	−0023	−0.1	−0.1	+0.1	0.0	2.25
	Cork Harbour											
751	COBH (RINGASKIDDY)	51 50	8 18	STANDARD PORT				See Table v				2.35
752	Passage West	51 52	8 20	+0005	+0005	+0005	+0005	+0.3	+0.3	+0.2	+0.2	⊙
753	Cork	51 54	8 27	+0020	+0020	+0020	+0020	+0.3	+0.1	0.0	0.0	⊙
754	Ballycotton	51 50	8 01	−0011	+0001	+0001	−0009	0.0	0.0	−0.1	0.0	⊙
755	Youghal	51 57	7 50	−0002	+0014	+0014	0000	0.0	0.0	−0.1	0.0	⊙
756	Dungarvan Harbour	52 05	7 34	+0004	+0012	+0012	−0001	0.0	+0.1	−0.2	0.0	⊙
	Waterford Harbour											
761	Dunmore East	52 09	6 59	+0013	+0013	+0013	+0001	0.0	0.0	−0.2	0.0	⊙
761a	Cheekpoint	52 16	7 00	+0022	+0022	+0022	+0022	+0.4	+0.2	+0.2	0.0	2.49
762	Waterford	52 16	7 07	+0057	+0057	+0057	+0046	+0.4	+0.3	−0.1	−0.1	⊙
763	New Ross	52 24	6 57	+0100	+0030	+0030	+0055	+0.4	+0.5	+0.1	+0.1	2.59
765	Great Saltee	52 07	6 38	+0010	+0009	+0009	−0004	−0.3	−0.4	⊙	⊙	⊙
766	Carnsore Point	52 10	6 22	+0029	+0019	+0019	−0002	−1.1	−1.0	⊙	⊙	⊙
819	REYKJAVIK	(see page 130)		—	—	—	—	4.0	2.9	1.3	0.2	
	Faeroe Islands											
	Vidoy											
780	Hvannesund	62 18	6 31	⊙	⊙	⊙	⊙	−2.7	−2.1	−0.8	−0.2	⊙
	Bordoy											
782	Klaksvik	62 14	6 35	+0345	+0345	+0345	+0345	−2.6	−2.1	−0.8	−0.2	⊙
	Eysturoy											
784	Nordskali	62 13	7 00	⊙	⊙	⊙	⊙	−2.0	−1.5	−0.7	−0.2	⊙
785	Oyndarfjordur	62 17	6 50	⊙	⊙	⊙	⊙	−2.3	−1.9	−0.6	−0.2	⊙
786	Fugalgjordur	62 14	6 48	⊙	⊙	⊙	⊙	−2.3	−1.9	−0.6	−0.2	⊙
787	Gotuvik (Nordragota)	62 12	6 44	⊙	⊙	⊙	⊙	−3.0	−2.3	−1.0	−0.2	⊙
788	Skalafjordur	62 06	6 43	No appreciable tide								
	Streymay											
790	Hvalvik	62 12	7 01	No appreciable tide								
791	Kollafjordur	62 06	6 56									
792	Torshavn	62 00	6 45	−0035	−0035	+0149	+0149	−3.7	−2.7	−1.2	−0.2	0.16
793	Vestmanna	62 09	7 10	+0145	+0145	+0145	+0145	−2.0	−1.5	−0.7	−0.2	⊙
	Vagar											
795	Sorvagur	62 04	7 18	+0058	+0058	+0058	+0058	−2.2	−1.6	−0.8	−0.2	⊙
796	Midvagur	62 03	7 11	⊙	⊙	⊙	⊙	−2.3	−1.7	−0.9	−0.2	⊙
	Sandoy											
798	Skopun	61 54	6 52	⊙	⊙	⊙	⊙	−2.2	−1.7	−0.7	−0.2	0.91
799	Sandur	61 50	6 48	+0100	+0100	+0100	+0100	−1.8	−1.5	−0.5	−0.2	⊙
	Suduroy											
800	Trongisvagur	61 33	6 49	+0040	+0040	+0040	+0040	−2.7	−2.1	−0.9	−0.2	⊙
801	Vagur	61 28	6 48	+0050	+0050	+0050	+0050	−3.0	−2.3	−0.9	−0.2	⊙

SEASONAL CHANGES IN MEAN LEVEL

No.	Jan 1	Feb 1	Mar 1	Apr 1	May 1	June 1	July 1	Aug 1	Sep 1	Oct 1	Nov 1	Dec 1	Jan 1
696–712	+0.1	0.0	−0.1	−0.1	0.0	0.0	0.0	−0.1	+0.1	0.0	+0.1	+0.1	+0.1
713–719	Negligible												
721–737	+0.1	0.0	0.0	−0.1	−0.1	0.0	0.0	0.0	0.0	0.0	0.0	+0.1	+0.1
741–766	0.0	0.0	0.0	−0.1	−0.1	−0.1	−0.1	0.0	+0.1	+0.1	+0.1	+0.1	0.0
780–801	+0.1	+0.1	0.0	0.0	0.0	−0.1	−0.1	−0.1	0.0	0.0	0.0	0.0	+0.1
819	+0.1	+0.1	+0.1	0.0	0.0	0.0	−0.1	−0.1	−0.1	−0.1	0.0	0.0	+0.1

No.	PLACE	Lat. N.		Long. E.		TIME DIFFERENCES				HEIGHT DIFFERENCES (IN METERS)				M.L.
						High Water	Low	Water		MHWS	MHWN	MLWN	MLWS	Z_0 m.
						(Zone -0070)								
1431	HEGOLAND	(see page 158)				—	—	—	—	2.6	2.3	0.4	0.0	
	Gulf of Yensiey													
985	Dikson Harbour	73	32	80	17	−0615	−0615	−0640	−0640	−2.2	−2.0	−0.2	+0.1	0.3
986	Sopochnaya Korga Point	71	45	82	45	−0155	−0155	−0220	−0220	−1.9	−1.7	−0.1	+0.2	0.5 x
987	Nassonovsky Island	70	52	83	22	+0415	+0415	+0350	+0350	−2.1	−1.9	−0.1	+0.2	0.37 x
	Gulf of Ob					(Zone −0600)								
989	Sabuleyaga River	72	10	75	00	−0550	−0550	−0615	−0615	−1.7	−1.7	−0.1	0.0	0.46
990	Kharsye Point	70	06	73	43	+0100	+0100	+0035	+0035	−1.9	−1.8	−0.2	0.0	0.37
992	Mys Kamyenny	68	30	73	35	+0435	+0435	+0315	+0315	−2.0	−1.9	−0.1	+0.1	0.35
	Kara Sea													
995	Mys Ragozin	73	23	70	07	+0010	+0010	−0015	−0015	−1.9	−1.7	−0.3	0.0	0.37
998	Cap Morrasale	69	43	66	48	−0510	−0510	−0535	−0535	−2.0	−1.8	−0.1	+0.2	0.4 x
1101	OSTROV YEKATERININSKAYA (KOL'SKIY ZALIU)	(see page 142)				—	—	—	—	3.7	3.0	1.3	0.5	
						(Zone −0500)								
	Novaya Zemlya													
1000	Guba Litke	72	26	55	30	−0404	−0404	⊙	⊙	−3.2	−2.7	−1.1	−0.4	⊙
1001	Mys Bvk	73	14	56	24	−0448	−0448	⊙	⊙	−3.1	−2.5	−1.0	−0.3	0.4 x
1002	Mys Zhelaniya	76	57	68	35	+0346	+0346	⊙	⊙	−3.1	−2.5	−1.0	−0.3	0.4 x
1003	Foki Bight	75	58	59	55	+0342	+0342	+0145	+0145	−3.1	−2.5	−1.0	−0.3	0.4 x
1004	Russkaya Gavan	76	14	62	32	+0320	+0320	⊙	⊙	−3.1	−2.5	−1.0	−0.3	0.4 x
1005	Krestovaya Bav	74	07	55	30	+0326	+0326	⊙	⊙	−2.9	−2.3	−0.9	−0.3	0.51
1006	Mityushika Bay	73	39	54	48	+350	+350	+0117	+0117	−2.7	−2.1	−1.0	−0.3	0.55
1007	Seryebryany Point	73	21	54	04	+0243	+0243	⊙	⊙	−3.0	−2.4	−1.1	−0.4	0.39
1008	Zaliv Pukhovy	72	39	52	42	+0328	+0328	+0052	+0052	−2.8	−2.3	−1.0	−0.3	0.55
1009	Guba Belushya	71	32	52	19	+0339	+0339	⊙	⊙	−3.2	−2.6	−1.1	−0.4	0.36
1010	Guba Nekhyatova	71	16	53	30	+0343	+0343	⊙	⊙	−3.4	−2.8	−1.2	−0.4	⊙
	Kara Strait													
1011	Guba Kamyenka	70	36	57	27	+0932	+0932	⊙	⊙	−3.0	−2.5	−1.0	−0.4	0.4 x
1012	Guba Dolgaya	70	15	58	43	−0252	−0252	−0456	−0456	−3.2	−2.6	−1.1	−0.4	⊙
	Yugorski Strait													
1013	Bukhta Varneka	69	42	60	04	+0026	+0026	+0022	+0022	−2.9	−2.4	−1.0	−0.3	0.46
1014	Ostrov Sokoliy	69	50	60	44	−0305	−0305	−0242	−0242	−2.9	−2.4	−1.0	−0.3	0.47
1015	Khabarovo	69	39	60	25	−0226	−0226	−0153	−0153	−3.0	−2.5	−1.0	−0.5	0.36
1017	Ostrov Dolgiy	69	12	59	10	−0131	−0131	⊙	⊙	−2.7	−2.2	−0.9	−0.3	⊙
1021	Ostrov Varandyey	68	49	58	00	−0129	−0129	⊙	⊙	−2.7	−2.2	−0.9	−0.3	⊙
1023	Mys Russki Zavorot	68	59	54	34	−0315	−0315	⊙	⊙	−2.7	−2.2	−0.9	−0.3	⊙
1024	Reka Pechora	68	23	54	26	+0004	+0004	−0008	−0008	−2.8	−2.3	−0.9	−0.3	0.56
						(Zone −0400)								
	Kolguyev Island													
1025	Bugrino	68	48	49	16	+0605	+0605	+0732	+0732	−2.1	−1.7	−0.7	−0.3	0.93 c
1026	Indiga River	67	42	48	46	+0021	+0021	+0921	+0929	−1.7	−1.2	−0.5	+0.1	1.31
1026a	Mys Mikulkin	67	48	46	41	+0833	+0833	+0833	+0821	−0.2	+0.3	−0.2	+0.4	2.2 x
1027	Ostrov Korga	68	23	46	10	+0646	+0646	+0646	+0634	−1.3	−1.1	−0.3	−0.1	1.45
1028	Kanin Nos.	68	40	43	17	+0410	+0410	+0358	+0358	−0.6	−0.5	−0.2	−0.1	⊙

SEASONAL CHANGES IN MEAN LEVEL

No.	Jan 1	Feb 1	Mar 1	Apr 1	May 1	June 1	July 1	Aug 1	Sep 1	Oct 1	Nov 1	Dec 1	Jan 1
885–996	+0.1	+0.1	0.0	0.0	0.0	0.0	−0.1	−0.1	−0.1	−0.1	0.0	+0.1	+0.1
910–1101	Negligible												
1431	+0.1	+0.1	0.0	0.0	0.0	0.0	−0.1	−0.1	−0.1	0.0	0.0	+0.1	+0.1

No.	PLACE		Lat. N.	Long. E.	TIME DIFFERENCES (Zone -0100)				HEIGHT DIFFERENCES (IN METERS)				M.L.
					High Water		Low Water		MHWS	MHWN	MLWN	MLWS	Z_0 m.
1539	ANTWERP (PROSPERPOLDER) Belgium		(see page 178)		0000 and 1200	0500 and 1700	0000 and 1200	0600 and 1800	5.8	4.2	1.0	0.0	2.66
1539a	Boudewijnsluis		51 17	4 20	+0013	+0005	+0025	+0020	0.0	+0.1	0.0	0.0	2.68 c
1539b	Royersluis	B	51 14	4 24	+0030	+0015	+0045	+0041	+0.3	+0.3	0.0	0.0	2.68 c
1539c	Boom		51 05	4 22	+0125	+0110	+0155	+0150	-0.2	+0.6	-0.4	+0.6	2.85
1539d	Gbent		51 03	3 44	+0430	+0415	+0630	+0600	-3.5	-2.4	-0.8	+0.1	1.0 x
1534	VLISSINGEN (FLUSHING)		(see page 174)		0300 and 1500	0900 and 2100	0400 and 1600	1000 and 2200	4.9	4.0	1.0	0.5	
1540	Cadzand (weilingen Sluis)		51 23	3 23	-0030	-0025	-0020	-0025	-0.2	-0.2	-0.1	-0.1	⊙
1562	Zeebrugge	BN	51 21	3 12	-0035	-0015	-0020	-0035	-0.1	-0.2	+0.1	-0.1	2.36
1564	Oostende	B	51 14	2 55	-0055	-0040	-0030	-0045	+0.2	+0.2	+0.1	-0.1	2.36
1565	Nieuwpoort		51 09	2 43	-0110	-0050	-0035	-0045	+0.4	+0.3	+0.2	-0.1	2.37
1568	DUNKERQUE France		(see page 182)		0200 and 1400	0800 and 2000	0200 and 1400	0900 and 2100	5.8	4.8	1.5	0.6	3.20
1569	Gravelines		51 01	2 06	-0010	-0010	-0020	0000	+0.1	+0.1	-0.1	-0.1	3.19
1570	CALAIS		50 58	1 51	STANDARD PORT				See Table V				4.02
1571	Wissant		50 33	1 40	-0030	⊙	⊙	⊙	+1.7	+1.5	+0.6	+0.6	⊙
1572	BOULOGNE		50 44	1 35	STANDARD PORT				See Table V				5.01
1579	DIEPPE		(see page 194)		0100 and 1300	0600 and 1800	0000 and1200	0700 and 1900	9.3	7.2	2.6	0.7	
1573	Le Touquet, Etaples		50 31	1 35	+0012	⊙	⊙	⊙	-0.3	0.0	+0.2	+0.3	⊙
1574	Berck La Somme		50 24	1 34	+0080	⊙	⊙	⊙	0.0	+0.1	+0.3	+0.3	⊙
1575	Le Hourdel		50 13	1 34	+0021	+0026	⊙	⊙	+0.7	+0.7	⊙	⊙	⊙
1576	St. Valery		50 11	1 37	+0028	+0040	⊙	⊙	+0.7	+0.8	⊙	⊙	⊙
1577	Caveux		50 11	1 29	+0007	+0010	-0008	+0013	+0.9	+0.7	+0.2	+0.3	5.50
1578	Le Treport		50 04	1 22	+0001	+0005	+0005	+0011	+0.1	+0.2	-0.1	0.0	5.02
1579	DIEPPE		49 56	1 05	STANDARD PORT				See Table V				4.97
1580	St. Valery-en-Caux		49 52	0 42	-0018	-0016	-0007	-0013	-0.4	-0.1	-0.1	+03	4.88
1581	Fecamp	F	49 46	0 22	-0022	-0018	-0034	-0043	-1.4	-0.7	0.0	+0.1	4.47
1581a	Antifer		49 39	0 09	-0046	-0039	-0051	-0100	-1.3	-0.6	+0.4	+0.5	4.73
1582	LE HAVRE La Seine		(see page 198)		0000 and 1200	0500 and 1700	0000 and 1200	0700 and 1900	7.9	6.6	3.0	1.2	4.87
1583	Honfluer		49 25	0 14	-0140	-0015	+0005	+0040	+0.2	+0.1	0.0	0.0	⊙ *
1584	Tancarville		49 28	0 28	-0150	+0025	+0105	+0140	0.0	0.0	+0.3	+1.0	⊙ *
1585	Quillebeouf		49 28	0 32	-0045	+0030	+0120	+0200	0.0	+0.1	+0.5	+1.4	⊙ *
1586	Vatteville		49 29	0 40	+0004	+0100	+0225	+0250	+0.1	0.0	+1.1	+2.4	⊙ *
1587	Caudebee		49 32	0 44	+0020	+0115	+0230	+0300	-0.2	0.0	+1.2	+2.5	⊙ *
1588	Duclair		49 29	0 53	+0225	+0300	+0355	+0410	-0.2	-0.1	+1.7	+3.3	⊙
1589	Rouen		49 27	1 66	+0440	+0415	+0525	+0525	-0.1	0.0	+1.9	+3.6	⊙
1590	Trouville		49 22 N	0 05 W	-0035	-0015	0000	-0010	-0.1	-0.2	-0.2	-0.1	4.50
1591	Dives		49 18	0 06	-0055	⊙	⊙	-0115	-0.4	-0.5	-0.6	-0.3	⊙
1592	Ouistreham		49 17	0 15	-0020	-0010	-0005	-0010	-0.2	-0.2	-0.2	-0.2	4.44
1593	Courseulles		49 20	0 27	-0010	⊙	⊙	-0040	-0.8	-1.0	-0.7	-0.3	3.95
1594	Port-en-Bessin		49 21	0 45	-0045	-0040	-0040	-0045	-0.7	-0.6	-0.3	-0.1	4.22

⊙   No data
*   See notes on page 344
B   Tides predicted in Belgian Tide Tables
F   Tides predicted in French Tide Tables.
N   Tides predicted in Netherlands Tide Tables
x   M.L. inferred

FRANCE, WEST COAST

No.	PLACE		Lat. N.		Long. W.		TIME DIFFERENCES				HEIGHT DIFFERENCES (IN METERS)				M.L.
							High Water	Water	Low	Water	MHWS	MHWN	MLWN	MLWS	Z_0 m.
							(Zone -0100)								
1638	BREST		(see page 210)				0000 and 1200	0600 and 1800	0000 and 1200	0600 and 1800	7.5	5.9	3.0	1.4	
1659	Ile de Hoedie		47	20	2	52	−0005	−0025	−0030	−0020	−2.3	−1.9	−1.1	−0.8	2.95
1660	Penerf		47	31	2	37	−0010	−0020	−0025	−0020	−2.0	−1.7	−1.0	−0.7	3.09
1661	Le Croisic		47	18	2	31	+0030	−0030	−0015	−0010	−2.3	−1.8	−1.1	−0.8	2.97
	La Loire														
1662	Le Pouliguen		47	17	2	25	+0015	−0040	0000	−0020	−2.1	−1.8	−1.2	−0.8	2.95
1663	Le Grand− Charpentier	F	47	13	2	19	+0010	−0030	−0020	−0020	−2.2	−2.0	−1.3	−0.9	2.87
1664	St. Nazaire		47	16	2	12	+0030	−0025	−0005	−0010	−2.0	−1.7	−1.1	−0.8	3.06
1665	Paimboeuf		47	17	2	02	+0015	−0005	+0120	+0030	−1.9	−1.6	−1.3	−0.5	3.15
1666	Le Pellerin		47	12	1	45	+0140	+0100	+0300	+0210	−1.7	−1.4	−1.2	+0.1	3.32
1667	Nantes(Chantenay)		47	12	1	35	+0155	+0140	+0330	+0245	−1.7	−1.3	−1.1	+0.2	3.74
1638	BREST		(see page 210)				0500 and 1700	1100 and 2300	0500 and 1700	1100 and 2300	7.5	5.9	3.0	1.4	
1668	Pornic		47	06	2	07	−0035	−0015	+0005	+0005	−2.1	−1.9	−1.4	−1.1	2.85
	Ile de Noirmoutier														
1669	Bois de la Chaise		47	01	2	13	−0030	−0020	0000	−0005	−2.1	−1.9	−1.4	−1.0	2.82
1670	Fromentine		46	54	2	10	−0025	−0020	−0005	+0015	−2.2	−1.9	−1.3	−0.9	2.89
	Ile d' Yeu														
1671	Port Joinville		46	42	2	20	−0035	−0010	−0035	−0035	−2.2	−1.8	−0.9	−0.6	3.09
1672	St. Gilles.sur.vie		46	41	1	56	−0030	−0015	−0030	−0035	−2.2	−1.7	−0.9	−0.6	3.12
1673	Les Sabies d' Olonne		46	30	1	47	−0030	+0015	−0035	−0030	−2.2	−1.7	−0.9	−0.6	3.12
1681	POINTE DE GRAVE		(see page 214)				0000 and 1200	0600 and 1800	0500 and 1700	1200 and 2400	5.3	4.3	2.1	1.0	
	Ile de Re														
1674	St. Martin		46	12	1	22	−0025	−0045	−0005	−0005	+0.8	+0.4	+0.1	−0.3	3.43
1675	La Pallice	F	46	10	1	13	+0005	−0035	−0020	−0015	+0.8	+0.6	+0.4	0.0	3.84
1676	La Rochelle		46	09	1	09	+0005	−0035	−0020	−0015	+0.8	+0.6	+0.4	0.0	3.64
1677	Ile d' Aix		46	01	1	10	−0005	−0035	−0025	−0015	+0.9	+0.7	+0.4	0.0	3.67
	La Charente														
1678	Rochefort		45	57	0	57	+0015	−0020	+0045	+0120	+1.1	+0.9	+0.1	+0.4	⊙
1679	La Cayenne		45	47	1	08	−0015	−0035	−0020	0000	+0.5	+0.3	+0.3	+0.1	3.51
	La Gironde														
1680	Royan		45	37	1	01	0000	−0020	−0010	−0005	−0.2	−0.2	−0.2	−0.1	3.01
1681	POINTE DE GRAVE		45	34	1	04	STANDARD PORT				See Table V				3.21
1682	Richard		45	27	0	56	+0015	+0020	+0025	+0030	0.0	0.0	−0.2	−0.2	3.10
1683	Lamena		45	20	0	48	+0035	+0045	+0100	+0130	+0.2	+0.1	−0.4	−0.3	3.15
1684	Pauillac		45	12	0	45	+0045	+0110	+0140	+0220	+0.2	0.0	−0.8	−0.5	2.97
1685	La Reuille		45	03	0	36	+0120	+0155	+0230	+0320	−0.2	−0.4	−1.4	−0.9	2.49
	La Garonne														
1686	Le Marquis		45	00	0	33	+0130	+0205	+0250	+0340	−0.2	−0.4	−1.6	−1.0	2.46
1687	Bordeaux	F	44	52	0	33	+0155	+0235	+0330	+0425	−0.1	−0.3	−1.7	−1.0	2.45
	La Dordogne														
1689	Libourne		44	55	0	15	+0245	+0315	+0525	+0600	−0.6	−0.8	−2.0	−0.4	⊙

No.	PLACE		Lat. N.		Long. W.		TIME DIFFERENCES				HEIGHT DIFFERENCES (IN METERS)				M.L.
							High	Water	Low	Water	MHWS	MHWN	MLWN	MLWS	Z_0 m.
							(Zone −0100)								
1681	POINTE DE GRAVE		(see page 214)				0000 and 1200	0600 and 1800	0500 and 1700	1200 and 2400	5.3	4.3	2.1	1.0	
	Bassin d' Arcachon														
1690	Cap Ferret		44	37	1	15	−0015	0000	+0005	+0015	−1.2	−1.0	−0.7	−0.6	2.32
1691	Arcachon		44	40	1	10	+0005	+0030	+0015	+0040	−1.1	−1.1	−1.0	−0.8	2.17
	L' Adour														
1692	Boucau	F	43	32	1	31	−0035	−0030	−0010	−0030	−1.0	−0.9	−0.3	−0.2	2.56
	St. Jean de Luz														
1693	Socoa	F	43	23	1	40	−0050	−0045	−0025	−0040	−1.0	−1.0	−0.5	−0.5	2.49
	Spain														
1694	Pasajes	S	43	20	1	56	−0050	−0030	−0015	−0045	−1.2	−1.2	−0.6	−0.5	2.31
1695	San Sebastian		43	19	1	59	−0055	−0040	−0020	−0035	−1.2	−1.2	−0.6	−0.5	⊙
1696	Guetaria		43	18	2	12	−0050	−0040	−0015	−0035	−1.2	−1.2	−0.6	−0.5	⊙
1697	Lequeitio		43	22	2	30	−0100	−0050	−0025	−0045	−1.2	−1.2	−0.6	−0.5	⊙
1698	Bermeo		43	25	2	43	−0040	−0025	−0005	−0025	−0.8	−0.8	−0.6	−0.5	⊙
1699	Abra de Bilbao	S	43	20	3	02	−0045	−0030	−0010	−0030	−1.3	−1.2	−0.7	−0.5	2.25
1701	Castro Urdiales		43	23	3	13	−0050	−0110	−0020	−0110	−1.4	−1.3	−0.7	−0.6	⊙
1702	Ria de Santona		43	26	3	28	−0010	−0040	+0015	−0030	−1.3	−1.2	−0.6	−0.5	⊙
1703	Santander	S	43	28	3	47	−0020	−0100	0000	−0050	−0.9	−0.8	−0.2	−0.1	2.29
1704	Ria de Suances		43	27	4	03	0000	−0030	+0020	−0020	−1.4	−1.3	−0.7	−0.6	⊙
1705	San Vicente de Barquera		43	23	4	24	−0020	−0050	−0005	−0045	−1.3	−1.2	−0.6	−0.5	⊙
1706	Ria de Tina Mayor		43	24	4	31	−0030	−0050	−0005	−0045	−1.3	−1.2	−0.6	−0.5	⊙
1707	Ribadesella		43	28	5	04	−0040	−0030	−0005	−0035	−1.3	−1.2	−0.6	−0.5	⊙
1708	Gijon	S	43	34	5	42	−0050	−0040	−0015	−0045	−1.3	−1.2	−0.6	−0.5	2.27
1709	Luanco		43	37	5	47	−0055	−0045	−0020	−0050	−1.4	−1.3	−0.5	−0.5	⊙
1710	Aviles	S	43	35	5	56	−0100	−0041	−0015	−0050	−1.3	−1.2	−0.6	−0.4	2.21
1711	San Esteban de Pravia		43	34	6	05	−0050	−0041	−0015	−0050	−1.4	−1.3	−0.6	−0.5	⊙
1712	Luarca		43	33	6	32	−0030	−0020	+0005	−0025	−1.4	−1.3	−0.6	−0.5	⊙
1713	Ribadeo		43	33	7	02	−0035	−0025	0000	−0030	−1.3	−1.2	−06	−0.5	⊙
1714	Ria de Vivero		43	43	7	36	−0035	−0025	0000	−0035	−1.4	−1.3	−0.7	−0.6	⊙
1715	Santa Marta de Ortigueira		43	41	7	51	−0020	+0010	+0020	0000	−1.7	−1.5	−0.8	−0.5	⊙
1716	El Ferrol del Caudillo	S	43	28	8	16	−0100	−0050	−0020	−0055	−1.6	−1.4	−0.8	−0.4	2.13
1717	La Coruna	S	43	22	8	24	−0110	−0050	−0030	−0100	−1.7	−1.5	−0.8	−0.5	2.04
1718	Ria de Corme		43	16	8	58	−0030	−0010	0000	−0030	−2.1	−1.9	−1.0	−0.7	⊙
1719	Ria de Camarinas		43	08	9	11	−0120	−0055	−0030	−0100	−2.0	−1.8	−0.9	−0.6	⊙
1741	LISBON		(see page 218)		0500 and 1700		1000 and 2200	0300 and 1500	0800 and 2000	3.8	3.0	1.4	0.5		
1720	Corcubion		42	57	9	12	+0055	+0110	+0120	+0135	−0.4	−0.4	−0.2	−0.2	⊙
1721	Muros		42	46	9	03	+0050	+0105	+0115	+0130	−0.3	−0.3	−0.1	−0.1	⊙
	Ria de Arosa														
1722	Villagarcia	S	42	36	8	47	+0040	+0100	+0110	+0120	−0.2	−0.2	0.0	0.0	2.05
	Ria de Pontevedra														
1723	Marin	S	42	24	8	42	+0050	+0110	+0120	+0130	−0.4	−0.4	−0.1	0.0	1.91
1724	Vigo	S	42	15	8	43	+0040	+0100	+0105	+0125	−0.3	−0.3	−0.1	−0.1	1.96
1725	Bayona		42	07	8	51	+0035	+0050	+0100	+0115	−0.3	−0.3	−0.1	−0.1	⊙
1726	La Guardia		41	54	8	53	+0040	+0055	+0105	+0120	−0.5	−0.5	−0.2	0.2	⊙

⊙ No data

F Tides predicted in French Tide Tables

P Tides predicted in Portuguese Tide Tables

S Tides predicted in Spanish Tide Tables

c For intermediate heights, use harmonic constants (see Part III) and N.P. 159

APPENDIX 2
Pacific tides

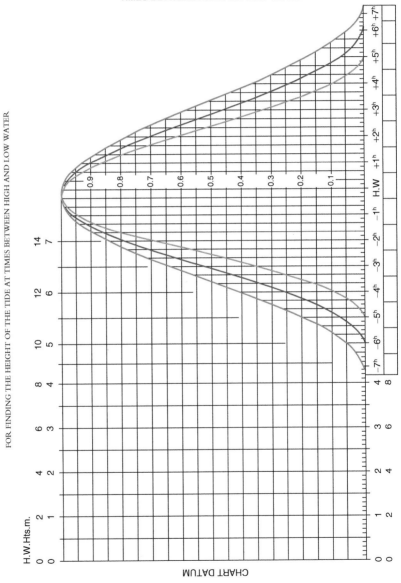

FOR FINDING THE HEIGHT OF THE TIDE AT
TIMES BETWEEN HIGH AND LOW WATER

PHILIPPINE ISLANDS – DAVAO
LAT 7°05'N   LONG 125°38'E
TIMES AND HEIGHTS OF HIGH AND LOW WATERS

TIME ZONE −0800   YEAR 1987

JANUARY

Day	TIME	M	Day	TIME	M
1 TH	0034	−0.2	16 F	0052	0.0
	0634	1.3		0650	1.3
	1212	0.1		1226	0.2
	1835	1.9		1843	1.7
2 F	0116	−0.2	17 SA	0120	−0.1
	0716	1.3		0719	1.3
	1256	0.1		1258	0.2
	1918	1.9		1915	1.7
3 SA	0158	−0.2	18 SU	0148	−0.1
	0758	1.3		0749	1.3
	1339	0.1		1330	0.2
	2002	1.8		1947	1.7
4 SU	0239	−0.1	19 M	0216	0.0
	0840	1.3		0820	1.3
	1424	0.2		1404	0.2
	2046	1.7		2021	1.7
5 M	0320	0.0	20 TU	0246	0.0
	0924	1.3		0852	1.4
	1512	0.2		1441	0.2
	2131	1.5		2057	1.6
6 TU	0402	0.1	21 W	0317	0.1
	1012	1.3		0928	1.3
	1606	0.3		1523	0.2
	2220	1.4		2137	1.4
7 W	0447	0.3	22 TH	0352	0.2
	1106	1.2		1011	1.3
	1711	0.4		1614	0.3
	2318	1.2		2225	1.3
8 TH	0538	0.4	23 F	0434	0.3
	1210	1.2		1105	1.3
	1838	0.5		1724	0.4
				2329	1.1
9 F	0031	1.0	24 SA	0528	0.4
	0641	0.5		1217	1.3
	1323	1.2		1905	0.4
	2016	0.5			
10 SA	0201	1.0	25 SU	0105	1.0
	0754	0.5		0649	0.5
	1434	1.3		1345	1.3
	2136	0.4		2055	0.3
11 SU	0323	1.0	26 M	0253	1.0
	0901	0.5		0826	0.5
	1534	1.4		1507	1.5
	2230	0.3		2212	0.2
12 M	0424	1.0	27 TU	0412	1.1
	0956	0.4		0944	0.4
	1621	1.5		1612	1.6
	2315	0.2		2307	0.0
13 TU	0509	1.1	28 W	0507	1.2
	1040	0.4		1043	0.3
	1702	1.6		1705	1.8
	2350	0.1		2351	−0.1
14 W	0546	1.1	29 TH	0551	1.3
	1119	0.3		1131	0.1
	1738	1.7		1715 •	1.9
15 TH	0022	0.0	30 F	0031	−0.2
o	0619	1.2		0631	1.4
	1153	0.3		1215	0.0
	1811	1.7		1833	1.9
			31 SA	0108	−0.2
				0707	1.5
				1256	0.0
				1913	1.9

FEBRUARY

Day	TIME	M	Day	TIME	M
1 SU	0142	−0.2	16 M	0120	−0.1
	0742	1.5		0722	1.5
	1335	0.0		1315	0.0
	1950	1.9		1927	1.8
2 M	0215	−0.1	17 TU	0144	−0.1
	0817	1.5		0748	1.6
	1413	0.0		1345	0.0
	2026	1.7		1958	1.7
3 TU	0246	0.0	18 W	0209	−0.1
	0851	1.5		0817	1.6
	1451	0.1		1419	0.0
	2102	1.5		2030	1.6
4 W	0315	0.1	19 TH	0236	0.0
	0926	1.4		0849	1.6
	1531	0.2		1457	0.1
	2137	1.4		2106	1.4
5 TH	0344	0.2	20 F	0306	0.1
	1005	1.4		0927	1.5
	1617	0.4		1542	0.2
	2216	1.1		2147	1.2
6 F	0415	0.4	21 SA	0340	0.3
	1051	1.3		1014	1.4
	1720	0.5		1644	0.3
	2306	1.0		2244	1.0
7 SA	0452	0.5	22 SU	0425	0.4
	1158	1.2		1123	1.3
	1919	0.6		1833	0.5
8 SU	0046	0.8	23 M	0033	0.9
	0602	0.6		0549	0.6
	1340	1.2		1315	1.3
	2134	0.5		2059	0.4
9 M	0326	0.8	24 TU	0305	0.8
	0827	0.6		0821	0.6
	1513	1.3		1502	1.4
	2235	0.4		2215	0.2
10 TU	0436	0.9	25 W	0420	1.1
	0951	0.6		0951	0.4
	1612	1.4		1611	1.6
	2312	0.2		2302	0.0
11 W	0512	1.1	26 TH	0505	1.2
	1040	0.4		1046	0.2
	1654	1.5		1702	1.8
	2341	0.1		2340	−0.1
12 TH	0541	1.2	27 F	0541	1.4
	1116	0.3		1130	0.1
	1728	1.6		1744	1.9
13 F	0007	0.0	28 SA	0014	−0.2
	0606	1.3		0614	1.5
o	1148	0.2	•	1209	−0.1
	1759	1.7		1822	1.9
14 SA	0032	−0.1			
	0631	1.4			
	1217	0.1			
	1829	1.8			
15 SU	0056	−0.1			
	0656	1.5			
	1245	0.0			
	1858	1.8			

MARCH

Day	TIME	M	Day	TIME	M
1 SU	0046	−0.2	16 M	0022	−0.1
	0646	1.6		0624	1.6
	1245	−0.1		1225	−0.1
	1857	1.9		1834	1.8
2 M	0114	−0.2	17 TU	0045	−0.1
	0716	1.7		0649	1.7
	1319	−0.2		1253	−0.2
	1929	1.8		1903	1.8
3 TU	0141	−0.1	18 W	0108	−0.1
	0745	1.7		0715	1.7
	1352	−0.1		1324	−0.2
	2000	1.7		1933	1.7
4 W	0206	0.0	19 TH	0133	−0.1
	0814	1.6		0744	1.8
	1424	0.0		1358	−0.1
	2029	1.5		2005	1.5
5 TH	0229	0.1	20 F	0200	0.0
	0843	1.6		0817	1.7
	1457	0.1		1436	0.0
	2058	1.3		2040	1.4
6 F	0250	0.2	21 SA	0229	0.1
	0914	1.5		0854	1.6
	1533	0.3		1521	0.1
	2126	1.1		2122	1.2
7 SA	0310	0.3	22 SU	0302	0.3
	0948	1.3		0941	1.5
	1618	0.5		1625	0.3
	2159	0.9		2222	0.9
8 SU	0328	0.5	23 M	0345	0.5
	1037	1.2		1053	1.3
	1800	0.6		1829	0.4
	2316	0.7			
9 M	0339	0.6	24 TU	0040	0.8
	1229	1.1		0528	0.6
	2138	0.5		1304	1.3
				2055	0.3
10 W	0438	0.8	25 W	0312	0.9
	0820	0.8		0833	0.6
	1455	1.2		1457	1.4
	2224	0.4		2159	0.2
11 W	0440	1.0	26 TH	0408	1.1
	0952	0.6		0949	0.4
	1558	1.3		1602	1.6
	2251	0.2		2241	0.0
12 TH	0458	1.1	27 F	0445	1.3
	1032	0.5		1038	0.1
	1636	1.4		1648	1.7
	2315	0.1		2316	−0.1
13 F	0518	1.2	28 SA	0518	1.5
	1103	0.3		1118	0.0
	1708	1.6		1727	1.8
	2338	0.0		2346	−0.2
14 SA	0539	1.4	29 SU	0548	1.7
	1131	0.1		1154	−0.2
	1737	1.7	•	1802	1.8
15 SU	0000	−0.1	30 M	0015	−0.2
o	0601	1.5		0617	1.7
	1157	0.0		1227	−0.2
	1805	1.7		1834	1.8
			31 TU	0040	−0.1
				0645	1.8
				1259	−0.2
				1904	1.7

APRIL

Day	TIME	M	Day	TIME	M
1 W	0104	−0.1	16 TH	0034	−0.1
	0713	1.8		0646	1.9
	1329	−0.1		1306	−0.2
	1932	1.5		1912	1.6
2 TH	0127	0.0	17 F	0101	0.0
	0740	1.7		0718	1.9
	1359	0.0		1343	−0.2
	1959	1.4		1947	1.4
3 F	0147	0.1	18 SA	0131	0.1
	0807	1.7		0754	1.8
	1430	0.1		1425	0.0
	2026	1.2		2026	1.3
4 SA	0207	0.2	19 SU	0204	0.2
	0835	1.5		0835	1.7
	1503	0.3		1515	0.1
	2054	1.1		2114	1.1
5 SU	0225	0.4	20 M	0243	0.3
	0906	1.4		0927	1.5
	1545	0.4		1626	0.3
	2128	0.9		2229	0.9
6 M	0241	0.5	21 TU	0339	0.5
	0947	1.2		1047	1.3
	1708	0.6		1826	0.4
	2251	0.7			
7 TU	0246	0.7	22 W	0047	0.9
	1121	1.1		0557	0.6
	2046	0.5		1252	1.3
				2020	0.3
8 W	0419	0.8	23 TH	0240	1.0
	0800	0.8		0821	0.5
	1411	1.1		1434	1.4
	2140	0.4		2123	0.2
9 TH	0404	1.0	24 F	0334	1.2
	0928	0.6		0930	0.3
	1522	1.2		1538	1.5
	2209	0.3		2206	0.1
10 F	0420	1.2	25 SA	0413	1.4
	1006	0.4		1018	0.1
	1604	1.4		1625	1.6
	2230	0.1		2241	0.0
11 SA	0440	1.3	26 SU	0446	1.6
	1036	0.3		1058	0.0
	1637	1.5		1704	1.6
	2258	0.1		2312	0.0
12 SU	0502	1.5	27 M	0518	1.7
	1104	0.1		1134	−0.1
	1708	1.6		1738	1.6
	2302	0.0		2340	0.0
13 M	0525	1.6	28 TU	0547	1.8
	1132	0.0		1208	−0.2
	1738	1.7	•	1810	1.6
	2344	−0.1			
14 TU			29 W		
o					
15 W	0008	−0.1	30 TH	0030	0.0
	0617	1.8		0644	1.8
	1233	−0.2		1310	−0.1
	1839	1.6		1909	1.4

JANUARY

Day	TIME	M	TIME	M	TIME	M	TIME	M
1 TH	0415	0.8	0936	0.5	1408	0.8	2142	0.1
2 F	0503	0.8	1035	0.5	1514	0.8	2236	0.1
3 SA	0542	0.9	1127	0.5	1629	0.9	2327	0.2
4 SU	0618	0.9	1218	0.5	1739	0.9		
5 M	0016	0.2	0653	0.9	1308	0.4	1839	0.9
6 TU	0103	0.3	0728	0.9	1358	0.4	1934	0.9
7 W	0146	0.3	0804	0.9	1448	0.3	2027	0.8
8 TH	0227	0.3	0840	0.9	1535	0.3	2122	0.8
9 F	0308	0.3	0919	0.8	1621	0.2	2223	0.7
10 SA	0351	0.3	1000	0.8	1708	0.1	2304	0.7
11 SU	0442	0.3	1045	0.7	1759	0.1		
12 M	0050	0.7	0553	0.4	1134	0.7	1851	0.1
13 TU	0153	0.7	0712	0.4	1224	0.7	1924	0.1
14 W	0256	0.7	0833	0.4	1316	0.7	2033	0.1
15 TH o	0354	0.7	0942	0.4	1408	0.7	2122	0.1
16 F	0441	0.8	1033	0.4	1503	0.7	2208	0.2
17 SA	0519	0.8	1113	0.4	1559	0.7	2253	0.2
18 SU	0550	0.9	1148	0.4	1651	0.8	2334	0.2
19 M	0616	0.9	1222	0.4	178	0.8		
20 TU	0012	0.3	0638	0.9	1256	0.4	1822	0.8
21 W	0049	0.3	0703	0.8	1332	0.3	1906	0.8
22 TH	0123	0.3	0730	0.8	1408	0.3	1951	0.8
23 F	0154	0.3	0758	0.8	1446	0.2	2038	0.7
24 SA	0223	0.3	0826	0.8	1524	0.1	2132	0.6
25 SU	0254	0.3	0855	0.8	1607	0.1	2236	0.6
26 M	0334	0.4	0934	0.8	1659	0.1		
27 TU	0009	0.6	0424	0.4	1026	0.8	1808	0.1
28 W	0138	0.5	0539	0.5	1133	0.8	1921	0.1
29 TH •	0252	0.7	0750	0.5	1243	0.8	2026	0.1
30 F	0351	0.8	0921	0.5	1357	0.8	2128	0.1
31 SA	0437	0.9	1025	0.5	1526	0.8	2228	0.2

FEBRUARY

Day	TIME	M	TIME	M	TIME	M	TIME	M
1 SU	0514	0.9	1121	0.4	1641	0.9	2323	0.2
2 M	0548	0.9	1211	0.4	1747	0.9		
3 TU	0014	0.3	0623	0.9	1258	0.3	1834	0.9
4 W	0058	0.3	0657	0.9	1342	0.3	1924	0.8
5 TH	0138	0.3	0734	0.9	1423	0.2	2015	0.8
6 F	0215	0.3	0810	0.8	1502	0.1	2108	0.7
7 SA	0249	0.3	0849	0.8	1539	0.1	2203	0.7
8 SU	0324	0.3	0927	0.8	1621	0.1	2304	0.7
9 M	0404	0.3	1008	0.7	1707	0.1		
10 TU	0010	0.7	0452	0.4	1053	0.7	1801	0.1
11 W	0117	0.7	0601	0.5	1141	0.7	1900	0.1
12 TH	0223	0.7	0754	0.5	1236	0.7	1959	0.2
13 F o	0323	0.7	0923	0.5	1338	0.7	2057	0.2
14 SA	0407	0.8	1010	0.5	1453	0.7	2153	0.2
15 SU	0439	0.8	1047	0.4	1556	0.7	2239	0.3
16 M	0507	0.8	1120	0.4	1645	0.8	2322	0.3
17 TU	0530	0.9	1154	0.3	1729	0.8		
18 W	0000	0.3	0556	0.8	1230	0.3	1811	0.8
19 TH	0036	0.3	0624	0.8	1307	0.2	1854	0.8
20 F	0108	0.3	0653	0.8	1343	0.2	1939	0.7
21 SA	0138	0.3	0721	0.8	1421	0.1	2027	0.7
22 SU	0204	0.3	0748	0.8	1457	0.1	2123	0.7
23 M	0232	0.4	0820	0.8	1538	0.1	2235	0.7
24 TU	0310	0.4	0902	0.8	1630	0.1	2358	0.7
25 W	0403	0.5	1003	0.8	1740	0.1		
26 TH	0108	0.7	0530	0.5	1120	0.8	1857	0.2
27 F	0211	0.8	0742	0.5	1238	0.8	2006	0.2
28 SA •	0308	0.8	0909	0.5	1415	0.8	2117	0.3

MARCH

Day	TIME	M	TIME	M	TIME	M	TIME	M
1 SU	0356	0.8	1015	0.4	1542	0.8	2226	0.3
2 M	0437	0.9	1108	0.3	1644	0.9	2322	0.3
3 TU	0513	0.8	1153	0.3	1738	0.9		
4 W	0008	0.3	0549	0.8	1235	0.2	1827	0.8
5 TH	0047	0.3	0624	0.8	1313	0.2	1914	0.8
6 F	0122	0.3	0701	0.8	1348	0.1	2001	0.8
7 SA	0153	0.3	0737	0.8	1422	0.1	2047	0.8
8 SU	0223	0.3	0812	0.8	1455	0.1	2135	0.7
9 M	0254	0.3	0847	0.8	1531	0.1	2224	0.7
10 TU	0329	0.4	0922	0.8	1611	0.1	2320	0.7
11 W	0409	0.4	1002	0.7	1659	0.2		
12 TH	0025	0.7	0501	0.5	1052	0.7	1758	0.3
13 F	0131	0.7	0659	0.5	1155	0.7	1912	0.3
14 SA	0224	0.7	0834	0.5	1316	0.7	2021	0.3
15 SU o	0305	0.8	0921	0.4	1440	0.7	2121	0.3
16 M	0336	0.8	1000	0.4	1538	0.8	2212	0.3
17 TU	0406	0.8	1038	0.3	1626	0.8	2256	0.3
18 W	0437	0.8	1116	0.2	1710	0.8	2336	0.3
19 TH	0508	0.8	1154	0.1	1755	0.8		
20 F	0013	0.3	0539	0.8	1234	0.1	1841	0.8
21 SA	0048	0.3	0611	0.8	1312	0.1	1931	0.8
22 SU	0122	0.3	0642	0.8	1352	0.1	2027	0.8
23 M	0152	0.4	0715	0.8	1431	0.1	2129	0.8
24 TU	0225	0.4	0755	0.8	1514	0.2	228	0.8
25 W	0308	0.4	0847	0.8	1606	0.2	2323	0.7
26 TH	0408	0.5	1003	0.8	1711	0.3		
27 F	0021	0.8	0557	0.5	1129	0.8	1829	0.3
28 SA	0121	0.8	0737	0.5	1251	0.8	1943	0.3
29 SU •	0219	0.8	0855	0.4	1435	0.8	2108	0.3
30 M	0310	0.8	0955	0.3	1546	0.8	2217	0.3
31 TU	0354	0.8	1043	0.2	1643	0.8	2307	0.3

APRIL

Day	TIME	M	TIME	M	TIME	M	TIME	M
1 W	0434	0.8	1124	0.2	1732	0.8		
2 TH	0510	0.8	1201	0.1	1817	0.8		
3 F	0023	0.3	0546	0.8	1235	0.1	1859	0.8
4 SA	0055	0.3	0622	0.8	1308	0.1	1941	0.8
5 SU	0124	0.3	0654	0.8	1339	0.1	2023	0.8
6 M	0153	0.3	0726	0.8	1411	0.1	2103	0.8
7 TU	0223	0.4	0756	0.8	1446	0.2	2142	0.8
8 W	0257	0.5	0831	0.8	1523	0.2	2223	0.8
9 TH	0338	0.5	0913	0.8	1605	0.3	2308	0.8
10 F	0427	0.5	1008	0.7	1654	0.3	2359	0.8
11 SA	0539	0.5	1123	0.7	1759	0.4		
12 SU	0053	0.8	0726	0.4	1257	0.7	1922	0.4
13 M	0140	0.8	0820	0.3	1416	0.7	2032	0.4
14 TU	0224	0.8	0906	0.2	1514	0.8	2132	0.4
15 W	0304	0.7	0950	0.2	1606	0.8	2223	0.3
16 TH	0341	0.7	1033	0.1	1654	0.8	2308	0.3
17 F	0418	0.8	1116	0.1	1745	0.8	2349	0.4
18 SA	0454	0.8	1159	0.1	1838	0.8		
19 SU	0029	0.4	0532	0.8	1242	0.1	1934	0.9
20 M	0108	0.4	0612	0.8	1325	0.1	2027	0.9
21 TU	0149	0.5	0657	0.9	1408	0.2	2114	0.9
22 W	0232	0.5	0752	0.9	1453	0.3	2157	0.9
23 TH	0322	0.5	0902	0.9	1541	0.3	2238	0.9
24 F	0429	0.5	1014	0.8	1640	0.4	2324	0.9
25 SA	0605	0.5	1128	0.8	1756	0.4		
26 SU	0019	0.8	0723	0.4	1312	0.8	1918	0.4
27 M	0122	0.8	0828	0.3	1438	0.8	2047	0.4
28 TU •	0219	0.8	0923	0.2	1541	0.8	2153	0.4
29 W	0308	0.8	1008	0.1	1634	0.8	2240	0.4
30 TH	0350	0.8	1047	0.1	1720	0.8	2321	0.3

FOR INTERMEDIATE HEIGHTS USE HARMONIC CONSTANTS (SEE PART III) AND NP 159.

MAY

	TIME	M		TIME	M
1 F	0428	0.8	16 SA	0331	0.8
	1123	0.1		1043	0.1
	1802	0.9		1749	0.9
	2355	0.3		2329	0.5
2 SA	0504	0.8	17 SU	0417	0.8
	1154	0.1		1131	0.1
	1841	0.9		1839	0.9
3 SU	0027	0.4	18 M	0014	0.5
	0537	0.8		0506	0.9
	1225	0.1		1219	0.1
	1920	0.9		1924	1.0
4 M	0056	0.4	19 TU	0100	0.5
	0609	0.8		0601	0.9
	1258	0.2		1305	0.2
	1955	0.9		2006	1.0
5 TU	0125	0.5	20 W	0147	0.5
	0642	0.8		0706	0.9
	1331	0.2		1350	0.3
	2028	0.9		2044	1.0
6 W	0157	0.5	21 TH	0237	0.5
	0718	0.8		0812	0.9
	1406	0.3		1435	0.4
	2028	0.9		2122	1.0
7 TH	0234	0.5	22 F	0332	0.5
	0718	0.8		0914	0.9
	1442	0.3		1520	0.4
	2129	0.9		2159	0.9
8 F	0317	0.5	23 SA	0441	0.5
	0847	0.8		1016	0.9
	1523	0.3		1612	0.5
	2206	0.9		2240	0.9
9 SA	0408	0.5	24 SU	0552	0.4
	0949	0.8		1133	0.8
	1609	0.4		1722	0.5
	2249	0.8		2328	0.9
10 SU	0511	0.4	25 M	0653	0.3
	1108	0.7		1314	0.8
	1706	0.4		1845	0.5
	2336	0.8			
11 M	0624	0.3	26 TU	0023	0.8
	1233	0.7		0750	0.2
	1817	0.4		1423	0.8
				2008	0.5
12 TU	0025	0.8	27 W	0121	0.8
	0724	0.2		0841	0.1
	1349	0.7		1523 ●	0.8
	1932	0.4		2119	0.4
13 W ○	0114	0.7	28 TH	0215	0.8
	0817	0.1		0926	0.1
	1453	0.8		1616	0.8
	2044	0.4		2210	0.4
14 TH	0202	0.7	29 F	0302	0.8
	0906	0.1		1007	0.1
	1553	0.9		1703	0.9
	2150	0.4		2253	0.4
15 F	0247	0.8	30 SA	0343	0.8
	0954	0.1		1043	0.1
	1653	0.8		1745	0.9
	2243	0.5		2331	0.4
			31 SU	0422	0.8
				1117	0.1
				1823	0.9

JUNE

	TIME	M		TIME	M
1 M	0005	0.5	16 TU	0005	0.6
	0500	0.8		0502	0.9
	1151	0.2		1202	0.3
	1859	0.9		1859	1.0
2 TU	0037	0.5	17 W	0054	0.5
	0538	0.8		0614	0.9
	1226	0.3		1250	0.3
	1930	0.9		1936	1.1
3 W	0109	0.5	18 TH	0145	0.5
	0619	0.9		0717	0.9
	1301	0.3		1337	0.4
	1957	1.0		2012	1.0
4 TH	0145	0.5	19 F	0238	0.5
	0701	0.9		0815	0.9
	1338	0.3		1421	0.4
	2022	1.0		2048	1.0
5 F	0224	0.5	20 SA	0334	0.5
	0749	0.8		0911	0.9
	1414	0.3		1505	0.4
	2049	1.0		2125	1.0
6 SA	0307	0.5	21 SU	0429	0.4
	0842	0.8		1013	0.8
	1453	0.4		1551	0.5
	2122	1.0		2207	0.9
7 SU	0353	0.4	22 M	0523	0.3
	0943	0.8		1136	0.8
	1536	0.4		1648	0.5
	2158	0.9		2252	0.9
8 M	0444	0.3	23 TU	0614	0.2
	1051	0.7		1253	0.7
	1624	0.4		1803	0.5
	2238	0.8		2341	0.8
9 TU	0540	0.2	24 W	0706	0.1
	1207	0.7		1356	0.8
	1723	0.5		1921	0.5
	2323	0.8			
10 W	0639	0.1	25 TH	0034	0.8
	1326	0.7		0756	0.1
	1835	0.5		1457	0.8
				2038	0.5
11 TH ○	0013	0.8	26 F	0126	0.8
	0738	0.1		0845	0.1
	144	0.8		1555 ●	0.8
	1954	0.5		2143	0.5
12 F	0107	0.8	27 SA	0218	0.8
	0833	0.1		0930	0.1
	1559	0.8		1646	0.9
	2118	0.6		2234	0.5
13 SA	0201	0.8	28 SU	0306	0.8
	0927	0.1		1012	0.2
	1657	0.9		1728	0.9
	2223	0.6		2317	0.5
14 SU	0256	0.8	29 M	0354	0.8
	1021	0.1		1052	0.2
	1742	0.9		1804	0.9
	2315	0.6		2355	0.5
15 M	0355	0.9	30 TU	0443	0.8
	1111	0.2		1131	0.3
	1822	1.0		1835	1.0

JULY

	TIME	M		TIME	M
1 W	0031	0.5	16 TH	0049	0.5
	0531	0.9		0617	1.0
	1209	0.3		1242	0.4
	1902	1.0		1905	1.1
2 TH	0105	0.5	17 F	0139	0.5
	0618	0.9		0713	0.9
	1248	0.3		1329	0.4
	1925	1.0		1940	1.0
3 F	0139	0.5	18 SA	0229	0.4
	0703	0.8		0808	0.9
	1323	0.4		1411	0.4
	1950	1.0		2018	1.0
4 SA	0216	0.4	19 SU	0316	0.3
	0749	0.8		0905	0.8
	1359	0.4		1452	0.4
	2016	0.9		2056	0.9
5 SU	0254	0.4	20 M	0400	0.3
	0838	0.8		1007	0.8
	1433	0.4		1532	0.4
	2045	0.9		2137	0.9
6 M	0334	0.3	21 TU	0445	0.2
	0930	0.7		1119	0.8
	1507	0.4		1617	0.4
	2115	0.9		2220	0.9
7 TU	0416	0.2	22 W	0532	0.2
	1031	0.7		1223	0.8
	1544	0.4		1713	0.5
	2149	0.8		2307	0.8
8 W	0504	0.2	23 TH	0623	0.1
	1145	0.7		1325	0.8
	1631	0.5		1830	0.5
	2231	0.8		2356	0.8
9 TH	0601	0.1	24 F	0716	0.1
	1317	0.7		1430	0.8
	1734	0.6		1958	0.5
	2325	0.8			
10 F	0707	0.1	25 SA	0049	0.8
	1443	0.8		0809	0.2
	1915	0.6		1533 ●	0.8
				2124	0.6
11 SA ○	0027	0.8	26 SU	0143	0.8
	0809	0.1		0902	0.2
	1552	0.8		1624	0.9
	2054	0.6		2223	0.6
12 SU	0131	0.9	27 M	0244	0.8
	0908	0.1		0953	0.3
	1640	0.9		1703	0.9
	2208	0.6		2307	0.6
13 M	0238	0.9	28 TU	0350	0.8
	1004	0.2		1040	0.3
	1719	1.0		1734	0.9
	2305	0.6		2342	0.5
14 TU	0357	0.9	29 W	0445	0.8
	1059	0.3		1123	0.3
	1754	1.0		1801	1.0
	2357	0.6			
15 W	0514	1.0	30 TH	0015	0.5
	1152	0.3		0532	0.8
	1829	1.1		1201	0.4
				1825	0.9
			31 F	0047	0.5
				0614	0.8
				1238	0.4
				1849	0.9

AUGUST

	TIME	M		TIME	M
1 SA	0121	0.4	16 SU	0204	0.3
	0655	0.8		0758	0.8
	1311	0.3		1357	0.4
	1913	0.9		1948	0.9
2 SU	0156	0.3	17 M	0244	0.2
	0738	0.8		0853	0.8
	1342	0.3		1433	0.4
	1939	0.9		2026	0.9
3 M	0232	0.3	18 TU	0323	0.2
	0823	0.8		0950	0.8
	1409	0.4		1507	0.4
	2006	0.9		2106	0.9
4 TU	0308	0.2	19 W	0402	0.2
	0914	0.7		1048	0.8
	1434	0.4		1543	0.4
	2033	0.8		2146	0.9
5 W	0346	0.2	20 TH	0445	0.2
	1014	0.7		1149	0.7
	1506	0.5		1627	0.5
	2104	0.8		2231	0.8
6 TH	0430	0.2	21 F	0537	0.2
	1138	0.7		1252	0.7
	1550	0.5		1726	0.6
	2149	0.9		2320	0.8
7 F	0529	0.2	22 SA	0636	0.2
	1308	0.7		1357	0.8
	1649	0.6		1926	0.6
	2252	0.9			
8 SA	0641	0.2	23 SU	0015	0.8
	1419	0.8		0738	0.3
	1854	0.7		1458	0.8
				2105	0.6
9 SU ○	0005	0.9	24 M	0120	0.8
	0749	0.2		0839	0.3
	1518	0.9		1545 ●	0.9
	2038	0.7		2157	0.6
10 M	0120	0.9	25 TU	0240	0.8
	0850	0.2		0936	0.3
	1605	0.9		1620	0.9
	2154	0.6		2235	0.5
11 TU	0247	0.9	26 W	0345	0.8
	0951	0.3		1023	0.3
	1643	1.0		1648	0.9
	2254	0.6		2308	0.5
12 W	0412	0.9	27 TH	0433	0.8
	1053	0.3		1105	0.3
	1720	1.0		1712	0.9
	2348	0.5		2339	0.4
13 TH	0515	0.9	28 F	0515	0.8
	1148	0.4		1143	0.3
	1755	1.0		1738	0.9
14 F	0036	0.4	29 SA	0014	0.3
	0611	0.9		0556	0.8
	1236	0.4		1218	0.3
	1832	1.0		1804	0.9
15 SA	0122	0.3	30 SU	0049	0.3
	0705	0.9		0638	0.8
	1319	0.4		1251	0.3
	1909	1.0		1831	0.8
			31 M	0125	0.2
				0721	0.8
				1319	0.4
				1858	0.8

FOR INTERMEDIATE HEIGHTS USE HARMONIC CONSTANTS (SEE PART III) AND NP 159.

JANUARY

	TIME	M		TIME	M
1	0528	1.2	16	0556	1.6
TH	1152	6.2	F	1210	6.1
	1740	2.1		1804	2.3
	2349	6.9			
2	0614	1.0	17	0007	6.5
F	1237	6.4	SA	0626	1.5
	1827	1.9		1241	6.2
				1834	2.1
3	0038	6.9	18	0038	6.6
SA	0657	1.0	SU	0653	1.4
	1319	6.5		1309	6.3
	1912	1.8		1904	2.0
4	0122	6.8	19	0108	6.6
SU	0738	1.2	M	0721	1.4
	1358	6.4		1338	6.4
	1954	1.9		1933	2.0
5	0203	6.6	20	0137	6.5
M	0817	1.4	TU	0748	1.5
	1436	6.3		1404	6.4
	2037	2.1		2003	2.0
6	0242	6.2	21	0206	6.3
TU	0854	1.8	W	0815	1.7
	1512	6.2		1432	6.3
	2119	2.4		2035	2.2
7	0322	5.8	22	0237	6.1
W	0932	2.3	TH	0845	2.0
	1552	6.0		1502	6.2
	2207	2.7		2112	2.4
8	0405	5.3	23	0311	5.7
TH	1014	2.7	F	0918	2.4
	1638	5.7		1537	6.0
	2309	3.0		2157	2.7
9	0502	4.9	24	0355	5.3
F	1109	3.2	SA	0958	2.9
	1743	5.4		1624	5.8
				2304	3.0
10	0040	3.1	25	0506	4.8
SA	0646	4.6	SU	1058	3.4
	1238	3.5		1743	5.5
	1910	5.3			
11	0210	3.0	26	0051	3.0
SU	0835	4.7	M	0715	4.6
	1413	3.5		1259	3.6
	2027	5.4		1926	5.5
12	0322	2.7	27	0223	2.8
M	0945	5.0	TU	0910	4.9
	1529	3.3		1436	3.5
	2128	5.7		2053	5.8
13	0412	2.4	28	0350	2.4
TU	1030	5.3	W	1023	5.4
	1620	3.0		1601	3.0
	2216	5.9		2208	6.2
14	0452	2.1	29	0446	1.8
W	1106	5.6	TH	1110	2.9
	1659	2.8	•	1658	2.5
	2256	6.2		2306	6.6
15	0526	1.8	30	0530	1.3
TH	1139	5.9	F	1152	6.3
○	1733	2.5		1743	1.9
	2333	6.4		2353	7.0
			31	0609	1.0
			SA	1230	6.7
				1823	1.5

FEBRUARY

	TIME	M		TIME	M
1	0036	7.2	16	0029	6.8
SU	0646	0.8	M	0937	1.2
	1306	6.9		1253	6.8
	1901	1.3		1848	1.6
2	0113	7.1	17	0056	6.9
M	0720	0.9	TU	0702	1.1
	1338	7.0		1317	6.9
	1937	1.3		1915	1.4
3	0147	6.9	18	0123	6.8
TU	0752	1.1	W	0726	1.2
	1408	6.9		1340	6.9
	2010	1.5		1942	1.5
4	0217	6.6	19	0148	6.6
W	0821	1.5	TH	0751	1.4
	1436	6.7		1403	6.8
	2043	1.8		2011	1.6
5	0247	6.1	20	0214	6.3
TH	0850	1.9	F	0816	1.8
	1504	6.4		1427	6.7
	2116	2.3		2041	1.9
6	0316	5.7	21	0241	6.0
F	0917	2.5	SA	0843	2.2
	1534	6.0		1454	6.4
	2153	2.7		2117	2.4
7	0348	5.1	22	0313	5.4
SA	0946	3.1	SU	0911	2.8
	1608	5.6		1528	6.0
	2247	3.2		2204	2.9
8	0433	4.6	23	0356	4.8
SU	1023	3.6	M	0944	3.4
	1706	5.1		1622	5.5
				2356	3.4
9	0057	3.5	24	0642	4.3
M	0750	4.3	TU	1210	4.0
	1309	4.0		1904	5.1
	1947	4.9			
10	0320	3.3	25	0241	3.3
TU	1006	4.7	W	0954	4.8
	1536	3.8		1509	3.8
	2123	5.2		2117	5.5
11	0417	2.8	26	0408	2.6
W	1038	5.2	TH	1034	5.5
	1625	3.3		1617	3.0
	2217	5.7		2223	6.1
12	0452	2.4	27	0446	1.9
TH	1106	5.6	F	1106	6.1
	1658	2.8		1657	2.2
	2255	6.1		2307	6.7
13	0520	1.9	28	0520	1.3
F	1132	6.0	SA	1138	6.7
○	1726	2.4	•	1733	1.6
	2328	6.4		2345	7.1
14	0546	1.6			
SA	1159	6.3			
	1753	2.0			
	2359	6.7			
15	0611	1.4			
SU	1226	6.6			
	1821	1.8			

MARCH

	TIME	M		TIME	M
1	0553	0.9	16	0545	1.3
SU	1210	7.1	M	1200	6.9
	1808	1.1		1758	1.4
2	0021	7.3	17	0009	7.0
M	0623	0.7	TU	0609	1.1
	1240	7.3		1224	7.1
	1840	0.9		1824	1.1
3	0053	7.3	18	0037	7.0
TU	0653	0.8	W	0635	1.1
	1308	7.4		1248	7.2
	1911	0.9		1852	1.0
4	0122	7.1	19	0102	6.9
W	0721	1.0	TH	0659	1.2
	1334	7.3		1310	7.2
	1939	1.1		1919	1.1
5	0148	6.7	20	0126	6.7
TH	0747	1.4	F	0724	1.4
	1358	7.0		1333	7.1
	2008	1.5		1946	1.3
6	0212	6.3	21	0151	6.4
F	0810	1.8	SA	0750	1.8
	1421	6.7		1356	6.9
	2035	1.9		2016	1.7
7	0236	5.9	22	0217	6.0
SA	0833	2.4	SU	0815	2.3
	1443	6.2		1424	6.5
	2102	2.5		2048	2.3
8	0258	5.3	23	0247	5.4
SU	0853	3.0	M	0841	2.9
	1507	5.7		1457	5.9
	2133	3.1		2131	3.0
9	0323	4.8	24	0327	4.8
M	0902	3.6	TU	0909	3.6
	1534	5.2		1548	5.3
	2228	3.7		2345	3.6
10	0402	4.2	25	0743	4.3
TU	0838	4.1	W	1253	4.2
	1901	4.6		1932	4.9
11	0333	3.6	26	0308	3.3
W	1039	4.7	TH	0949	5.1
	1548	3.9		1527	3.5
	2130	5.0		2132	5.5
12	0412	3.0	27	0354	2.6
TH	1036	5.2	F	1016	5.8
	1619	3.4		1608	2.7
	2213	5.6		2217	6.2
13	0437	2.5	28	0426	1.9
F	1051	5.7	SA	1044	6.4
	1642	2.8		1641	1.9
	2243	6.1		2253	6.7
14	0458	2.0	29	0457	1.4
SA	1111	6.2	SU	1113	6.9
	1707	2.2	•	1713	1.3
	2312	6.5		2326	7.0
15	0522	1.6	30	0527	1.1
SU	1135	6.6	M	1142	7.3
○	1732	1.8		1745	0.9
	2341	6.8		2358	7.2
			31	0556	0.9
			TU	1209	7.4
				1814	0.8

APRIL

	TIME	M		TIME	M
1	0027	7.1	16	0011	6.9
W	0624	1.0	TH	0606	1.2
	1236	7.4		1215	7.3
	1842	0.8		1826	0.8
2	0053	6.9	17	0039	6.8
TH	0650	1.2	F	0633	1.4
	1259	7.3		1241	7.3
	1909	1.0		1856	1.0
3	0118	6.6	18	0108	6.6
F	0714	1.6	SA	0701	1.6
	1323	7.0		1308	7.1
	1935	1.4		1927	1.3
4	0142	6.3	19	0135	6.2
SA	0738	2.0	SU	0730	2.0
	1345	6.7		1338	6.8
	2001	1.9		2000	1.8
5	0205	2.9	20	0206	5.8
SU	0800	2.5	M	0801	2.6
	1408	6.2		1411	6.3
	2027	2.4		2039	2.4
6	0229	5.4	21	0245	5.3
M	0820	3.0	TU	0837	3.2
	1432	5.7		1455	5.7
	2053	3.0		2138	3.1
7	0254	4.9	22	0352	4.8
TU	0833	3.5	W	0948	3.8
	1458	5.2		1621	5.1
	2128	3.6			
8	0329	4.4	23	0012	3.5
W	0832	4.1	TH	0729	4.7
	1540	4.6		1322	3.8
				1938	5.0
9	0230	3.8	24	0223	3.1
TH	1008	4.6	F	0903	5.3
	1520	4.0		1458	3.1
	2104	4.8		2110	5.5
10	0337	3.3	25	0320	2.6
F	0958	5.2	SA	0940	5.9
	1548	3.4		1541	2.4
	2146	5.4		2154	6.1
11	0400	2.8	26	0355	2.1
SA	1014	5.7	SU	1010	6.5
	1610	2.8		1616	1.8
	2216	5.9		2229	6.5
12	0423	2.3	27	0428	1.7
SU	1036	6.2	M	1040	6.8
	1635	2.2		1648	1.3
	2245	6.3		2302	6.7
13	0446	1.8	28	0459	1.5
M	1059	6.6	TU	1109	7.1
	1701	1.6	•	1719	1.0
	2313	6.7		2333	6.7
14	0511	1.5	29	0528	1.4
TU	1124	7.0	W	1137	7.2
	1729	1.2	○	1749	0.9
	2342	6.8			
15	0538	1.3	30	0002	6.7
W	1150	7.2	TH	0555	1.5
	1757	0.9		1204	7.1
				1816	1.0

AUSTRALIA, NORTH COAST – DARWIN
LAT 12°28'S   LONG 130°51'E
TIMES AND HEIGHTS OF HIGH AND LOW WATERS

TIME ZONE −0930

YEAR 1987

JANUARY

Day	TIME	M	Day	TIME	M
1 TH	0047	2.8	16 F	0104	2.8
	0556	6.5		0629	6.1
	1251	0.4		1306	1.3
	1941	7.7		1950	7.0
2 F	0130	2.6	17 SA	0134	2.6
	0648	6.6		0705	6.2
	1335	0.4		1337	1.3
	2022	7.7		2018	7.1
3 SA	0214	2.4	18 SU	0206	2.5
	0738	6.6		0739	6.3
	1418	0.7		1403	1.4
	2059	7.5		2044	7.1
4 SU	0300	2.3	19 M	0239	2.4
	0831	6.5		0814	6.2
	1459	1.1		1424	1.5
	2137	7.3		2109	7.0
5 M	0349	2.3	20 TU	0316	2.4
	0925	6.2		0851	6.1
	1539	1.6		1443	1.7
	2213	7.0		2136	6.9
6 TU	0442	2.3	21 W	0354	2.4
	1023	5.8		0932	6.0
	1619	2.3		1506	2.1
	2250	6.6		2201	6.7
7 W	0539	2.4	22 TH	0437	2.3
	1124	5.4		1022	5.7
	1701	3.0		1539	2.6
	2326	6.2		2229	6.4
8 TH	0641	2.5	23 F	0523	2.3
	1237	5.1		1120	5.5
	1756	3.6		1630	3.2
				2302	6.1
9 F	0006	5.7	24 SA	0616	2.3
	0745	2.4		1233	5.3
	1408	5.1		1743	3.8
	1922	4.0		2347	5.7
10 SA	0059	5.3	25 SU	0723	2.2
	0847	2.3		1423	5.3
	1540	5.4		1928	4.2
	2108	4.2			
11 SU	0210	5.1	26 M	0056	5.4
	0944	2.2		0847	2.0
	1642	5.8		1615	5.8
	2246	4.0		2202	4.2
12 M	0324	5.1	27 TU	0238	5.3
	1033	2.0		1008	1.6
	1729	6.2		1721	6.4
	2337	3.6		2318	3.7
13 TU	0423	5.3	28 W	0408	5.6
	1116	1.8		1115	1.2
	1809	6.6		1812	7.0
14 W	0009	3.3	29 TH	0009	3.1
	0509	5.6		0516	6.0
	1155	1.6	•	1211	0.8
	1845	6.8		1858	7.5
15 TH	0037	3.0	30 F	0054	2.6
o	0551	5.8		0613	6.5
	1232	1.4		1259	0.5
	1919	7.0		1938	7.8
			31 SA	0137	2.1
				0705	6.8
				1341	0.4
				2015	7.9

FEBRUARY

Day	TIME	M	Day	TIME	M
1 SU	0218	1.8	16 M	0203	2.0
	0753	7.0		0742	6.7
	1419	0.6		1401	1.2
	2047	7.9		2026	7.4
2 M	0258	1.6	17 TU	0233	1.8
	0839	7.0		0815	6.7
	1453	1.0		1423	1.4
	2117	7.7		2047	7.4
3 TU	0338	1.5	18 W	0305	1.6
	0924	6.7		0850	6.7
	1524	1.5		1445	1.6
	2144	7.4		2107	7.2
4 W	0418	1.6	19 TH	0337	1.6
	1008	6.3		0928	6.6
	1551	2.2		1510	2.0
	2208	6.9		2128	7.0
5 TH	0457	1.8	20 F	0410	1.6
	1051	5.9		1009	6.4
	1612	2.8		1542	2.6
	2228	6.4		2151	6.6
6 F	0538	2.1	21 SA	0446	1.7
	1140	5.4		1056	6.0
	1641	3.5		1622	3.2
	2249	5.8		2217	6.2
7 SA	0625	2.4	22 SU	0529	1.9
	1248	5.1		1155	5.6
	1738	4.1		1715	3.8
	2323	5.2		2252	5.6
8 SU	0729	2.6	23 M	0631	2.2
	1457	5.1		1348	5.3
	1945	4.5		1843	4.4
				2354	5.1
9 M	0051	4.7	24 TU	0818	2.4
	0848	2.7		1607	5.6
	1636	5.5		2218	4.2
	2332	4.1			
10 TU	0308	4.6	25 W	0249	4.9
	1018	2.6		1003	2.1
	1729	5.9		1713	6.2
				2323	3.5
11 W	0003	3.6	26 TH	0432	5.5
	0433	5.0		1117	1.6
	1126	2.2		1804	6.8
	1809	6.3			
12 TH	0026	3.3	27 F	0009	2.8
	0523	5.4		0536	6.2
	1209	1.9		1211	1.1
	1843	6.6		1846	7.4
13 F	0047	2.9	28 SA	0051	2.1
	0603	5.8		0628	6.8
o	1243	1.6	•	1255	0.7
	1913	6.9		1923	7.8
14 SA	0109	2.6			
	0638	6.2			
	1311	1.4			
	1940	7.1			
15 SU	0135	2.3			
	0710	6.5			
	138	1.2			
	2005	7.3			

MARCH

Day	TIME	M	Day	TIME	M
1 SU	0129	1.5	16 M	0111	1.9
	0714	7.2		0702	6.8
	1333	0.6		1317	1.4
	1953	7.9		1931	7.3
2 M	0206	1.1	17 TU	0139	1.4
	0755	7.4		0734	7.0
	1405	0.8		1341	1.4
	2021	7.9		1951	7.4
3 TU	0240	0.9	18 W	0209	1.1
	0834	7.3		0807	7.2
	1435	1.2		1407	1.5
	2045	7.7		2009	7.3
4 W	0314	1.0	19 TH	0238	0.9
	0910	7.1		0840	7.2
	1501	1.7		1434	1.8
	2107	7.3		2030	7.2
5 TH	0345	1.2	20 F	0310	0.9
	0945	6.7		0917	7.1
	1523	2.3		1506	2.2
	2124	6.8		2054	6.9
6 F	0414	1.5	21 SA	0343	1.1
	1019	6.3		0956	6.8
	1545	2.9		1541	2.8
	2138	6.3		2121	6.5
7 SA	0441	1.9	22 SU	0420	1.4
	1055	5.8		1040	6.3
	1612	3.4		1621	3.3
	2154	5.7		2149	5.9
8 SU	0516	2.3	23 M	0504	1.9
	1141	5.3		1138	5.7
	1659	4.0		1713	3.9
	2205	5.1		2226	5.3
9 M	0617	2.8	24 TU	0615	2.4
	1305	5.0		1333	5.3
	1833	4.4		1911	4.3
	2107	4.6			
10 TU	0751	3.1	25 W	0009	4.7
	1615	5.2		0809	2.6
	2328	4.0		1545	5.6
				2217	3.8
11 W	0326	4.4	26 TH	0323	5.0
	0959	3.0		0953	2.4
	1709	5.6		1651	6.1
	2347	3.5		2310	3.0
12 TH	0447	5.0	27 F	0441	5.7
	1113	2.5		1105	1.9
	1746	6.0		1738	6.7
				2353	2.2
13 F	0005	3.1	28 SA	0537	6.4
	0525	5.5		1154	1.5
	1153	2.1		1816	7.1
	1815	6.4			
14 SA	0023	2.7	29 SU	0031	1.6
	0558	6.0		0623	7.0
	1223	1.7	•	1235	1.2
	1843	6.8		1848	7.4
15 SU	0046	2.3	30 M	0107	1.1
o	0631	6.4		0704	7.3
	1251	1.5		1308	1.2
	1908	7.1		1915	7.5
			31 TU	0138	0.8
				0741	7.5
				1338	1.4
				1938	7.4

APRIL

Day	TIME	M	Day	TIME	M
1 W	0209	0.7	16 TH	0135	0.6
	0815	7.4		0751	7.4
	1405	1.7		1343	1.9
	2000	7.2		1924	7.1
2 TH	0237	0.8	17 F	0206	0.5
	0847	7.2		0826	7.4
	1431	2.1		1416	2.2
	2019	6.9		1952	6.9
3 F	0303	1.0	18 SA	0239	0.6
	0918	6.8		0903	7.2
	1455	2.5		1453	2.5
	2037	6.5		2023	6.6
4 SA	0327	1.4	19 SU	0317	0.9
	0947	6.5		0944	6.8
	1523	3.0		1533	2.9
	2055	6.0		2058	6.1
5 SU	0353	1.8	20 M	0359	1.4
	1019	6.1		1031	6.3
	1555	3.4		1619	3.4
	2117	5.5		2138	5.6
6 M	0426	2.3	21 TU	0453	2.0
	1058	5.6		1133	5.7
	1641	3.8		1726	3.8
	2136	5.0		2244	5.0
7 TU	0525	2.9	22 W	0613	2.5
	1158	5.1		1311	5.4
	1755	4.1		1946	3.8
	2316	4.4			
8 W	0700	3.2	23 TH	0118	4.7
	1447	5.0		0754	2.7
	2238	3.9		1455	5.5
				2143	3.2
9 TH	0253	4.4	24 F	0319	5.2
	0858	3.2		0924	2.6
	1610	5.3		1602	5.9
	2254	3.4		2239	2.5
10 F	0414	4.9	25 SA	0428	5.8
	1022	2.8		1033	2.3
	1651	5.7		1650	6.3
	2314	3.0		2322	1.9
11 SA	0453	5.5	26 SU	0520	6.4
	1107	2.4		1123	2.1
	1723	6.2		1726	6.6
	2338	2.5		2359	1.3
12 SU	0441	5.7	27 M	0603	6.8
	1142	2.1		1201	2.0
	1751	6.5		1755	6.7
	2353	2.2			
13 M	0006	1.9	28 TU	0031	1.0
	0606	6.5		0641	7.1
	1214	1.9	•	1233	2.0
	1817	6.8		1821	6.8
14 TU	0035	1.4	29 W	0100	0.8
	0641	6.9		0717	7.2
o	1244	1.7		1302	2.1
	1839	7.0		1844	6.8
15 W	0105	0.9	30 TH	0126	1.0
	0716	7.3		0749	7.2
	1313	1.8		1330	2.3
	1900	71		1908	6.7

NEW ZEALAND, SOUTH ISLAND – WESTPORT
LAT 41°45'S   LONG 171°36'E
TIMES AND HEIGHTS OF HIGH AND LOW WATERS

TIME ZONE −1200

YEAR 1987

SEPTEMBER

Date	TIME	M	Date	TIME	M
1 TU	0309	2.7	16 W	0434	2.3
	0951	0.8		1115	1.1
	1537	2.6		1726	2.2
	2222	0.8		2352	1.1
2 W	0413	2.5	17 TH	0626	2.2
	1103	0.9		1223	1.1
	1653	2.5		1914	2.3
	2339	0.8			
3 TH	0550	2.4	18 F	0059	1.1
	1217	0.8		0751	2.3
	1837	2.5		1326	1.1
				2016	2.4
4 F	0054	0.7	19 SA	0157	1.0
	0732	2.6		0840	2.5
	1326	0.7		1418	0.9
	2001	2.7		2058	2.6
5 SA	0200	0.5	20 SU	0243	0.8
	0840	2.8		0917	2.7
	1426	0.4		1500	0.8
	2101	3.0		2132	2.8
6 SU	0258	0.3	21 M	0321	0.7
	0932	3.1		0949	2.8
	1520	0.2		1535	0.7
	2150	3.3		2203	3.0
7 M ○	0349	0.1	22 TU	0356	0.6
	1017	3.3		1020	3.0
	1609	0.1		1609	0.6
	2235	3.5		2234	3.1
8 TU	0438	0.0	23 W	0429	0.5
	1101	3.4		1051	3.1
	1657	0.0		1642	0.5
	2318	3.6		2305	3.2
9 W	0525	0.0	24 TH ●	0501	0.4
	1142	3.5		1122	3.1
	1744	0.1		1716	0.5
				2337	3.2
10 TH	0000	3.6	25 F	0539	0.4
	0611	0.1		1153	3.1
	1223	3.4		1751	0.5
	1831	0.1			
11 F	0041	3.5	26 SA	0009	3.2
	0657	0.2		0615	0.5
	1303	3.3		1226	3.1
	1917	0.3		1829	0.5
12 SA	0121	3.3	27 SU	0043	3.1
	0743	0.4		0655	0.5
	1342	3.1		1301	3.0
	2004	0.5		1911	0.5
13 SU	0201	3.1	28 M	0120	3.0
	0829	0.6		0739	0.6
	1422	2.9		1338	2.9
	2053	0.7		1959	0.6
14 M	0242	2.8	29 TU	0201	2.9
	0917	0.8		0830	0.7
	1505	2.6		1422	2.8
	2145	0.9		2058	0.7
15 TU	0329	2.5	30 W	0251	2.7
	1012	1.0		0933	0.8
	1558	2.4		1518	2.6
	2245	1.0		2208	0.8

OCTOBER

Date	TIME	M	Date	TIME	M
1 TH	0400	2.5	16 F	0533	2.2
	1046	0.9		1142	1.2
	1641	2.5		1822	2.3
	2326	0.8			
2 F	0549	2.4	17 SA	0019	1.1
	1201	0.8		0704	2.3
	1832	2.6		1246	1.1
				1932	2.4
3 SA	0041	0.7	18 SU	0117	1.0
	0725	2.6		0759	2.5
	1310	0.7		1340	1.0
	1950	2.8		2018	2.6
4 SU	0146	0.5	19 M	0206	0.9
	0826	2.9		0839	2.7
	1410	0.5		1424	0.8
	2044	3.1		2055	2.8
5 M	0241	0.3	20 TU	0246	0.7
	0913	3.1		0914	2.8
	1501	0.3		1501	0.7
	2130	3.3		2129	3.0
6 TU	0330	0.2	21 W	0322	0.6
	0955	3.3		0946	3.0
	1548	0.2		1536	0.6
	2212	3.5		2201	3.1
7 W ○	0415	0.1	22 TH ●	0358	0.5
	1035	3.4		1018	3.1
	1633	0.1		1611	0.5
	2252	3.5		2234	3.2
8 TH	0458	0.1	23 F	0434	0.4
	1114	3.4		1051	3.1
	1717	0.1		1648	0.4
	2331	3.5		2309	3.2
9 F	0542	0.2	24 SA	0512	0.4
	1153	3.4		1126	3.1
	1802	0.2		1728	0.4
				2345	3.2
10 SA	0011	3.4	25 SU	0553	0.4
	0625	0.3		1203	3.1
	1231	3.2		1811	0.4
	1847	0.3			
11 SU	0050	3.2	26 M	0024	3.1
	0709	0.5		0637	0.5
	1309	3.1		1243	3.0
	1933	0.5		1859	0.5
12 M	0129	3.1	27 TU	0106	3.0
	0754	0.6		0727	0.6
	1349	2.9		1326	3.0
	2020	0.7		1953	0.6
13 TU	0209	2.7	28 W	0153	2.9
	0842	0.8		0822	0.7
	1431	2.7		1415	2.8
	2112	0.9		2053	0.7
14 W	0255	2.5	29 TH	0249	2.7
	0935	1.0		0924	0.8
	1521	2.5		1516	2.7
	2210	1.0		2201	0.8
15 TH	0356	2.3	30 F	0403	2.5
	1036	1.1		1033	0.9
	1638	2.3		1640	2.6
	2314	1.1		2313	0.8
			31 SA	0542	2.5
				1143	0.8
				1815	2.7

NOVEMBER

Date	TIME	M	Date	TIME	M
1 SU	0023	0.7	16 M	0027	1.0
	0704	2.7		0659	2.5
	1250	0.7		1250	1.0
	1926	2.9		1923	2.6
2 M	0126	0.6	17 TU	0119	0.9
	0802	2.9		0750	2.6
	1349	0.6		1339	0.9
	2021	3.1		2009	2.7
3 TU	0220	0.4	18 W	0205	0.7
	0849	3.0		0831	2.8
	1440	0.4		1422	0.7
	2106	3.2		2049	2.9
4 W	0307	0.3	19 TH	0246	0.6
	0931	3.2		0909	2.9
	1526	0.3		1503	0.6
	2147	3.3		2127	3.0
5 TH ○	0350	0.3	20 F	0327	0.5
	1009	3.3		0947	3.0
	1609	0.2		1543	0.5
	2226	3.3		2206	3.1
6 F	0432	0.3	21 SA ●	0407	0.4
	1047	3.3		1025	3.1
	1652	0.3		1626	0.4
	2305	3.3		2247	3.1
7 SA	0513	0.3	22 SU	0451	0.4
	1125	3.2		1106	3.1
	1735	0.3		1712	0.4
	2344	3.2		2330	3.1
8 SU	0556	0.4	23 M	0537	0.4
	1204	3.1		1150	3.1
	1819	0.4		1801	0.4
9 M	0024	3.0	24 TU	0015	3.1
	0639	0.6		0627	0.4
	1243	3.0		1236	3.1
	1905	0.6		1854	0.4
10 TU	0104	2.9	25 W	0104	3.0
	0724	0.7		0720	0.5
	1324	2.9		1324	3.1
	1952	0.7		1950	0.5
11 W	0146	2.7	26 TH	0155	2.9
	0811	0.8		0816	0.6
	1407	2.7		1417	3.0
	2041	0.9		2049	0.5
12 TH	0231	2.6	27 F	0253	2.8
	0901	1.0		0914	0.7
	1455	2.6		1517	2.9
	2134	1.0		2150	0.6
13 F	0324	2.5	28 SA	0359	2.7
	0956	1.1		1016	0.7
	1554	2.5		1626	2.8
	2230	1.1		2253	0.7
14 SA	0434	2.4	29 SU	0515	2.6
	1054	1.1		1120	0.8
	1711	2.4		1743	2.8
	2329	1.1		2357	0.7
15 SU	0554	2.4	30 M	0629	2.7
	1154	1.1		1223	0.7
	1826	2.5		1853	2.8

DECEMBER

Date	TIME	M	Date	TIME	M
1 TU	0059	0.6	16 W	0027	0.9
	0731	2.8		0647	2.5
	1323	0.7		1250	0.9
	1952	2.9		1914	2.6
2 W	0154	0.6	17 TH	0121	0.8
	0822	2.9		0743	2.6
	1416	0.6		1342	0.8
	2042	3.0		2007	2.7
3 TH	0243	0.5	18 F	0211	0.6
	0907	3.0		0832	2.8
	1504	0.5		1432	0.6
	2125	3.0		2057	2.9
4 F	0327	0.5	19 SA	0259	0.5
	0947	3.0		0920	2.9
	1548	0.4		1520	0.4
	2205	3.0		2144	3.0
5 SA ○	0409	0.5	20 SU ●	0346	0.4
	1026	3.0		1006	3.0
	1631	0.4		1609	0.3
	2245	3.0		2233	3.1
6 SU	0450	0.5	21 M	0435	0.3
	1105	3.0		1054	3.1
	1714	0.5		1700	0.3
	2325	3.0		2322	3.1
7 M	0532	0.5	22 TU	0525	0.3
	1145	3.0		1143	3.2
	1757	0.5		1753	0.2
8 TU	0005	2.9	23 W	0012	3.2
	0615	0.6		0618	0.2
	1225	2.9		1232	3.2
	1842	0.6		1847	0.2
9 W	0027	2.9	24 TH	0102	3.1
	0659	0.7		0711	0.3
	1306	2.9		1322	3.2
	1926	0.7		1942	0.3
10 TH	0127	2.8	25 F	0152	3.1
	0743	0.8		0804	0.4
	1347	2.8		1412	3.2
	2011	0.8		2036	0.3
11 F	0209	2.7	26 SA	0244	3.0
	0827	0.9		0858	0.5
	1429	2.7		1504	3.1
	2056	0.9		2131	0.5
12 SA	0253	2.6	27 SU	0338	2.9
	0913	1.0		0953	0.6
	1514	2.6		1600	2.9
	2144	0.9		2227	0.6
13 SU	0341	2.5	28 M	0437	2.7
	1003	1.0		1052	0.7
	1605	2.6		1703	2.8
	2236	1.0		2327	0.7
14 M	0438	2.5	29 TU	0545	2.6
	1057	1.0		1153	0.8
	1707	2.5		1813	2.7
	2331	1.0			
15 TU	0543	2.5	30 W	0028	0.7
	1154	1.0		0654	2.6
	1813	2.5		1255	0.8
				1921	2.6
			31 TH	0126	0.7
				0756	2.6
				1353	0.7
				2021	2.7

TIME ZONE −1200

YEAR 1987

SEPTEMBER

Day	Time1	M	Time2	M	Time3	M	Time4	M
1 TU	0557	0.4	1217	1.3	1804	0.5		
16 W	0103	1.4	0731	0.5	1359	1.2	1935	0.7
2 W	0034	1.5	0702	0.4	1328	1.3	1910	0.5
17 TH	0202	1.4	0837	0.5	1456	1.3	2040	0.6
3 TH	0138	1.5	0811	0.4	1436	1.3	2020	0.5
18 F	0256	1.4	0934	0.4	1543	1.3	2136	0.6
4 F	0242	1.6	0918	0.3	1536	1.4	2127	0.5
19 SA	0345	1.4	1020	0.4	1627	1.4	2223	0.6
5 SA	0339	1.6	1017	0.2	1630	1.5	2226	0.4
20 SU	0428	1.5	1101	0.4	1705	1.4	2305	0.5
6 SU	0435	1.7	1109	0.1	1720	1.6	2320	0.3
21 M	0508	1.5	1136	0.3	1739	1.5	2343	0.5
7 M ○	0527	1.8	1158	0.1	1808	1.7		
22 TU	0544	1.5	1208	0.3	1811	1.5		
8 TU	0014	0.2	0619	1.8	1245	0.0	1855	1.7
23 W •	0019	0.4	0621	1.5	1239	0.3	1843	1.6
9 W	0106	0.2	0712	1.8	1331	0.1	1942	1.7
24 TH	0056	0.4	0656	1.5	1312	0.3	1916	1.6
10 TH	0157	0.2	0802	1.7	1416	0.1	2029	1.7
25 F	0134	0.3	0734	1.5	1345	0.3	1951	1.6
11 F	0247	0.2	0854	1.6	1501	0.3	2117	1.6
26 SA	0213	0.3	0815	1.5	1422	0.4	2029	1.6
12 SA	0338	0.3	0949	1.5	1548	0.4	2207	1.6
27 SU	0257	0.3	0901	1.4	1504	0.4	2114	1.6
13 SU	0430	0.3	1047	1.4	1635	0.5	2302	1.5
28 M	0343	0.3	0955	1.4	1550	0.5	2206	1.5
14 M	0526	0.4	1151	1.3	1729	0.6		
29 TU	0437	0.4	1057	1.3	1645	0.6	2308	1.5
15 TU	0001	1.4	0627	0.5	1257	1.2	1829	0.6
30 W	0537	0.4	1207	1.3	1750	0.6		

OCTOBER

Day	Time1	M	Time2	M	Time3	M	Time4	M
1 TH	0015	1.5	0645	0.4	1317	1.3	1902	0.6
16 F	0117	1.4	0751	0.5	1415	1.3	2006	0.7
2 F	0123	1.5	0754	0.4	1422	1.4	2013	0.5
17 SA	0213	1.4	0847	0.5	1504	1.3	2103	0.6
3 SA	0225	1.6	0858	0.3	1518	1.5	2117	0.5
18 SU	0305	1.4	0934	0.5	1546	1.4	2150	0.6
4 SU	0322	1.6	0955	0.2	1609	1.6	2214	0.4
19 M	0350	1.4	1013	0.4	1623	1.5	2233	0.5
5 M	0417	1.7	1044	0.2	1657	1.7	2306	0.3
20 TU	0431	1.4	1049	0.4	1657	1.5	2311	0.5
6 TU	0511	1.7	1132	0.2	1743	1.7	2358	0.2
21 W	0511	1.5	1123	0.4	1729	1.6	2349	0.4
7 W ○	0601	1.7	1217	0.2	1827	1.8		
22 TH •	0549	1.5	1157	0.4	1803	1.6		
8 TH	0048	0.2	0652	1.7	1302	0.2	1912	1.7
23 F	0027	0.3	0629	1.5	1232	0.4	1838	1.7
9 F	0135	0.2	0742	1.6	1345	0.3	1955	1.7
24 SA	0107	0.3	0712	1.5	1312	0.4	1919	1.7
10 SA	0223	0.2	0834	1.5	1427	0.4	2040	1.6
25 SU	0149	0.2	0759	1.5	1354	0.4	2004	1.7
11 SU	0310	0.3	0925	1.4	1511	0.5	2128	1.5
26 M	0236	0.2	0850	1.4	1440	0.5	2054	1.6
12 M	0357	0.3	1020	1.3	1557	0.6	2219	1.5
27 TU	0327	0.3	0948	1.4	1534	0.5	2152	1.6
13 TU	0448	0.4	1116	1.3	1651	0.7	2315	1.4
28 W	0421	0.3	1049	1.4	1634	0.6	2254	1.5
14 W	0546	0.5	1218	1.2	1751	0.7		
29 TH	0523	0.3	1156	1.4	1743	0.6		
15 TH	0017	1.4	0648	0.5	1319	1.2	1900	0.7
30 F	0000	1.5	0628	0.3	1300	1.4	1856	0.6
31 SA	0104	1.5	0733	0.3	1401	1.5	2004	0.5

NOVEMBER

Day	Time1	M	Time2	M	Time3	M	Time4	M
1 SU	0206	1.5	0832	0.3	1456	1.6	2104	0.5
16 M	0215	1.3	0836	0.5	1457	1.4	2108	0.6
2 M	0305	1.6	0927	0.3	1546	1.6	2200	0.4
17 TU	0305	1.3	0918	0.5	1535	1.5	2153	0.5
3 TU	0402	1.6	1017	0.3	1633	1.7	2252	0.3
18 W	0350	1.4	0957	0.5	1612	1.6	2235	0.4
4 W	0455	1.6	1105	0.3	1718	1.7	2343	0.3
19 TH	0435	1.4	1037	0.5	1648	1.6	2318	0.3
5 TH	0549	1.6	1150	0.3	1801	1.7		
20 F ○	0520	1.4	1118	0.4	1727	1.7		
6 F	0031	0.2	0639	1.5	1234	0.4	1843	1.7
21 SA •	0001	0.3	0608	1.4	1201	0.4	1810	1.7
7 SA	0116	0.2	0728	1.5	1317	0.4	1927	1.6
22 SU	0046	0.2	0657	1.5	1246	0.4	1857	1.7
8 SU	0201	0.2	0816	1.4	1358	0.5	2011	1.6
23 M	0133	0.2	0749	1.5	1335	0.4	1948	1.7
9 M	0244	0.3	0904	1.4	1442	0.6	2054	1.5
24 TU	0223	0.2	0843	1.5	1427	0.5	2044	1.7
10 TU	0328	0.3	0950	1.3	1527	0.6	2142	1.5
25 W	0315	0.2	0939	1.5	1525	0.5	2141	1.6
11 W	0416	0.4	1040	1.3	1617	0.7	2234	1.4
26 TH	0412	0.2	1037	1.5	1628	0.5	2241	1.6
12 TH	0506	0.5	1133	1.3	1716	0.7	2329	1.4
27 F	0509	0.3	1137	1.5	1736	0.5	2342	1.6
13 F	0601	0.5	1229	1.3	1821	0.7		
28 SA	0608	0.3	1238	1.5	1842	0.5		
14 SA	0025	1.3	0657	0.5	1326	1.3	1924	0.7
29 SU	0043	1.5	0706	0.3	1337	1.6	1947	0.5
15 SU	0121	1.3	0749	0.5	1415	1.4	2019	0.7
30 M	0147	1.5	0802	0.4	1432	1.6	2047	0.4

DECEMBER

Day	Time1	M	Time2	M	Time3	M	Time4	M
1 TU	0250	1.5	0857	0.4	1522	1.7	2145	0.4
16 W	0218	1.3	0820	0.5	1446	1.5	2111	0.5
2 W	0349	1.5	0949	0.4	1610	1.7	2240	0.3
17 TH	0312	1.3	0907	0.5	1529	1.6	2200	0.4
3 TH	0445	1.5	1040	0.4	1657	1.7	2330	0.3
18 F	0406	1.3	0956	0.5	1613	1.7	2251	0.3
4 F	0540	1.4	1127	0.5	1740	1.7		
19 SA	0459	1.4	1047	0.5	1701	1.7	2340	0.2
5 SA	0017	0.3	0629	1.4	1212	0.5	1824	1.6
20 SU •	0551	1.4	1137	0.5	1751	0.7		
6 SU	0100	0.2	0716	1.4	1255	0.5	1906	1.6
21 M	0031	0.2	0645	1.5	1231	0.4	1845	1.8
7 M	0142	0.3	0759	1.4	1335	0.6	1948	1.6
22 TU	0121	0.1	0738	1.5	1324	0.4	1938	1.8
8 TU	0223	0.3	0840	1.4	1418	0.6	2032	1.5
23 W	0212	0.1	0830	1.6	1419	0.4	2033	1.8
9 W	0305	0.3	0922	1.4	1503	0.6	2114	1.5
24 TH	0304	0.1	0924	1.6	1517	0.4	2128	1.7
10 TH	0348	0.4	1004	1.4	1552	0.7	2159	1.5
25 F	0356	0.1	1017	1.6	1617	0.4	2223	1.7
11 F	0433	0.4	1051	1.4	1645	0.7	2245	1.4
26 SA	0448	0.2	1113	1.6	1719	0.5	2322	1.6
12 SA	0519	0.5	1140	1.4	1740	0.7	2334	1.4
27 SU	0543	0.3	1211	1.6	1822	0.5		
13 SU	0605	0.5	1229	1.4	1836	0.7		
28 M	0024	1.5	0636	0.3	1310	1.6	1924	0.5
14 M	0027	1.3	0650	0.5	1319	1.4	1928	0.7
29 TU	0130	1.5	0733	0.4	1406	1.6	2027	0.4
15 TU	0121	1.3	0735	0.5	1404	1.5	2020	0.6
30 W	0236	1.4	0829	0.5	1500	1.7	2128	0.4
31 TH	0339	1.4	0925	0.5	1550	1.6	2226	0.4

SEPTEMBER

Day		TIME	M	Day		TIME	M
1	TU	0323	1.2	16	W	0349	1.2
		1055	3.2			1838	3.4
		1508	2.9				
		2002	3.4				
2	W	0408	1.1	17	TH	0437	1.4
		1937	3.4			1720	3.5
3	TH	0511	1.2	18	F	0555	1.6
		1727	3.5			1705	3.6
4	F	0644	1.2	19	SA	0757	1.7
		1713	3.7			1711	3.6
5	SA	0830	1.2	20	SU	0920	1.6
		1729	3.8			1721	3.6
6	SU	0948	1.1	21	M	0017	2.4
		1748	3.8			0346	2.6
						1012	1.6
						1732	3.6
7	M (o)	0002	2.7	22	TU	0001	2.2
		0335	2.9			0440	2.8
		1045	1.0			1053	1.6
		1807	3.8			1743	3.6
8	TU	0003	2.4	23	W (•)	0002	2.1
		0447	3.2			0520	3.0
		1133	1.1			1129	1.6
		1824	3.7			1754	3.5
9	W	0019	2.1	24	TH	0011	1.8
		0542	3.4			0556	3.3
		1214	1.2			1203	1.7
		1839	3.6			1806	3.5
10	TH	0040	1.8	25	F	0026	1.6
		0632	3.6			0634	3.5
		1251	1.5			1236	1.9
		1853	3.5			1817	3.4
11	F	0106	1.5	26	SA	0044	1.4
		0718	3.6			0713	3.6
		1324	1.8			1309	2.1
		1906	3.5			1829	3.4
12	SA	0134	1.3	27	SU	0108	1.1
		0805	3.6			0756	3.7
		1354	2.1			1341	2.4
		1917	3.4			1840	3.3
13	SU	0204	1.1	28	M	0135	1.0
		0855	3.5			1414	2.7
		1420	2.5			1849	3.3
		1925	3.4				
14	M	0236	1.1	29	TU	0208	0.9
		0953	3.3			0945	3.6
		1438	2.8			1449	3.0
		1928	3.4			1855	3.3
15	TU	0311	1.1	30	W	0247	0.9
		1121	3.2			1112	3.5
		1438	3.1			1533	3.2
		1919	3.4			1847	3.4

OCTOBER

Day		TIME	M	Day		TIME	M
1	TH	0335	1.0	16	F	0344	1.4
		1336	3.5			1443	3.5
2	F	0440	1.2	17	SA	0439	1.7
		1515	3.6			1515	3.5
3	SA	0617	1.4	18	SU	0626	1.9
		1553	3.6			1536	3.5
						2338	2.3
4	SU	0808	1.5	19	M	0250	2.4
		1620	3.6			0818	2.0
		2258	2.5			1554	3.5
						2259	2.1
5	M	0243	2.8	20	TU	0400	2.6
		0930	1.5			0928	2.0
		1641	3.6			1609	3.4
		2254	2.2			2255	1.9
6	TU	0406	3.1	21	W	0443	2.9
		1029	1.6			1021	2.1
		1659	3.5			1624	3.4
		2310	1.9			2303	1.7
7	W (o)	0504	3.4	22	TH (•)	0521	3.2
		1118	1.7			1105	2.2
		1716	3.4			1638	3.3
		2333	1.5			2319	1.4
8	TH	0554	3.6	23	F	0559	3.5
		1159	1.9			1147	2.0
		1731	3.4			1652	3.3
		2359	1.2			2339	1.1
9	F	0640	3.8	24	SA	0638	3.7
		1237	2.2			1226	2.5
		1745	3.3			1706	3.2
10	SA	0028	1.0	25	SU	0004	0.9
		0726	3.8			0721	3.8
		1311	2.4			1306	2.7
		1758	3.3			1721	3.2
11	SU	0058	1.0	26	M	0034	0.7
		0813	3.7			0808	3.9
		1344	2.7			1348	2.9
		1809	3.3			1735	3.3
12	M	0129	0.8	27	TU	0107	0.6
		0903	3.6			0901	3.9
		1415	2.9			1434	3.1
		1817	3.3			1748	3.3
13	TU	0200	0.8	28	W	0145	0.6
		1003	3.5			1002	3.8
		1449	3.1			1534	3.2
		1816	3.3			1755	3.3
14	W	0232	1.0	29	TH	0228	0.7
		1125	3.4			1115	3.7
		1545	3.3				
		1745	3.3				
15	TH	0305	1.2	30	F	0319	0.9
		1328	3.4			1232	3.7
				31	SA	0423	1.2
						1335	3.6

NOVEMBER

Day		TIME	M	Day		TIME	M
1	SU	0548	1.6	16	M	0454	2.0
		1420	3.5			1340	3.4
		2135	2.5			2144	2.2
2	M	0102	2.6	17	TU	0236	2.4
		0729	1.8			0633	2.3
		1452	3.4			1406	3.3
		2134	2.1			2136	1.9
3	TU	0305	2.9	18	W	0354	2.7
		0857	2.0			0820	2.5
		1517	3.3			1429	3.3
		2156	1.7			2148	1.6
4	W	0418	3.2	19	TH	0444	3.1
		1005	2.2			1450	3.2
		1539	3.3			2208	1.3
		2224	1.4				
5	TH (o)	0516	3.5	20	F	0528	3.4
		1101	2.4			1050	2.7
		1558	3.2			1510	3.2
		2254	1.1			2303	1.0
6	F	0607	3.7	21	SA (•)	0611	3.7
		1149	2.6			1148	2.8
		1617	3.2			1530	3.2
		2326	0.8			2304	0.8
7	SA	0655	3.8	22	SU	0654	3.9
		1233	2.8			1241	3.0
		1634	3.2			1552	3.2
		2359	0.6			2338	0.6
8	SU	0742	3.9	23	M	0739	4.0
		1316	2.9			1332	3.1
		1650	3.3			1615	3.2
9	M	0032	0.6	24	TU	0015	0.4
		0829	3.9			0826	4.1
		1401	3.1			1424	3.1
		1704	3.3			1641	3.2
10	TU	0104	0.6	25	W	0055	0.4
		0917	3.8			0913	4.1
		1453	3.1			1516	3.2
		1712	3.2			1712	3.2
11	W	0135	0.8	26	TH	0138	0.5
		1007	3.7			1000	4.0
						1613	3.1
						1750	3.1
12	TH	0207	0.9	27	F	0224	0.7
		1057	3.7			1046	3.9
						1712	3.0
						1843	3.0
13	F	0239	1.2	28	SA	0313	1.0
		1147	3.6			1129	3.7
						1808	2.8
						2013	2.8
14	SA	0312	1.4	29	SU	0408	1.4
		1232	3.5			1207	3.6
						1856	2.5
						2252	2.4
15	SU	0353	1.7	30	M	0511	1.8
		1310	3.5			1240	3.4
						1940	2.1

DECEMBER

Day		TIME	M	Day		TIME	M
1	TU	0118	2.7	16	W	0145	2.5
		0630	2.2			0446	2.4
		1310	3.3			1205	3.3
		2022	1.8			2003	1.7
2	W	0315	3.0	17	TH	0401	2.8
		0805	2.6			0615	2.8
		1336	3.2			1224	3.2
		2104	1.4			2039	1.4
3	TH	0438	3.3	18	F	0512	3.2
		0940	2.8			0916	3.1
		1401	3.2			1242	3.2
		2144	1.1			2119	1.1
4	F	0542	3.6	19	SA	0556	3.5
		1425	3.2			1145	3.1
		2224	0.8			1255	3.1
						2200	0.9
5	SA (o)	0634	3.8	20	SU (•)	0636	3.8
		1210	3.1			2243	0.6
		1450	3.2				
		2303	0.6				
6	SU	0721	3.9	21	M	0714	4.0
		1310	3.1			2326	0.4
		1515	3.2				
		2340	0.6				
7	M	0802	4.0	22	TU	0751	4.2
		1405	3.1			1417	3.1
		1541	3.2			1534	3.1
8	TU	0016	0.6	23	W	0010	0.4
		0839	1.0			0826	4.2
		1457	3.1			1436	3.1
		1608	3.1			1644	3.1
9	W	0050	0.6	24	TH	0054	0.4
		0912	4.0			0859	4.1
		1552	3.0			1500	3.0
		1633	3.0			1749	3.1
10	TH	0122	0.8	25	F	0137	0.6
		0942	3.9			0930	4.0
						1529	2.8
						1855	3.1
11	F	0153	1.0	26	SA	0221	0.8
		1010	3.8			0958	3.8
						1602	2.6
						2005	3.0
12	SA	0224	1.2	27	SU	0303	1.2
		1036	3.7			1023	3.7
						1640	2.3
						2123	2.9
13	SU	0255	1.5	28	M	0345	1.6
		1100	3.5			1045	3.5
		1831	2.5			1725	2.0
		2028	2.5			2301	2.8
14	M	0327	1.8	29	TU	0427	2.1
		1122	3.5			1104	3.4
		1859	2.3			1817	1.7
		2252	2.4				
15	TU	0402	2.1	30	W	0111	2.8
		1144	3.4			0511	2.6
		1929	2.0			1119	3.3
						1915	1.5
				31	TH	0355	3.0
						0616	3.0
						1127	3.3
						2015	1.2

FOR INTERMEDIATE HEIGHTS USE HARMONIC CONSTANTS (SEE PART III) AND NP 159

JANUARY

Days 1–15

Day	Time	M
1 TH	0559	5.5
	1543	−0.3
2 F	0650	5.6
	1635	−0.2
3 SA	0743	5.5
	1725	0.0
4 SU	0836	5.3
	1811	0.5
5 M	0931	4.9
	1851	1.0
6 TU	1028	4.3
	1918	1.5
7 W	1130	3.7
	1922	2.0
8 TH	0139	2.8
	0555	2.4
	1244	3.1
	1853	2.2
9 F	0200	3.2
	0938	2.2
	1423	2.5
	1805	2.3
10 SA	0233	3.7
	1150	1.7
11 SU	0311	4.0
	1300	1.2
12 M	0351	4.3
	1351	0.8
13 TU	0431	4.5
	1433	0.5
14 W ●	0510	4.7
	1510	0.3
15 TH ○	0547	4.7
	1542	0.2

Days 16–31

Day	Time	M
16 F	0623	4.8
	1610	0.2
17 SA	0659	4.8
	1638	0.3
18 SU	0737	4.7
	1704	0.5
19 M	0815	4.6
	1729	0.8
20 TU	0859	4.3
	1750	1.1
21 W	0948	4.0
	1806	1.5
22 TH	0117	2.4
	0255	2.3
	1046	3.6
	1812	1.9
23 F	0038	2.8
	0550	2.3
	1205	2.9
	1803	2.2
24 SA	0055	3.3
	0840	2.0
	1414	2.4
	1724	2.3
25 SU	0131	3.8
	1046	1.4
26 M	0217	4.3
	1202	0.8
27 TU	0309	
	1301	
28 W ○	0404	5.1
	1355	−0.1
29 TH ●	0500	5.3
	1445	−0.2
30 F	0556	5.3
	1533	−0.1
31 SA	0653	5.2
	1618	0.2

FEBRUARY

Days 1–15

Day	Time	M
1 SU	0748	5.0
	1657	0.6
2 M	0844	4.6
	1729	1.2
	2310	1.8
3 TU	0102	1.8
	0940	4.1
	1745	1.7
	2302	2.3
4 W	0302	1.9
	1338	3.5
	1735	2.1
	2323	2.8
5 TH	0450	2.0
	1144	2.9
	1703	2.3
	2358	3.2
6 F	0708	2.0
	1315	2.4
	1618	2.3
7 SA	0042	3.6
	1023	1.7
8 SU	0131	4.0
	1156	1.3
9 M	0222	4.2
	1251	1.0
10 TU	0311	4.3
	1333	0.7
11 W	0358	4.4
	1407	0.5
12 TH	0442	4.5
	1436	0.4
13 F ○	0524	4.5
	1501	0.4
14 SA	0605	4.4
	1525	0.5
15 SU	0647	4.4
	1548	0.7

Days 16–28

Day	Time	M
16 M	0731	4.2
	1608	1.0
17 TU	0818	4.0
	1625	1.3
	2210	2.0
18 W	0119	1.9
	0910	3.7
	1635	1.7
	2156	2.4
19 TH	0309	1.9
	1010	3.3
	1634	2.1
	2213	2.9
20 F	0452	1.8
	1130	2.8
	1616	2.4
	2247	3.5
21 SA	0649	1.6
	1402	2.4
	1501	2.4
	2334	4.0
22 SU	0858	1.3
23 M	0031	4.4
	1037	0.8
24 TU	0135	4.7
	1147	0.4
25 W	0243	4.9
	1245	0.2
26 TH	0350	5.0
	1336	0.1
27 F	0455	5.0
	1423	0.3
28 SA ●	0558	
	1504	

MARCH

Days 1–15

Day	Time	M
1 SU	0700	4.5
	1538	1.0
	2045	1.7
	2333	1.6
2 M	0759	4.2
	1601	1.5
	2039	2.1
3 TU	0119	1.5
	0859	3.7
	1605	2.0
	2053	2.5
4 W	0245	1.5
	1544	2.3
	2119	3.0
5 TH	0408	1.5
	1113	2.7
	1507	2.5
	2155	3.5
6 F	0542	1.6
	2238	3.8
7 SA	0752	1.5
	2328	4.0
8 SU	1007	1.3
9 M	0023	4.2
	1120	1.1
10 TU	0121	4.2
	1207	0.9
11 W	0219	4.2
	1244	0.8
12 TH	0315	4.2
	1313	0.7
13 F	0408	4.1
	1338	0.7
14 SA	0459	4.1
	1401	0.8
15 SU ○	0550	3.9
	1421	1.0

Days 16–31

Day	Time	M
16 M	0641	3.8
	1438	1.3
	2011	2.0
17 TU	0007	1.8
	0736	3.6
	1450	1.7
	1956	2.4
18 W	0135	1.6
	0835	3.3
	1453	2.1
	2006	2.9
19 TH	0253	1.4
	0945	3.0
	1443	2.4
	2030	3.4
20 F	0412	1.3
	1126	2.6
	1402	2.5
	2106	3.9
21 SA	0539	1.1
	2151	4.4
22 SU	0716	0.9
	2246	4.7
23 M	0854	0.7
	2351	4.9
24 TU	1015	0.5
25 W	0105	4.9
	1120	0.4
26 TH	0223	4.8
	1215	0.5
27 F	0340	4.7
	1301	0.7
28 SA	0454	4.4
	1340	1.1
	1920	1.9
	2129	1.9
29 SU ●	0604	4.1
	1408	1.5
	1855	2.2
	2345	1.6
30 M	0712	3.7
	1421	2.0
	1903	2.6
31 TU	0112	1.4
	0821	3.2
	1411	2.3
	1922	3.1

APRIL

Days 1–15

Day	Time	M
1 W	0226	1.2
	0936	2.8
	1335	2.5
	1950	3.5
2 TH	0338	1.1
	1146	2.5
	1204	2.5
	2023	3.9
3 F	0452	1.1
	2100	4.2
4 SA	0615	1.1
	2141	4.4
5 SU	0752	1.1
	2225	4.4
6 M	0918	1.0
	2313	4.4
7 TU	1017	1.0
8 W	0008	4.3
	1059	1.0
9 TH	0112	4.1
	1132	1.0
10 F	0220	4.0
	1200	1.0
11 SA	0328	3.8
	1223	1.2
12 SU	0434	3.6
	1242	1.4
	1907	2.3
	2238	2.2
13 M	0540	3.4
	1256	1.7
	1920	1.9
	2129	1.9
14 TU ○	0014	1.8
	0647	3.1
	1302	2.0
	1831	2.9
15 W	0125	1.4
	0802	2.8
	1255	2.3
	1847	3.4

Days 16–30

Day	Time	M
16 TH	0232	1.1
	0940	2.6
	1218	2.5
	1914	3.9
17 F	0338	0.8
	1950	4.5
18 SA	0449	0.6
	2033	4.9
19 SU	0604	0.5
	2122	5.1
20 M	0722	0.4
	2219	5.2
21 TU	0839	0.4
	2325	5.1
22 W	0946	0.5
23 TH	0043	4.8
	1043	0.7
24 F	0208	4.5
	1130	1.1
25 SA	0334	4.1
	1206	1.5
	1742	2.4
	2128	2.2
26 SU	0456	3.6
	1227	1.9
	1734	2.7
	2336	1.8
27 M	0618	3.2
	1226	2.3
	1748	3.2
28 TU ●	0059	1.4
	0746	2.8
	1150	2.5
	1811	3.5
29 W	0208	1.0
	1841	4.0
30 TH	0312	0.8
	1914	4.4

FOR INTERMEDIATE HEIGHTS USE HARMONIC CONSTANTS (SEE PART III) AND NP 159

KOREA, WEST COAST – INCH' ON (CHEMULPHO)
LAT 37°28'N LONG 126°37'E
TIMES AND HEIGHTS OF HIGH AND LOW WATERS

TIME ZONE – 0900 YEAR 1987

(Heights in metres. ○ = open-circle moon symbol, ● = filled moon symbol, as printed beside the day.)

JANUARY

Day		TIME	M	TIME	M	TIME	M	TIME	M
1	TH	0540	7.6	1143	−0.4	1822	9.0		
2	F	0030	0.7	0631	7.6	1230	−0.4	1909	8.9
3	SA	0116	0.7	0720	7.6	1315	−0.2	1954	8.7
4	SU	0159	0.8	0807	7.5	1359	0.1	2036	8.4
5	M	0239	0.9	0852	7.4	1443	0.5	2114	8.0
6	TU	0322	1.1	0935	7.1	1526	1.1	2151	7.5
7	W	0403	1.4	1022	6.8	1614	1.9	2230	6.9
8	TH	0449	1.7	1115	6.5	1715	2.5	2321	6.4
9	F	0550	2.0	1222	6.3	1836	3.0		
10	SA	0027	5.9	0702	2.2	1343	6.4	2003	3.0
11	SU	0147	5.9	0811	2.0	1459	6.8	2115	2.6
12	M	0256	6.1	0912	1.7	1554	7.3	2208	2.2
13	TU	0354	6.4	1002	1.3	1640	7.6	2251	1.8
14	W	0439	6.7	1045	1.0	1719	7.9	2330	1.6
15	TH	0522	6.9	1124	0.7	○1759	8.0		
16	F	0007	1.4	0601	7.0	1159	0.6	1832	8.0
17	SA	0039	1.3	0639	7.1	1234	0.5	1905	8.0
18	SU	0111	1.2	0715	7.1	1307	0.4	1936	8.0
19	M	0141	1.0	0747	7.2	1341	0.4	2005	8.0
20	TU	0210	0.9	0818	7.3	1414	0.5	2029	8.0
21	W	0239	0.8	0850	7.3	1450	0.8	2103	7.8
22	TH	0314	0.8	0927	7.3	1532	1.2	2140	7.4
23	F	0356	1.0	1014	7.0	1625	1.8	2228	6.9
24	SA	0451	1.4	1121	6.8	1739	2.5	2342	6.3
25	SU	0606	1.7	1252	6.7	1921	2.7		
26	M	0115	6.1	0734	1.6	1427	7.1	2048	2.3
27	TU	0240	6.1	0853	1.2	1541	7.8	2157	1.7
28	W	0349	6.9	0954	0.5	1640	8.4	2250	1.1
29	TH	0447	7.3	1047	0.0	●1730	8.8	2337	0.7
30	F	0537	7.7	1136	−0.4	1815	9.0		
31	SA	0019	0.4	0624	7.9	1221	−0.5	1857	9.0

FEBRUARY

Day		TIME	M	TIME	M	TIME	M	TIME	M
1	SU	0059	0.2	0709	8.0	1303	−0.4	1936	8.8
2	M	0136	0.2	0751	8.0	1343	−0.1	2009	8.5
3	TU	0213	0.3	0829	7.9	1420	0.4	2039	8.1
4	W	0244	0.5	0906	7.7	1458	1.0	2108	7.6
5	TH	0319	0.9	0940	7.3	1537	1.7	2140	7.0
6	F	0351	1.3	1020	6.9	1622	2.5	2217	6.4
7	SA	0433	1.9	1113	6.4	1729	3.2	2318	5.8
8	SU	0539	2.5	1237	6.1	1915	3.5		
9	M	0051	5.4	0720	2.7	1424	6.2	2053	3.2
10	TU	0230	5.6	0845	2.3	1541	6.8	2154	2.6
11	W	0339	6.2	0946	1.7	1629	7.3	2239	2.0
12	TH	0429	6.7	1031	1.2	1708	7.8	2316	1.5
13	F	0511	7.1	1109	0.7	○1741	8.0	2348	1.1
14	SA	0547	7.4	1144	0.4	1811	8.2		
15	SU	0017	0.9	0619	7.6	1217	0.2	1840	8.3
16	M	0045	0.6	0654	7.7	1249	0.2	1908	8.4
17	TU	0114	0.4	0722	7.9	1320	0.2	1934	8.4
18	W	0141	0.3	0752	8.0	1354	0.2	2002	8.3
19	TH	0210	0.2	0823	8.1	1429	0.5	2031	8.0
20	F	0244	0.3	0858	8.0	1508	1.1	2108	7.6
21	SA	0322	0.6	0940	7.7	1556	1.8	2151	6.9
22	SU	0412	1.2	1041	7.1	1709	2.7	2302	6.2
23	M	0526	1.9	1219	6.7	1857	3.1		
24	TU	0052	5.8	0710	2.1	1416	6.9	2043	2.7
25	W	0235	6.2	0843	1.6	1538	7.7	2152	1.8
26	TH	0349	6.9	0949	0.8	1632	8.4	2239	1.0
27	F	0443	7.6	1039	0.1	1717	8.9	2316	0.5
28	SA	0527	8.1	1124	−0.2	●1756	9.0		

MARCH

Day		TIME	M	TIME	M	TIME	M	TIME	M
1	SU	0000	0.1	0609	8.4	1206	−0.3	1832	9.0
2	M	0035	0.0	0649	8.5	1244	−0.2	1904	8.7
3	TU	0107	0.0	0725	8.5	1322	0.2	1935	8.4
4	W	0138	0.1	0759	8.4	1356	0.6	2003	8.0
5	TH	0207	0.3	0829	8.1	1431	1.1	2029	7.6
6	F	0237	0.7	0858	7.8	1503	1.8	2058	7.1
7	SA	0306	1.2	0931	7.3	1540	2.5	2132	6.5
8	SU	0342	1.9	1013	6.7	1633	3.3	2224	5.9
9	M	0436	2.6	1131	6.1	1820	3.8		
10	TU	0005	5.3	0621	3.1	1337	6.0	2027	3.5
11	W	0200	5.5	0816	2.8	1515	6.5	2133	2.8
12	TH	0320	6.1	0925	2.1	1603	7.2	2215	2.0
13	F	0408	6.8	1008	1.4	1639	7.7	2247	1.4
14	SA	0447	7.4	1047	0.8	1708	8.1	2316	0.9
15	SU	0520	7.8	1120	0.5	○1737	8.4	2345	0.6
16	M	0553	8.1	1153	0.2	1806	8.4		
17	TU	0012	0.3	0624	8.3	1226	0.2	1834	8.5
18	W	0043	0.1	0655	8.5	1301	0.2	1904	8.4
19	TH	0112	−0.1	0725	8.7	1335	0.4	1936	8.3
20	F	0144	0.5	0759	8.7	1412	0.7	2009	8.0
21	SA	0220	0.6	0837	8.4	1452	1.3	2047	7.5
22	SU	0257	0.7	0922	7.9	1543	2.1	2135	6.8
23	M	0351	1.4	1025	7.3	1657	2.9	2249	6.1
24	TU	0505	2.1	1208	6.7	1850	3.2		
25	W	0046	5.8	0657	2.3	1406	7.0	2032	2.7
26	TH	0232	6.3	0832	1.8	1523	7.7	2133	1.8
27	F	0320	7.1	0936	1.1	1613	8.3	2219	0.9
28	SA	0429	7.9	1026	0.5	1653	8.7	2258	0.4
29	SU ●	0511	8.4	1108	0.1	1727	8.8	2332	0.0
30	M	0548	8.7	1146	0.1	1759	8.7		
31	TU	0005	−0.1	0624	8.8	1223	0.3	1830	8.4

APRIL

Day		TIME	M	TIME	M	TIME	M	TIME	M
1	W	0034	0.0	0656	8.7	1258	0.6	1859	8.1
2	TH	0106	0.2	0728	8.6	1332	1.0	1928	7.8
3	F	0134	0.5	0757	8.3	1405	1.5	1957	7.4
4	SA	0205	0.9	0829	8.0	1439	2.0	2031	7.0
5	SU	0234	1.3	0858	7.5	1514	2.6	2106	6.6
6	M	0311	1.9	0943	6.9	1559	3.2	2159	6.0
7	TU	0359	2.6	1052	6.3	1729	3.7	2331	5.5
8	W	0531	3.1	1242	6.1	1935	3.6		
9	TH	0121	5.5	0731	3.0	1416	6.4	2051	2.9
10	F	0243	6.2	0845	2.4	1515	7.1	2133	2.1
11	SA	0333	6.9	0934	1.7	1554	7.6	2208	1.4
12	SU	0413	7.6	1015	1.1	1626	8.0	2237	0.9
13	M	0447	8.1	1050	0.7	1658	8.3	2308	0.4
14	TU	0521	8.5	1127	0.5	○1730	8.4	2338	0.2
15	W	0555	8.8	1203	0.4	1802	8.4		
16	TH	0012	0.0	0628	9.0	1241	0.5	1837	8.3
17	F	0045	−0.1	0704	9.0	1320	0.7	1914	8.0
18	SA	0124	0.5	0746	8.9	1402	1.1	1954	7.7
19	SU	0204	0.7	0829	8.6	1447	1.6	2041	7.3
20	M	0250	0.9	0919	8.0	1540	2.3	2137	6.7
21	TU	0343	1.5	1028	7.4	1654	2.8	2257	6.2
22	W	0502	2.2	1158	7.0	1831	3.0		
23	TH	0038	6.1	0640	2.4	1334	7.1	2000	2.5
24	F	0211	6.6	0808	2.0	1448	7.5	2104	1.7
25	SA	0317	7.4	0912	1.5	1538	8.0	2149	1.0
26	SU	0405	8.1	1002	1.0	1618	8.2	2226	0.5
27	M	0445	8.6	1045	0.8	1653	8.3	2300	0.2
28	TU	0522	8.8	1124	0.8	●1724	8.1	2332	0.2
29	W	0556	8.9	1200	0.9	1756	7.9		
30	TH	0004	0.4	0628	8.7	1236	1.2	1830	7.7

SEPTEMBER

Day	TIME / M		Day	TIME / M
1 TU	0327 2.2 / 0922 7.4 / 1540 1.5 / 2207 7.6		16 W	0421 3.6 / 1009 6.2 / 1617 2.8 / 2321 6.5
2 W	0428 3.0 / 1020 6.7 / 1645 2.1 / 2333 7.1		17 TH	0603 4.1 / 1144 5.7 / 1800 3.3
3 TH	0612 3.5 / 1205 6.2 / 1823 2.5		18 F	0118 6.4 / 0808 3.8 / 1339 5.8 / 1955 3.1
4 F	0131 7.2 / 0803 3.3 / 1355 6.4 / 2003 2.1		19 SA	0253 6.9 / 0917 3.1 / 1459 6.4 / 2104 2.4
5 SA	0301 7.9 / 0917 2.4 / 1515 −7.1 / 2117 1.3		20 SU	0344 7.6 / 0957 2.3 / 1549 7.1 / 2152 1.7
6 SU	0402 8.6 / 1013 1.6 / 1610 7.9 / 2211 0.5		21 M	0418 8.1 / 1031 1.6 / 1629 7.7 / 2229 1.1
7 M ○	0447 9.2 / 1055 0.8 / 1700 8.5 / 2258 0.1		22 TU	0450 8.5 / 1100 1.1 / 1703 8.1 / 2303 0.8
8 TU	0527 9.5 / 1133 0.4 / 1742 8.8 / 2340 −0.1		23 W ●	0519 8.7 / 1128 0.8 / 1735 8.4 / 2335 0.7
9 W	0604 9.4 / 1210 0.2 / 1824 9.0		24 TH	0545 8.4 / 1154 0.6 / 1805 8.6
10 TH	0021 0.0 / 0640 9.2 / 1246 0.1 / 1903 9.0		25 F	0007 0.6 / 0614 8.6 / 1223 0.5 / 1834 8.7
11 F	0101 0.4 / 0712 8.8 / 1319 0.3 / 1940 8.8		26 SA	0039 0.7 / 0639 8.5 / 1251 0.4 / 1903 8.7
12 SA	0138 0.9 / 0744 8.4 / 1349 0.6 / 2013 8.5		27 SU	0114 0.9 / 0709 8.4 / 1320 0.4 / 1935 8.7
13 SU	0215 1.5 / 0813 7.9 / 1421 1.0 / 2047 8.1		28 M	0149 1.2 / 0741 8.1 / 1354 0.5 / 2009 8.5
14 M	0250 2.2 / 0844 7.4 / 1452 1.5 / 2121 7.7		29 TU	0225 1.6 / 0818 7.6 / 1433 0.9 / 2052 8.2
15 TU	0329 2.9 / 0919 6.8 / 1529 2.1 / 2207 7.0		30 W	0313 2.3 / 0903 7.1 / 1519 1.5 / 2151 7.6

OCTOBER

Day	TIME / M		Day	TIME / M
1 TH	0417 3.0 / 1009 6.4 / 1628 2.2 / 2323 7.0		16 F	0515 3.9 / 1113 5.6 / 1713 3.2
2 F	0600 3.5 / 1203 6.0 / 1811 2.5		17 SA	0024 6.3 / 0715 3.7 / 1302 5.7 / 1908 3.1
3 SA	0121 7.1 / 0750 3.0 / 1350 6.4 / 1953 2.1		18 SU	0158 6.6 / 0835 3.0 / 1424 6.2 / 2027 2.6
4 SU	0245 7.8 / 0902 2.1 / 1504 7.2 / 2104 1.3		19 M	0256 7.2 / 0917 2.2 / 1517 7.0 / 2117 1.9
5 M	0341 8.5 / 0952 1.2 / 1557 8.1 / 2157 0.6		20 TU	0336 7.7 / 0952 1.5 / 1557 7.7 / 2157 1.3
6 TU	0423 9.0 / 1031 0.5 / 1642 8.7 / 2242 0.2		21 W	0409 8.1 / 1023 0.9 / 1631 8.2 / 2234 0.9
7 W ○	0500 9.1 / 1108 0.1 / 1722 9.1 / 2322 0.2		22 TH ●	0439 8.3 / 1050 0.5 / 1703 8.6 / 2308 0.7
8 TH	0533 9.0 / 1142 −0.1 / 1759 9.2		23 F	0508 8.4 / 1119 0.3 / 1735 8.8 / 2343 0.7
9 F	0001 0.4 / 0607 8.7 / 1214 0.0 / 1833 9.1		24 SA	0540 8.3 / 1149 0.2 / 1804 8.9
10 SA	0038 0.7 / 0637 8.4 / 1246 0.3 / 1909 8.8		25 SU	0018 0.8 / 0612 8.2 / 1222 0.2 / 1840 8.9
11 SU	0114 1.2 / 0709 7.9 / 1316 0.6 / 1941 8.5		26 M	0056 1.0 / 0647 8.0 / 1259 0.2 / 1916 8.8
12 M	0149 1.7 / 0741 7.5 / 1349 1.0 / 2013 8.1		27 TU	0135 1.3 / 0724 7.7 / 1336 0.4 / 1958 8.5
13 TU	0226 2.3 / 0813 7.1 / 1421 1.5 / 2049 7.6		28 W	0218 1.7 / 0810 7.3 / 1421 0.8 / 2050 8.1
14 W	0303 2.9 / 0852 6.6 / 1458 2.0 / 2132 7.1		29 TH	0308 2.2 / 0903 6.9 / 1614 1.3 / 2151 7.6
15 TH	0348 3.4 / 0943 6.1 / 1548 2.7 / 2241 6.6		30 F	0417 2.7 / 1017 6.3 / 1622 2.0 / 2218 7.2
			31 SA	0547 3.0 / 1155 6.1 / 1755 2.3

NOVEMBER

Day	TIME / M		Day	TIME / M
1 SU	0052 7.1 / 0723 2.6 / 1331 6.5 / 1931 2.1		16 M	0042 6.4 / 0720 2.9 / 1327 6.1 / 1929 2.7
2 M	0211 7.5 / 0832 1.8 / 1445 7.2 / 2043 1.6		17 TU	0150 6.7 / 0819 2.2 / 1430 6.7 / 2033 2.2
3 TU	0309 7.9 / 0920 1.0 / 1538 8.0 / 2138 1.0		18 W	0243 7.1 / 0901 1.6 / 1517 7.4 / 2123 1.7
4 W	0354 8.3 / 1002 0.4 / 1623 8.6 / 2223 0.7		19 TH	0325 7.5 / 0938 1.0 / 1557 8.0 / 2205 1.2
5 TH ○	0431 8.4 / 1039 0.1 / 1700 9.0 / 2303 0.6		20 F	0402 7.7 / 1013 0.5 / 1631 8.4 / 2245 1.0
6 F	0504 8.3 / 1113 0.0 / 1736 9.0 / 2341 0.8		21 SA ●	0437 7.9 / 1048 0.2 / 1708 8.7 / 2324 0.9
7 SA	0538 8.0 / 1145 0.2 / 1812 8.8		22 SU	0514 7.9 / 1125 0.1 / 1747 8.9
8 SU	0019 1.1 / 0610 7.7 / 1219 0.4 / 1844 8.6		23 M	0004 0.9 / 0553 7.8 / 1205 0.0 / 1829 8.8
9 M	0055 1.5 / 0645 7.4 / 1251 0.7 / 1919 8.3		24 TU	0046 1.0 / 0638 7.7 / 1247 0.0 / 1914 8.7
10 TU	0133 1.8 / 0722 7.1 / 1326 1.0 / 1954 7.8		25 W	0130 1.2 / 0722 7.4 / 1330 0.2 / 2005 8.5
11 W	0207 2.2 / 0759 6.8 / 1402 1.4 / 2031 7.6		26 TH	0218 1.4 / 0815 7.2 / 1418 0.5 / 2055 8.2
12 TH	0247 2.6 / 0841 6.5 / 1439 1.8 / 2116 7.2		27 F	0308 1.7 / 0914 6.9 / 1511 0.9 / 2151 7.8
13 F	0332 2.9 / 0934 6.1 / 1527 2.2 / 2212 6.7		28 SA	0407 2.0 / 1017 6.6 / 1612 1.5 / 2256 7.3
14 SA	0433 3.2 / 1043 5.8 / 1632 2.7 / 2323 6.4		29 SU	0518 2.2 / 1134 6.4 / 1729 2.0
15 SU	0557 3.2 / 1208 5.7 / 1801 2.9		30 M	0007 7.0 / 0634 2.1 / 1255 6.5 / 1857 2.2

DECEMBER

Day	TIME / M		Day	TIME / M
1 TU	0123 6.9 / 0747 1.7 / 1411 7.0 / 2014 2.0		16 W	0033 6.4 / 0657 2.2 / 1327 6.4 / 1936 2.6
2 W	0225 7.1 / 0843 1.2 / 1512 7.7 / 2115 1.7		17 TH	0140 6.5 / 0800 1.8 / 1432 6.9 / 2043 2.2
3 TH	0317 7.3 / 0930 0.8 / 1600 8.1 / 2205 1.4		18 F	0238 6.7 / 0853 1.3 / 1523 7.8 / 2136 1.7
4 F	0402 7.4 / 1011 0.5 / 1642 8.4 / 2247 1.2		19 SA	0328 7.0 / 0941 0.8 / 1608 8.1 / 2223 1.3
5 SA ○	0439 7.4 / 1047 0.4 / 1720 8.5 / 2327 1.2		20 SU	0415 7.3 / 1024 0.3 / 1655 8.5 / 2310 1.1
6 SU	0516 7.3 / 1125 0.4 / 1756 8.4		21 M ●	0500 7.5 / 1109 0.0 / 1740 8.7 / 2356 0.9
7 M	0005 1.4 / 0556 7.2 / 1201 0.6 / 1832 8.2		22 TU	0548 7.5 / 1154 −0.2 / 1827 8.8
8 TU	0043 1.5 / 0634 7.0 / 1237 0.7 / 1909 8.0		23 W	0041 0.8 / 0638 7.6 / 1240 −0.3 / 1917 8.8
9 W	0120 1.7 / 0714 6.9 / 1314 0.9 / 1946 7.8		24 TH	0125 0.8 / 0728 7.5 / 1326 −0.2 / 2004 8.7
10 TH	0157 1.9 / 0754 6.8 / 1349 1.1 / 2021 7.7		25 F	0210 0.9 / 0818 7.5 / 1413 0.0 / 2050 8.4
11 F	0233 2.0 / 0834 6.6 / 1428 1.3 / 2058 7.4		26 SA	0258 1.0 / 0908 7.3 / 1502 0.5 / 2135 8.0
12 SA	0311 2.1 / 0916 6.4 / 1506 1.6 / 2104 7.1		27 SU	0345 1.2 / 0959 7.1 / 1553 1.1 / 2222 7.5
13 SU	0351 2.3 / 1007 6.2 / 1553 2.0 / 2227 6.8		28 M	0436 1.5 / 1059 6.8 / 1654 1.9 / 2315 6.9
14 M	0444 2.4 / 1105 6.0 / 1654 2.4 / 2323 6.6		29 TU	0538 1.7 / 1208 6.6 / 1812 2.4
15 TU	0549 2.4 / 1217 6.1 / 1814 2.7		30 W	0020 6.4 / 0648 1.8 / 1327 6.6 / 1936 2.6
			31 TH	0135 6.2 / 0758 1.7 / 1443 7.0 / 2051 2.4

JAPAN, HONSHU, WEST COAST – SHIMONOSEKI
LAT 33°58'N   LONG 130°57'E

TIME ZONE – 0900

TIMES AND HEIGHTS OF HIGH AND LOW WATERS

YEAR 1987

JANUARY

Day	TIME	M	Day	TIME	M
1 TH	0348	-0.4	16 F	0355	-0.1
	1042	2.2		1029	2.0
	1619	0.6		1616	0.6
	2147	2.3		2149	2.1
2 F	0435	-0.4	17 SA	0428	-0.1
	1127	2.2		1056	2.0
	1704	0.5		1647	0.6
	2239	2.3		2223	2.1
3 SA	0519	-0.3	18 SU	0500	-0.1
	1207	2.2		1122	2.0
	1747	0.5		1718	0.5
	2329	2.2		2257	2.1
4 SU	0600	-0.1	19 M	0532	0.0
	1243	2.1		1146	2.0
	1831	0.5		1751	0.5
				2332	2.0
5 M	0017	2.1	20 TU	0605	0.1
	0641	0.1		1210	2.0
	1316	2.0		1828	0.5
	1917	0.6			
6 TU	0107	1.9	21 W	0012	2.0
	0721	0.4		0640	0.3
	1348	1.9		1236	2.0
	2009	0.6		1909	0.5
7 W	0202	1.8	22 TH	0059	1.9
	0803	0.7		0719	0.5
	1423	1.8		1309	1.9
	2109	0.6		2000	0.5
8 TH	0307	1.6	23 F	0157	1.7
	0855	0.9		0806	0.7
	1510	1.8		1352	1.9
	2220	0.6		2104	0.5
9 F	0434	1.6	24 SA	0316	1.6
	1012	1.1		0911	1.0
	1618	1.7		1453	1.8
	2331	0.5		2225	0.4
10 SA	0613	1.6	25 SU	0516	1.6
	1152	1.1		1056	1.1
	1732	1.8		1621	1.8
				2350	0.3
11 SU	0031	0.4	26 M	0703	1.8
	0726	1.7		1240	1.1
	1307	1.1		1753	1.9
	1833	1.8			
12 M	0122	0.8	27 TU	0103	0.1
	0815	1.8		0814	1.9
	1358	1.0		1349	0.9
	1921	1.9		1905	2.0
13 TU	0206	0.2	28 W	0204	-0.1
	0854	1.9		0908	2.1
	1439	0.9		1442	0.7
	2002	2.0		2007	2.1
14 W	0245	0.1	29 TH ●	0256	-0.3
	0928	2.0		0954	2.2
	1514	0.8		1528	0.5
	2040	2.1		2104	2.2
15 TH ○	0321	0.1	30 F	0343	-0.4
	0959	2.0		1034	2.2
	1545	0.7		1610	0.4
	2115	2.1		2155	2.3
			31 SA	0425	-0.4
				1109	2.2
				1649	0.3
				2242	2.3

FEBRUARY

Day	TIME	M	Day	TIME	M
1 SU	0503	-0.3	16 M	0442	-0.1
	1140	2.2		1058	2.1
	1727	0.2		1659	0.2
	2324	2.2		2254	2.1
2 M	0539	-0.2	17 TU	0513	-0.1
	1206	2.1		1118	2.1
	1804	0.2		1730	0.2
				2327	2.1
3 TU	0003	2.1	18 W	0544	0.1
	0611	0.1		1136	2.1
	1226	2.1		1803	0.2
	1840	0.3			
4 W	0040	2.0	19 TH	0002	2.0
	0641	0.4		0616	0.3
	1243	2.0		1157	2.1
	1919	0.4		1840	0.2
5 TH	0121	1.8	20 F	0041	1.9
	0710	0.7		0650	0.6
	1303	1.9		1224	2.0
	2003	0.5		1923	0.3
6 F	0210	1.6	21 SA	0129	1.8
	0740	0.9		0729	0.8
	1335	1.8		1300	1.9
	2103	0.6		2019	0.4
7 SA	0329	1.5	22 SU	0243	1.6
	0825	1.2		0826	1.1
	1433	1.6		1357	1.8
	2231	0.7		2146	0.5
8 SU	0541	1.5	23 M	0523	1.6
	1112	1.3		1044	1.2
	1641	1.6		1552	1.7
				2336	0.4
9 M	0000	0.6	24 TU	0718	1.7
	0725	1.6		1246	1.1
	1311	1.2		1753	1.8
	1811	1.7			
10 TU	0107	0.5	25 W	0101	0.2
	0815	1.7		0818	1.9
	1359	1.0		1348	0.8
	1909	1.8		1912	1.9
11 W	0156	0.3	26 TH	0200	0.0
	0849	1.9		0900	2.1
	1433	0.8		1433	0.6
	1955	1.9		2015	2.1
12 TH	0235	0.1	27 F	0248	-0.2
	0919	1.9		0937	2.2
	1503	0.7		1513	0.3
	2035	2.0		2108	2.3
13 F ○	0309	0.0	28 SA ●	0329	-0.3
	0946	2.0		1010	2.2
	1531	0.5		1551	0.1
	2112	2.1		2153	2.3
14 SA	0341	-0.1			
	1012	2.0			
	1559	0.4			
	2147	2.1			
15 SU	0412	-0.1			
	1036	2.1			
	1629	0.3			
	2221	2.2			

MARCH

Day	TIME	M	Day	TIME	M
1 SU	0406	-0.3	16 M	0348	-0.1
	1039	2.2		1003	2.1
	1626	0.0		1603	0.1
	2234	2.3		2213	2.2
2 M	0440	-0.2	17 TU	0419	0.0
	1104	2.2		1025	2.2
	1700	0.0		1634	0.0
	2310	2.2		2248	2.2
3 TU	0511	0.0	18 W	0451	0.1
	1122	2.2		1044	2.2
	1732	0.0		1706	-0.1
	2342	2.1		2322	2.2
4 W	0539	0.2	19 TH	0523	0.2
	1135	2.1		1102	2.2
	1803	0.1		1740	-0.1
				2355	2.1
5 TH	0013	2.0	20 F	0555	0.5
	0604	0.5		1122	2.1
	1149	2.1		1816	0.0
	1834	0.2			
6 F	0046	1.8	21 SA	0032	1.9
	0628	0.8		0630	0.7
	1208	2.0		1151	2.1
	1908	0.4		1858	0.2
7 SA	0126	1.7	22 SU	0119	1.8
	0652	1.0		0710	1.0
	1234	1.8		1230	1.9
	1952	0.6		1955	0.4
8 SU	0229	1.6	23 M	0245	1.6
	0722	1.2		0819	1.2
	1312	1.6		1336	1.7
				2129	0.6
9 M	0456	1.5	24 TU	0542	1.6
	1016	1.4		1105	1.3
	1549	1.5		1605	1.6
	2323	0.8		2331	0.5
10 TU	0710	1.6	25 W	0709	1.8
	1309	1.2		1239	1.0
	1750	1.6		1800	1.8
11 W	0047	0.6	26 TH	0051	0.3
	0755	1.7		0756	1.9
	1343	1.0		1331	0.7
	1854	1.7		1915	2.0
12 TH	0137	0.3	27 F	0146	0.1
	0824	1.8		0833	2.1
	1411	0.8		1412	0.4
	1943	1.9		2012	2.2
13 F	0213	0.3	28 SA	0229	0.0
	0850	1.9		0906	2.2
	1438	0.6		1450	0.2
	2024	2.0		2100	2.3
14 SA	0246	0.0	29 SU ●	0307	0.0
	0915	2.0		0935	2.2
	1505	0.4		1526	0.0
	2101	2.1		2141	2.3
15 SU ○	0317	0.0	30 M	0342	0.0
	0939	2.1		1000	2.2
	1534	0.2		1559	-0.1
	2138	2.2		2217	2.3
			31 TU	0413	0.1
				1020	2.2
				1630	-0.1
				2250	2.2

APRIL

Day	TIME	M	Day	TIME	M
1 W	0442	0.3	16 TH	0430	0.3
	1035	2.2		1009	2.3
	1700	-0.1		1644	-0.2
	2320	2.1		2315	2.2
2 TH	0509	0.5	17 F	0506	0.5
	1050	2.2		1033	2.2
	1729	0.0		1721	-0.1
	2348	2.0		2354	2.1
3 F	0535	0.7	18 SA	0543	0.7
	1109	2.1		1101	2.2
	1758	0.2		1802	0.0
4 SA	0019	1.9	19 SU	0036	2.0
	0600	0.9		0624	0.9
	1133	2.0		1138	2.1
	1830	0.4		1849	0.2
5 SU	0055	1.8	20 M	0132	1.8
	0627	1.1		0717	1.1
	1203	1.9		1230	1.9
	1909	0.6		1952	0.5
6 M	0149	1.6	21 TU	0315	1.7
	0703	1.2		0849	1.2
	1243	1.7		1404	1.7
	2010	0.8		2125	0.6
7 TU	0348	1.5	22 W	0513	1.7
	0843	1.4		1051	1.1
	1457	1.5		1615	1.7
	2216	0.9		2309	0.6
8 W	0553	1.6	23 TH	0624	1.9
	1220	1.2		1208	0.9
	1710	1.6		1749	1.9
9 TH	0000	0.8	24 F	0024	0.5
	0655	1.7		0712	2.0
	1300	1.0		1300	0.6
	1821	1.7		1900	2.0
10 F	0056	0.6	25 SA	0118	0.4
	0730	1.8		0750	2.1
	1330	0.8		1343	0.3
	1913	1.9		1956	2.2
11 SA	0137	0.4	26 SU	0201	0.3
	0759	2.0		0822	2.2
	1400	0.5		1421	0.1
	1959	2.0		2043	2.2
12 SU	0212	0.3	27 M	0239	0.3
	0827	2.1		0850	2.2
	1430	0.3		1457	0.0
	2041	2.1		2123	2.3
13 M	0246	0.2	28 TU ●	0314	0.4
	0854	2.2		0913	2.2
	1502	0.1		1530	-0.1
	2121	2.2		2158	2.2
14 TU ○	0320	0.4	29 W	0346	0.5
	0921	2.2		0933	2.2
	1535	-0.1		1601	-0.1
	2200	2.3		2230	2.2
15 W	0355	0.3	30 TH	0417	0.6
	0945	2.3		0953	2.2
	1609	-0.2		1631	0.0
	2238	2.3		2259	2.2

JANUARY

Day		TIME	M	Day		TIME	M
1	TH	0135	3.4	16	F	0150	3.2
		0655	1.7			0705	1.8
		1255	4.1			1300	3.6
		1945	0.1			1950	0.6
2	F	0220	3.5	17	SA	0220	3.2
		0750	1.6			0745	1.8
		1345	4.0			1335	3.5
		2035	0.2			2015	0.7
3	SA	0305	3.5	18	SU	0245	3.3
		0845	1.6			0825	1.7
		1440	3.7			1410	3.4
		2120	0.5			2050	0.8
4	SU	0355	3.5	19	M	0320	3.3
		0945	1.6			0905	1.6
		1535	3.4			1455	3.3
		2205	0.8			2120	0.9
5	M	0440	3.5	20	TU	0355	3.3
		1045	1.5			0950	1.6
		1630	3.1			1540	3.1
		2255	1.1			2155	1.1
6	TU	0530	3.5	21	W	0430	3.4
		1155	1.5			1045	1.5
		1740	2.9			1635	2.9
		2340	1.4			2230	1.4
7	W	0620	3.4	22	TH	0510	3.4
		1305	1.4			1145	1.4
		1900	2.7			1745	2.7
						2315	1.6
8	TH	0035	1.6	23	F	0555	3.4
		0710	3.4			1255	1.3
		1415	1.3			1910	2.6
		2025	2.7				
9	F	0140	1.9	24	SA	0015	1.9
		0805	3.4			0655	3.5
		1515	1.2			1410	1.1
		2140	2.7			2045	2.7
10	SA	0245	2.0	25	SU	0130	2.0
		0855	3.4			0800	3.6
		1605	1.0			1520	0.8
		2240	2.8			2200	2.8
11	SU	0340	2.1	26	M	0250	2.0
		0945	3.4			0905	3.7
		1655	0.9			1620	0.5
		2330	3.0			2300	3.0
12	M	0430	2.1	27	TU	0400	2.0
		1025	3.5			1015	3.9
		1735	0.8			1715	0.3
						2350	3.2
13	TU	0010	3.0	28	W	0500	1.8
		0515	2.0			1105	4.0
		1105	3.6			1805	0.2
		1810	0.7				
14	W	0045	3.1	29	TH •	0045	3.4
		0555	2.0			0555	1.6
		1145	3.6			1155	4.1
		1845	0.6			1845	0.1
15	TH o	0115	3.1	30	F	0115	3.5
		0630	1.9			0650	1.5
		1220	3.6			1250	4.0
		1915	0.6			1930	0.2
				31	SA	0200	3.6
						0740	1.3
						1340	3.9
						2010	0.4

FEBRUARY

Day		TIME	M	Day		TIME	M
1	SU	0240	3.6	16	M	0205	3.4
		0830	1.2			0800	1.3
		1425	3.7			1400	3.4
		2050	0.6			2015	0.9
2	M	0315	3.6	17	TU	0240	3.5
		0920	1.2			0845	1.2
		1515	3.4			1445	3.3
		2130	0.9			2045	1.0
3	TU	0355	3.6	18	W	0305	3.5
		1010	1.2			0925	1.1
		1605	3.1			1525	3.1
		2205	1.2			2115	1.3
4	W	0435	3.5	19	TH	0345	3.5
		1110	1.2			1015	1.1
		1705	2.8			1620	2.9
		2250	1.6			2155	1.5
5	TH	0515	3.4	20	F	0420	3.5
		1215	1.3			1110	1.1
		1820	2.6			1730	2.7
		2340	1.9			2240	1.9
6	F	0605	3.3	21	SA	0510	3.4
		1325	1.3			1225	1.1
		2000	2.6			1900	2.6
						2345	2.0
7	SA	0045	2.1	22	SU	0615	3.4
		0715	3.2			1350	1.0
		1440	1.2			2045	2.6
		2120	2.6				
8	SU	0200	2.2	23	M	0115	2.1
		0815	3.2			0735	3.4
		1540	1.2			1505	0.8
		2225	2.7			2155	2.8
9	M	0315	2.2	24	TU	0250	2.0
		0915	3.2			0855	3.5
		1630	1.0			1610	0.6
		2310	2.9			2245	3.0
10	TU	0410	2.1	25	W	0400	1.8
		1005	3.3			1000	3.7
		1715	0.9			1700	0.4
		2345	3.0			2330	3.2
11	W	0500	2.0	26	TH	0500	1.6
		1055	3.4			1100	3.8
		1745	0.8			1750	0.3
12	TH	0020	3.1	27	F	0010	3.4
		0540	1.8			0550	1.3
		1130	3.5			1155	3.8
		1820	0.7			1830	0.3
13	F o	0050	3.2	28	SA •	0050	3.5
		0615	1.7			0640	1.1
		1205	3.6			1240	3.8
		1850	0.6			1915	0.4
14	SA	0115	3.2				
		0650	1.5				
		1245	3.5				
		1920	0.6				
15	SU	0145	3.3				
		0725	1.4				
		1320	3.5				
		1945	0.7				

MARCH

Day		TIME	M	Day		TIME	M
1	SU	0125	3.6	16	M	0100	3.4
		0725	0.9			0705	0.9
		1325	3.7			1310	3.4
		1945	0.6			1910	0.9
2	M	0200	3.7	17	TU	0125	3.5
		0810	0.9			0745	0.8
		1410	3.5			1345	3.3
		2015	0.9			1945	1.0
3	TU	0235	3.7	18	W	0155	3.6
		0855	0.8			0820	0.7
		1500	3.3			1435	3.2
		2050	1.1			2015	1.2
4	W	0305	3.6	19	TH	0230	3.6
		0940	0.9			0905	0.7
		1550	3.0			1520	3.0
		2125	1.4			2050	1.4
5	TH	0340	3.4	20	F	0305	3.6
		1025	1.0			0955	0.7
		1640	2.8			1615	2.8
		2205	1.7			2130	1.7
6	F	0420	3.3	21	SA	0350	3.5
		1120	1.2			1055	0.8
		1750	2.6			1725	2.7
		2250	2.0			2230	1.9
7	SA	0500	3.1	22	SU	0445	3.4
		1230	1.3			1215	0.9
		1925	2.5			1855	2.6
		2355	2.1			2345	2.0
8	SU	0605	3.0	23	M	0555	3.2
		1350	1.3			1330	0.9
		2100	2.6			2025	2.7
9	M	0135	2.2	24	TU	0125	2.0
		0735	2.9			0730	3.2
		1500	1.2			1445	0.8
		2200	2.7			2130	2.8
10	TU	0255	2.1	25	W	0255	1.8
		0850	3.0			0850	3.3
		1555	1.1			1550	0.6
		2240	2.8			2220	3.0
11	W	0355	2.0	26	TH	0355	1.5
		0945	3.1			0955	3.4
		1640	0.9			1640	0.6
		2310	2.9			2300	3.3
12	TH	0440	1.7	27	F	0455	1.2
		1030	3.2			1055	3.5
		1715	0.8			1720	0.5
		2340	3.0			2340	3.4
13	F	0520	1.5	28	SA	0540	0.9
		1110	3.3			1145	3.5
		1745	0.7			1800	0.6
14	SA	0015	3.6	29	SU •	0010	3.6
		0555	0.7			0625	0.7
		1230	3.5			1230	3.5
		1835	0.8			1835	0.8
15	SU o	0030	3.3	30	M	0050	3.6
		0630	1.1			0715	0.6
		1230	3.4			1315	3.4
		1845	0.8			1910	0.9
				31	TU	0120	3.6
						0745	0.5
						1400	3.3
						1945	1.2

APRIL

Day		TIME	M	Day		TIME	M
1	W	0145	3.6	16	TH	0120	3.7
		0825	0.5			0805	0.3
		1445	3.1			1430	3.1
		2015	1.4			1955	1.4
2	TH	0220	3.5	17	F	0200	3.7
		0905	0.6			0845	0.3
		1525	2.9			1520	3.0
		2055	1.6			2035	1.6
3	F	0255	3.3	18	SA	0240	3.6
		0950	0.8			0945	0.4
		1615	2.7			1615	2.8
		2130	1.8			2130	1.7
4	SA	0330	3.1	19	SU	0330	3.4
		1040	1.0			1045	0.5
		1715	2.6			1725	2.7
		2215	2.0			2235	1.9
5	SU	0410	2.8	20	M	0435	3.2
		1140	1.1			1150	0.7
		1840	2.5			1845	2.7
		2325	2.1			2355	1.9
6	M	0510	2.8	21	TU	0555	3.0
		1250	1.2			1315	0.8
		2010	2.5			1955	2.8
7	TU	0055	2.1	22	W	0130	1.7
		0640	2.7			0725	3.0
		1405	1.2			1415	0.8
		2110	2.6			2055	3.0
8	W	0225	2.0	23	TH	0250	1.5
		0805	2.7			0845	3.1
		1505	1.1			1515	0.8
		2150	2.8			2145	3.1
9	TH	0325	1.7	24	F	0350	1.2
		0910	2.8			0950	3.1
		1550	1.0			1605	1.0
		2220	2.9			2225	3.3
10	F	0410	1.5	25	SA	0440	0.9
		1005	3.0			1045	3.1
		1625	0.9			1650	0.9
		2250	3.1			2300	3.4
11	SA	0450	1.2	26	SU	0525	0.6
		1050	3.1			1135	3.2
		1700	0.9			1730	1.0
		2315	3.2			2335	3.5
12	SU	0530	0.9	27	M	0605	0.5
		1130	3.2			1220	3.2
		1735	0.9			1805	1.1
		2345	3.4				
13	M	0605	0.7	28	TU •	0005	3.6
		1215	3.2			0645	0.4
		1815	1.0			1305	3.1
						1840	1.3
14	TU o	0045	3.5	29	W	0040	3.5
		0645	0.5			0725	0.3
		1255	3.2			1350	3.1
		1915	1.1			1915	1.5
15	W	0045	3.6	30	TH	0110	3.5
		0720	0.3			0800	0.4
		1340	3.2			1430	3.0
		1915	1.3			1950	1.6

CANADA, BRITISH COLUMBIA – VANCOUVER HARBOUR
LAT 49°17'N   LONG 123°07'W

TIME ZONE +0800   TIMES AND HEIGHTS OF HIGH AND LOW WATERS   YEAR 1987

JANUARY

Day	TIME	M	Day	TIME	M
1 TH	0005	0.0	16 F	0015	0.8
	0745	5.1		0750	4.7
	1255	3.7		1310	3.5
	1705	4.4		1720	4.0
2 F	0055	0.1	17 SA	0045	0.9
	0825	5.1		0815	4.7
	1350	3.5		1350	3.3
	1810	4.2		1755	3.8
3 SA	0140	0.4	18 SU	0115	1.0
	0905	5.1		0840	4.7
	1455	3.2		1435	3.1
	1920	3.9		1855	3.7
4 SU	0220	0.8	19 M	0145	1.2
	0945	5.0		0900	4.7
	1600	2.9		1515	2.9
	2035	3.6		1955	3.5
5 M	0305	1.3	20 TU	0220	1.6
	1020	4.9		0925	4.7
	1700	2.5		1600	2.6
	2155	3.4		2100	3.4
6 TU	0345	1.9	21 W	0255	2.0
	1055	4.8		0945	4.7
	1800	2.1		1650	2.2
	2330	3.3		2225	3.4
7 W	0440	2.5	22 TH	0345	2.5
	1125	4.7		1015	4.6
	1855	1.8		1740	1.8
				2355	3.4
8 TH	0125	3.4	23 F	0430	3.0
	0545	3.0		1045	4.5
	1155	4.5		1830	1.4
	1945	1.5			
9 F	0315	3.8	24 SA	0210	3.7
	0705	3.4		0545	3.5
	1225	4.3		1115	4.6
	2025	1.2		1925	1.1
10 SA	0420	4.1	25 SU	0345	4.1
	0830	3.7		0720	3.8
	1255	4.1		1200	4.5
	2110	1.1		2025	0.8
11 SU	0505	4.4	26 M	0445	4.4
	0950	3.7		0850	3.9
	1330	4.1		1255	4.5
	2145	1.0		2250	0.5
12 M	0545	4.6	27 TU	0520	4.7
	1045	3.7		0955	3.8
	1410	4.0		1400	4.5
	2230	0.9		2215	0.3
13 TU	0620	4.7	28 W	0605	4.8
	1120	3.7		1055	3.7
	1455	4.0		1515	4.5
	2305	0.8		2315	0.2
14 W	0650	4.7	29 TH	0640	4.9
	1200	3.7		1150	3.5
	1545	4.0	•	1625	4.4
	2345	0.8			
15 TH	0720	4.7	30 F	0715	4.9
	1230	3.6		1240	3.3
o	1630	4.0		1725	4.3
			31 SA	0040	0.4
				0745	4.9
				1330	3.0
				1830	4.1

FEBRUARY

Day	TIME	M	Day	TIME	M
1 SU	0120	0.8	16 M	0055	1.3
	0820	4.9		0745	4.6
	1420	2.6		1355	2.5
	1930	3.9		1910	3.8
2 M	0200	1.2	17 TU	0125	1.6
	0850	4.9		0800	4.6
	1515	2.3		1435	2.2
	2035	3.7		2005	3.7
3 TU	0240	1.8	18 W	0200	2.0
	0920	4.8		0825	4.6
	1605	2.0		1510	1.9
	2145	3.5		2110	3.7
4 W	0315	2.4	19 TH	0235	2.5
	0945	4.6		0845	4.5
	1700	1.7		1600	1.6
	2320	3.5		2225	3.7
5 TH	0400	3.0	20 F	0325	3.0
	1010	4.4		0910	4.5
	1750	1.6		1650	1.3
6 F	0120	3.7	21 SA	0010	3.7
	0500	3.4		0420	3.4
	1035	4.2		0940	4.4
	1845	1.5		1745	1.1
7 SA	0255	3.9	22 SU	0210	4.0
	0650	3.7		0545	3.7
	1100	4.0		1025	4.3
	1940	1.4		1850	1.0
8 SU	0355	4.2	23 M	0315	4.2
	0840	3.7		0735	3.8
	1145	3.9		1130	4.2
	2035	1.3		2000	0.8
9 M	0445	4.4	24 TU	0410	4.4
	0955	3.7		0900	3.7
	1245	3.8		1245	4.1
	2120	1.2		2105	0.7
10 TU	0515	4.5	25 W	0455	4.6
	1025	3.6		1000	3.5
	1400	3.8		1425	4.1
	2205	1.1		2200	0.6
11 W	0545	4.5	26 TH	0530	4.6
	1100	3.5		1050	3.2
	1505	3.9		1540	4.2
	2250	1.0		2255	0.6
12 TH	0615	4.5	27 F	0600	4.7
	1135	3.4		1135	2.9
	1600	3.9		1645	4.2
	2320	1.0		2335	0.7
13 F	0640	4.5	28 SA	0630	4.7
	1215	3.2		1220	2.5
	1650	4.0	•	1745	4.2
o	2355	1.0			
14 SA	0705	4.5			
	1240	3.0			
	1735	4.0			
15 SU	0025	1.1			
	0725	4.6			
	1315	2.7			
	1820	3.9			

MARCH

Day	TIME	M	Day	TIME	M
1 SU	0015	1.0	16 M	0620	4.4
	0700	4.7		1245	1.9
	1305	2.2		1835	4.0
	1840	4.1			
2 M	0055	1.4	17 TU	0030	1.9
	0725	4.7		0640	4.5
	1345	1.8		1315	1.6
	1940	4.0		1925	4.0
3 TU	0135	1.9	18 W	0105	2.3
	0750	4.6		0700	4.5
	1430	1.6		1355	1.3
	2040	3.9		2020	4.0
4 W	0210	2.4	19 TH	0145	2.7
	0815	4.5		0725	4.4
	1510	1.4		1435	1.1
	2145	3.8		2130	4.0
5 TH	0255	2.9	20 F	0230	3.0
	0830	4.3		0745	4.4
	1555	1.4		1520	0.9
	2305	3.8		2245	4.1
6 F	0345	3.3	21 SA	0325	3.4
	0855	4.1		0815	4.2
	1640	1.4		1610	0.9
7 SA	0040	3.9	22 SU	0010	4.1
	0455	3.5		0440	3.6
	0910	3.9		0850	4.1
	1735	1.4		1710	0.9
8 SU	0200	4.1	23 M	0135	4.2
	0715	3.6		0615	3.7
	0945	3.7		0945	3.9
	1835	1.5		1820	1.0
9 M	0305	4.2	24 TU	0240	4.4
	0910	3.5		0805	3.5
	1045	3.5		1130	3.7
	1945	1.5		1935	1.0
10 TU	0350	4.2	25 W	0325	4.4
	0945	3.4		0905	3.2
	1240	3.5		1315	3.7
	2045	1.4		2045	1.0
11 W	0425	4.3	26 TH	0405	4.5
	1010	3.2		0955	2.9
	1410	3.5		1445	3.8
	2135	1.4		2140	1.1
12 TH	0455	4.3	27 F	0440	4.5
	1035	3.0		1035	2.5
	1515	3.7		1600	3.9
	2215	1.3		2230	1.2
13 F	0520	4.3	28 SA	0505	4.5
	1105	2.8		1115	2.1
	1605	3.8		1700	4.1
	2255	1.3		2315	1.5
14 SA	0545	4.4	29 SU	0535	4.5
	1135	2.6		1155	1.7
	1655	3.9	•	1755	4.1
	2325	1.4		2355	1.9
15 SU	0605	4.4	30 M	0600	4.5
	1205	2.3		1235	1.4
o	1745	4.0		1850	4.2
	2355	1.6			
			31 TU	0035	2.3
				0625	4.4
				1310	1.1
				1945	4.2

APRIL

Day	TIME	M	Day	TIME	M
1 W	0110	2.7	16 TH	0055	2.9
	0645	4.3		0600	4.4
	1350	1.0		1320	0.6
	2040	4.2		2030	4.4
2 TH	0155	3.0	17 F	0140	3.2
	0705	4.1		0625	4.3
	1425	1.0		1405	0.5
	2140	4.2		2135	4.4
3 F	0240	3.3	18 SA	0230	3.4
	0725	4.0		0655	4.2
	1500	1.0		1455	0.5
	2245	4.2		2245	4.4
4 SA	0345	3.5	19 SU	0340	3.6
	0740	3.8		0740	4.0
	1540	1.2		1545	0.6
	2355	4.2		2355	4.4
5 SU	0515	3.5	20 M	0510	3.6
	0750	3.6		0830	3.8
	1625	1.4		1645	0.8
6 M	0100	4.2	21 TU	0100	4.5
	0515	3.6		0700	3.4
	1015	3.3		0950	3.5
	1725	1.5		1755	1.1
7 TU	0155	4.2	22 W	0150	4.5
	0815	3.2		0805	3.0
	1200	3.1		1200	3.4
	1840	1.6		1905	1.3
8 W	0245	4.2	23 TH	0230	4.5
	0915	3.0		0855	2.6
	1240	3.2		1350	3.4
	1950	1.7		2015	1.6
9 TH	0320	4.2	24 F	0305	4.5
	0935	2.8		0935	2.1
	1405	3.3		1515	3.6
	2045	1.7		2110	1.8
10 F	0345	4.2	25 SA	0340	4.4
	1000	2.5		1015	1.7
	1510	3.5		1610	3.8
	2130	1.8		2200	2.1
11 SA	0410	4.5	26 SU	0410	4.4
	1030	2.2		1055	1.3
	1605	3.7		1715	4.0
	2210	1.9		2255	2.4
12 SU	0435	4.2	27 M	0435	4.3
	1100	1.8		1130	1.0
	1700	3.9		1810	4.2
	2255	2.1		2335	2.7
13 M	0455	4.3	28 TU	0500	4.2
	1135	1.5		1200	0.8
	1745	4.1	•	1900	4.3
	2330	2.3			
14 TU	0515	4.3	29 W	0015	3.0
	1215	1.1		0520	4.1
o	1840	4.2		1240	0.7
				1950	4.4
15 W	0005	2.7	30 TH	0100	3.2
	0535	4.4		0540	4.1
	1245	0.8		1310	0.7
	1935	4.3		2035	4.5

MAY

Day	Time	M	Day	Time	M
1 F	0530	-0.1	16 SA	0542	-0.4
	1210	0.9		1231	1.1
	1555	0.8		1630	0.8
	2230	1.7		2259	2.0
2 SA	0619	0.0	17 SU	0644	-0.3
	1335	0.9		1351	1.1
	1609	0.9		1742	0.9
	2307	1.6			
3 SU	0722	0.1	18 M	0001	1.8
	2357	1.4		0750	-0.2
				1507	1.2
				1922	0.9
4 M	0838	0.1	19 TU	0120	1.6
				0857	-0.3
				1606	1.3
				2112	0.8
5 TU	0117	1.3	20 W	0249	1.5
	0945	0.2		0957	-0.1
	1737	1.1		1654	1.5
	2151	1.0		2239	0.6
6 W	0300	1.2	21 TH	0415	1.4
	1037	0.1		1048	0.0
	1751	1.2		1732	1.6
	2306	0.8		2345	0.3
7 TH	0423	1.2	22 F	0529	1.3
	1116	0.1		1133	0.1
	1807	1.3		1806	1.8
	2352	0.6			
8 F	0525	1.3	23 SA	0041	0.8
	1150	0.2		0631	1.3
	1825	1.5		1212	0.2
				1839	1.9
9 SA	0034	0.4	24 SU	0126	-0.1
	0620	1.3		0728	1.2
	1222	0.2		1247	0.5
	1847	1.6		1910	2.0
10 SU	0113	0.1	25 M	0205	-0.2
	0707	1.3		0817	1.2
	1251	0.2		1319	0.5
	1910	1.8		1939	2.0
11 M	0148	-0.1	26 TU	0244	-0.3
	0755	1.3		0903	1.2
	1320	0.3		1351	0.5
	1939	1.9		2009	2.0
12 TU	0227	-0.3	27 W ●	0322	-0.3
	0842	1.3		0948	1.1
	1352	0.4		1417	0.6
	2009	2.0		2037	2.0
13 W ○	0309	-0.4	28 TH	0358	-0.3
	0930	1.2		1033	1.1
	1426	0.5		1448	0.7
	2043	2.1		2108	1.9
14 TH	0356	-0.5	29 F	0435	-0.2
	1023	1.2		1115	1.0
	1502	0.5		1513	0.8
	2124	2.1		2140	1.8
15 F	0446	-0.5	30 SA	0515	-0.2
	1123	1.1		1208	1.0
	1542	0.7		1545	0.8
	2207	2.0		2212	1.7
			31 SU	0557	-0.1
				1307	1.0
				1624	0.9
				2250	1.6

JUNE

Day	Time	M	Day	Time	M
1 M	0642	0.0	16 TU	0717	-0.2
	1412	1.0		1411	1.4
	1706	0.9		1921	0.8
	2335	1.5			
2 TU	0732	0.1	17 W	0103	1.6
	1511	1.1		0809	-0.1
	1840	1.0		1506	1.5
				2055	0.7
3 W	0027	1.4	18 TH	0224	1.4
	0821	0.2		0903	-0.1
	1553	1.2		1557	1.5
	2039	0.9		2223	0.7
4 TH	0147	1.2	19 F	0354	1.2
	0907	0.2		0952	0.3
	1625	1.3		1644	1.7
	2210	0.8		2333	0.3
5 F	0314	1.1	20 SA	0521	1.1
	0953	0.3		1044	0.4
	1654	1.4		1726	1.8
	2316	0.5			
6 SA	0439	1.1	21 SU	0035	0.1
	1037	0.3		0635	1.1
	1721	1.6		1130	0.5
				1805	1.9
7 SU	0004	0.3	22 M	0118	-0.1
	0552	1.1		0739	1.1
	1116	0.4		1212	0.6
	1753	1.8		1840	2.0
8 M	0054	0.1	23 TU	0200	-0.2
	0655	1.1		0829	1.1
	1158	0.5		1254	0.7
	1828	1.9		1916	2.0
9 TU	0136	-0.2	24 W	0241	-0.2
	0750	1.2		0914	1.1
	1240	0.5		1329	0.7
	1903	2.1		1948	2.0
10 W	0218	-0.4	25 TH	0314	-0.2
	0845	1.2		0953	1.1
	1323	0.6		1405	0.8
	1945	2.2		2024	2.0
11 TH ○	0303	-0.5	26 F ●	0349	-0.2
	0936	1.2		1028	1.1
	1404	0.6		1437	0.8
	2027	2.3		2055	2.0
12 F	0352	-0.6	27 SA	0422	-0.2
	1027	1.2		1100	1.1
	1453	0.6		1509	0.8
	2115	2.3		2127	1.9
13 SA	0441	-0.6	28 SU	0457	-0.2
	1120	1.2		1139	1.1
	1545	0.7		1545	0.8
	2204	2.2		2159	1.8
14 SU	0531	-0.5	29 M	0529	-0.1
	1216	1.2		1211	1.2
	1643	0.7		1624	0.9
	2257	2.0		2234	1.7
15 M	0623	-0.4	30 TU	0601	0.0
	1312	1.3		1249	1.2
	1755	0.8		1710	0.9
	2355	1.8		2310	1.6

JULY

Day	Time	M	Day	Time	M
1 W	0633	0.1	16 TH	0042	1.5
	1328	1.2		0717	0.1
	1816	0.9		1402	1.6
	2355	1.4		2022	0.5
2 TH	0709	0.2	17 F	0158	1.2
	1407	1.3		0800	0.4
	1938	0.9		1455	1.7
				2152	0.4
3 F	0054	1.2	18 SA	0336	1.0
	0743	0.3		0853	0.5
	1452	1.4		1551	1.7
	2111	0.7		2318	0.3
4 SA	0215	1.1	19 SU	0529	1.0
	0825	0.4		0951	0.7
	1533	1.5		1648	1.8
	2234	0.5			
5 SU	0400	1.0	20 M	0023	0.1
	0918	0.5		0656	1.0
	1619	1.7		1056	0.8
	2343	0.3		1737	1.8
6 M	0545	1.0	21 TU	0112	0.0
	1017	0.6		0752	1.1
	1705	1.8		1157	0.9
				1822	1.9
7 TU	0035	0.0	22 W	0153	-0.1
	0659	1.1		0834	1.2
	1116	0.7		1246	0.9
	1756	2.0		1905	1.9
8 W	0124	-0.2	23 TH	0229	-0.2
	0759	1.1		0906	1.2
	1215	0.7		1324	0.8
	1844	2.2		1939	2.0
9 TH	0211	-0.2	24 F	0301	-0.2
	0845	1.2		0934	1.2
	1307	0.7		1402	0.8
	1933	2.3		2013	2.0
10 F	0255	-0.5	25 SA ●	0329	-0.2
	0931	1.3		1001	1.2
	1400	0.6		1434	0.6
	2021	2.3		2046	2.0
11 SA ○	0340	-0.6	26 SU	0357	-0.2
	1013	1.3		1024	1.3
	1453	0.6		1506	0.7
	2110	2.3		2115	2.0
12 SU	0425	-0.6	27 M	0422	-0.2
	1055	1.4		1049	1.3
	1546	0.6		1540	0.7
	2159	2.3		2147	1.9
13 M	0509	-0.5	28 TU	0448	-0.1
	1139	1.5		1115	1.3
	1643	0.6		1617	0.7
	2251	2.1		2219	1.8
14 TU	0549	-0.3	29 W	0512	0.0
	1226	1.5		1140	1.4
	1746	0.6		1657	0.7
	2343	1.8		2251	1.4
15 W	0634	-0.1	30 TH	0534	0.2
	1311	1.6		1212	1.4
	1900	0.6		1749	0.7
				2330	1.4
			31 F	0603	0.3
				1244	1.5
				1856	0.7

AUGUST

Day	Time	M	Day	Time	M
1 SA	0023	1.2	16 SU	0340	1.0
	0630	0.5		0737	0.8
	1322	1.6		1453	1.6
	2018	0.6		2259	0.3
2 SU	0145	1.0	17 M	0609	1.0
	0706	0.6		0906	0.9
	1418	1.6		1606	1.6
	2201	0.5			
3 M	0400	0.9	18 TU	0005	0.2
	0802	0.8		0718	1.1
	1527	1.7		1055	1.0
	2324	0.2		1716	1.7
4 TU	0606	1.0	19 W	0057	0.1
	0931	0.9		0753	1.2
	1634	1.9		1201	0.9
				1809	1.8
5 W	0023	0.0	20 TH	0133	0.0
	0710	1.1		0815	1.3
	1100	0.9		1246	0.9
	1741	2.0		1852	1.9
6 TH	0115	-0.2	21 F	0205	-0.1
	0752	1.2		0837	1.3
	1214	0.8		1323	0.8
	1835	2.2		1927	1.9
7 F	0200	-0.4	22 SA	0232	-0.1
	0830	1.3		0858	1.4
	1310	0.7		1355	0.7
	1927	2.3		1959	2.0
8 SA	0241	-0.5	23 SU	0257	-0.1
	0905	1.5		0917	1.4
	1401	0.5		1424	0.6
	2016	2.4		2031	2.0
9 SU ○	0320	-0.5	24 M ●	0319	-0.1
	0940	1.6		0938	1.5
	1453	0.4		1456	0.5
	2103	2.3		2100	1.9
10 M	0359	-0.5	25 TU	0342	0.1
	1016	1.6		0957	1.5
	1540	0.4		1528	0.5
	2150	2.2		2131	1.8
11 TU	0433	-0.3	26 W	0404	0.1
	1054	1.7		1019	1.6
	1633	0.3		1603	0.5
	2236	2.0		2200	1.7
12 W	0510	-0.1	27 TH	0422	0.2
	1131	1.7		1042	1.6
	1728	0.3		1641	0.4
	2326	1.7		2235	1.5
13 TH	0544	0.2	28 F	0444	0.3
	1211	1.7		1107	1.6
	1831	0.4		1728	0.4
				2318	1.3
14 F	0021	1.4	29 SA	0506	0.5
	0616	0.4		1134	1.7
	1253	1.7		1827	0.4
	1946	0.4			
15 SA	0135	1.1	30 SU	0016	1.1
	0653	0.6		0528	0.6
	1346	1.7		1218	1.7
	2123	0.4		1949	0.4
			31 M	0155	0.9
				0557	0.8
				1318	1.7
				2139	0.3

FOR IMMEDIATE HEIGHTS USE HARMONIC CONSTANTS (SEE PART III) AND NP 159.

PANAMA, PACIFIC COAST – BALBOA
LAT 8°57'N   LONG 79°34'W
TIMES AND HEIGHTS OF HIGH AND LOW WATERS

TIME ZONE +0500   YEAR 1987

MAY

Day	Time	M	Day	Time	M
1 F	0057	4.5	16 SA	0543	4.9
	1154	0.5		1147	0.2
	1806	4.6		1755	5.1
2 SA	0015	0.2	17 SU	0015	-0.2
	0636	4.3		0633	4.8
	1233	0.7		1236	0.3
	1843	4.4		1847	4.9
3 SU	0055	0.5	18 M	0108	0.1
	0713	4.1		0726	4.8
	1316	1.0		1332	0.5
	1921	4.1		1942	4.7
4 M	0139	0.8	19 TU	0209	0.3
	0755	3.9		0825	4.6
	1403	1.3		1435	0.7
	2001	3.8		2043	4.4
5 Tu	0027	1.0	20 W	0306	0.5
	0843	3.7		0925	4.5
	1502	1.5		1541	0.9
	2054	3.6		2152	4.2
6 W	0323	1.2	21 TH	0410	0.7
	0941	3.6		1036	4.4
	1604	1.6		1649	0.9
	2159	3.5		2308	4.1
7 TH	0424	1.3	22 F	0516	0.8
	1050	3.6		1144	4.4
	1708	1.5		1755	0.9
	2314	3.5	23 SA	0018	4.1
8 F	0526	1.3		0818	0.8
	1155	3.7		1245	4.5
	1809	1.3		1856	0.7
9 SA	0018	3.6	24 SU	0117	4.2
	0624	1.2		0717	0.8
	1245	4.0		1335	4.6
	1904	1.1		1951	0.5
10 SU	0109	3.9	25 M	0207	4.4
	0719	1.0		0810	0.7
	1327	4.3		1420	4.7
	1956	0.8		2041	0.4
11 M	0155	4.1	26 TU	0252	4.4
	0808	0.8		0857	0.6
	1410	4.6		1503	4.8
	2041	0.4		2123	0.3
12 TU	0236	4.4	27 W	0337	4.5
	0852	0.5		0937	0.6
	1450	4.8		1542	4.8
	2123	0.1		2203	0.2
13 W	0319	4.6	28 TH	0419	4.5
	0937	0.3		1018	0.6
	1532	5.0		1622	4.6
	2204	-0.2		2241	0.2
14 TH	0406	4.8	29 F	0458	4.5
	1019	0.2		1055	0.6
	1618	5.1		1702	4.6
	2246	-0.3		2317	0.3
15 F	0453	4.9	30 SA	0538	4.4
	1102	0.1		1134	0.8
	1705	5.1		1742	4.5
	2330	-0.3		2354	0.4
			31 SU	0617	4.3
				1212	0.9
				1820	4.4

JUNE

Day	Time	M	Day	Time	M
1 M	0031	0.6	16 TU	0054	-0.1
	0655	4.2		0718	5.1
	1255	1.1		1321	0.4
	1900	4.2		1936	4.9
2 TU	0113	0.8	17 W	0150	0.2
	0734	4.1		0811	4.9
	1340	1.2		1419	0.5
	1940	4.0		2032	4.6
3 W	0158	1.0	18 TH	0246	0.4
	0817	4.1		0907	4.8
	1430	1.4		1522	0.7
	2025	3.9		2133	4.4
4 TH	0248	1.1	19 F	0346	0.6
	0859	4.0		1007	4.6
	1525	1.4		1623	0.8
	2115	3.7		2239	4.2
5 F	0339	1.2	20 SA	0444	0.9
	0952	3.9		1110	4.5
	1621	1.4		1724	0.9
	2216	3.7		2347	4.1
6 SA	0433	1.3	21 SU	0544	1.0
	1049	4.0		1211	4.5
	1717	1.3		1826	0.9
	2320	3.7	22 M	0048	4.1
7 SU	0529	1.2		0642	1.1
	1148	4.1		1303	4.5
	1815	1.1		1922	0.8
8 M	0021	3.9	23 TU	0143	4.1
	0626	1.2		0741	1.1
	1240	4.4		1352	4.5
	1911	0.9		2015	0.7
9 TU	0112	4.1	24 W	0231	4.1
	0725	1.0		0833	1.0
	1327	4.6		1436	4.5
	2004	0.5		2103	0.6
10 W	0204	4.4	25 TH	0317	4.2
	0820	0.7		0917	1.0
	1415	4.8		1521	4.5
	2055	0.2		2142	0.5
11 TH	0253	4.6	26 F	0358	4.3
	0911	0.5		0958	0.9
	1506	5.0		1601	4.5
	2142	0.0		2221	0.4
12 F	0343	4.8	27 SA	0440	4.4
	0958	0.2		1038	0.9
	1558	5.1		1642	4.5
	2142	0.0		2258	0.4
13 SA	0439	5.0	28 SU	0521	4.4
	1046	0.2		1116	0.8
	1651	5.2		1723	4.5
	2228	-0.2		2335	0.4
14 SU	0531	5.1	29 M	0559	4.4
	1135	0.2		1154	0.9
	1702	5.2		1801	4.4
	2316	-0.3			
15 M	0003	-0.2	30 TU	0010	0.5
	0625	5.1		0636	4.4
	1227	0.2		1233	0.9
	1841	5.0		1839	4.3

JULY

Day	Time	M	Day	Time	M
1 W	0049	0.6	16 TH	0126	0.0
	0708	4.4		0750	5.2
	1315	1.0		1355	0.3
	1916	4.2		2009	4.8
2 TH	0129	0.8	17 F	0217	0.3
	0745	4.3		0838	4.9
	1358	1.1		1451	0.5
	1954	4.1		2104	4.5
3 F	0210	0.9	18 SA	0312	0.6
	0819	4.3		0931	4.7
	1446	1.2		1549	0.8
	2036	4.0		2202	4.1
4 SA	0256	1.1	19 SU	0409	0.9
	0902	4.2		1028	4.4
	1536	1.2		1648	0.9
	2125	3.9		2309	3.9
5 SU	0347	1.2	20 M	0507	1.2
	0954	4.2		1133	4.2
	1628	1.2		1747	1.1
	2224	3.8			
6 M	0441	1.2	21 TU	0018	3.8
	1051	4.2		0609	1.3
	1727	1.1		1233	4.1
	2332	3.9		1851	1.1
7 TU	0541	1.2	22 W	0119	3.8
	1155	4.3		0714	1.4
	1829	0.9		1327	4.1
				1951	1.0
8 W	0037	4.1	23 TH	0212	3.9
	0646	1.1		0810	1.3
	1256	4.5		1418	4.2
	1933	0.7		2041	0.9
9 TH	0135	4.3	24 F	0300	4.1
	0751	0.9		0900	1.2
	1351	4.7		1501	4.3
	2031	0.4		2124	0.7
10 F	0233	4.5	25 SA	0342	4.2
	0849	0.6		0942	1.0
	1449	4.9		1545	4.4
	2124	0.0		2201	0.5
11 SA	0329	4.8	26 SU	0422	4.3
	0943	0.3		1021	0.8
	1545	5.1		1625	4.5
	2214	-0.2		2329	-0.5
12 SU	0424	5.1	27 M	0459	4.5
	1035	0.1		1057	0.7
	1642	5.2		1704	4.5
	2301	-0.4		2313	0.3
13 M	0519	5.2	28 TU	0533	4.5
	1124	0.0		1134	0.6
	1736	5.2		1739	4.5
	2348	-0.4			
14 TU	0611	5.3	29 W	0607	4.6
	1212	0.0		1209	0.6
	1828	5.2		1815	4.5
15 W	0036	-0.2	30 TH	0022	0.4
	0700	5.3		0639	4.6
	1304	0.1		1246	0.7
	1920	5.0		1848	4.4
			31 F	0057	0.6
				0709	4.6
				1324	0.8
				1924	4.3

AUGUST

Day	Time	M	Day	Time	M
1 SA	0134	0.7	16 SU	0235	0.7
	0743	4.5		0853	4.6
	1406	0.9		1510	0.8
	2001	4.2		2123	4.1
2 SU	0217	0.9	17 M	0330	1.1
	0821	4.4		0947	4.2
	1453	1.0		1607	1.1
	2045	4.0		2226	3.7
3 M	0306	1.1	18 TU	0430	1.4
	0907	4.3		1049	4.0
	1548	1.1		1711	1.3
	2143	3.9		2344	3.6
4 TU	0400	1.2	19 W	0537	1.6
	1007	4.2		1203	3.8
	1649	1.1		1817	1.3
	2253	3.8			
5 W	0507	1.3	20 TH	0056	3.6
	1120	4.2		0645	1.6
	1758	1.0		1306	3.9
				1922	1.2
6 TH	0011	3.9	21 F	0154	3.8
	0621	1.2		0750	1.5
	1234	4.3		1359	4.0
	1909	0.8		2018	1.0
7 F	0119	4.2	22 SA	0239	4.0
	0733	1.0		0841	1.2
	1339	4.6		1444	4.1
	2012	0.5		2103	0.8
8 SA	0220	4.5	23 SU	0318	4.2
	0836	0.6		0922	0.9
	1439	4.8		1524	4.3
	2108	0.1		2140	0.6
9 SU	0316	4.9	24 M	0354	4.4
	0932	0.2		0958	0.7
	1535	5.1		1603	4.5
	2158	-0.2		2214	0.4
10 M	0411	5.2	25 TU	0427	4.6
	1020	-0.1		1035	0.5
	1630	5.3		1638	4.6
	2244	-0.4		2247	0.3
11 TU	0500	5.4	26 W	0500	4.7
	1107	-0.2		1108	0.4
	1721	5.3		1712	4.7
	2329	-0.5		2319	0.2
12 W	0550	5.5	27 TH	0532	4.8
	1153	-0.3		1141	0.3
	1810	5.3		1746	4.7
				2353	0.3
13 TH	0014	-0.3	28 F	0602	4.8
	0635	5.4		1215	0.4
	1239	-0.1		1819	4.6
	1857	5.1			
14 F	0100	-0.1	29 SA	0027	0.4
	0721	5.2		0636	4.8
	1326	0.1		1252	0.5
	1942	4.8		1855	4.5
15 SA	0147	0.3	30 SU	0102	0.6
	0806	4.9		0708	4.7
	1417	0.5		1332	0.6
	2030	4.5		1932	4.4
			31 M	0145	0.8
				0745	4.6
				1419	0.8
				2016	4.1

PANAMA, PACIFIC COAST-BALBOA
LAT 8°57'N   LONG 79°34'W

TIME ZONE +0500

TIMES AND HEIGHTS OF HIGH AND LOW WATERS

YEAR 1987

SEPTEMBER

Day		TIME	M	Day		TIME	M
1	TU	0234	1.1	16	W	0355	1.6
		0835	4.3			1005	3.7
		1515	1.0			1632	1.4
		2112	3.9			2305	3.5
2	W	0333	1.2o	17	TH	0505	1.7
		0936	4.1			1127	3.6
		1623	1.1			1741	1.5
		2229	3.9				
3	TH	0447	1.3	18	F	0026	3.6
		1100	4.1			0620	1.7
		1737	1.1			1242	3.7
		2357	3.9			1850	1.4
4	F	0605	1.2	19	SA	0125	3.8
		1224	4.2			0722	1.6
		1853	0.9			1335	3.9
						1948	1.2
5	SA	0109	4.2	20	SU	0210	4.0
		0722	1.0			0815	1.2
		1333	4.5			1420	4.1
		1957	0.5			2031	0.9
6	SU	0210	4.6	21	M	0247	4.3
		0824	0.5			0857	0.9
		1431	4.8			1457	4.4
		2053	0.1			2111	0.7
7	M o	0303	5.0	22	TU	0318	4.5
		1003	0.1			0932	0.6
		1613	5.1			1532	4.5
		2226	-0.2			2145	0.5
8	TU	0350	5.3	23	W •	0351	4.7
		1003	-0.2			1006	0.4
		1613	5.3			1607	4.7
		2225	-0.4			2219	0.3
9	W	0438	5.5	24	TH	0423	4.9
		1047	-0.4			1040	0.2
		1700	5.3			1641	4.8
		2307	-0.4			2251	0.2
10	TH	0523	5.5	25	F	0455	5.0
		1129	-0.4			1113	0.1
		1746	5.3			1716	4.8
		2348	-0.2			2325	0.3
11	F	0607	5.4	26	SA	0530	5.0
		1211	-0.2			1148	0.2
		1830	5.1			1753	4.8
						2359	0.4
12	SA	0031	0.0	27	SU	0604	4.9
		0649	5.2			1225	0.3
		1255	0.1			1830	4.7
		1913	4.8				
13	SU	0113	0.4	28	M	0039	0.6
		0728	4.8			0641	4.7
		1340	0.5			1307	0.5
		1955	4.4			1912	4.5
14	M	0200	0.9	29	TU	0121	0.8
		0814	4.5			0724	4.6
		1430	0.9			1355	0.7
		2043	4.0			1958	4.3
15	TU	0251	1.2	30	W	0216	1.1
		0902	4.1			0817	4.4
		1528	1.2			1454	1.0
		2144	3.7			2059	4.1

OCTOBER

Day		TIME	M	Day		TIME	M
1	TH	0323	1.3	16	F	0429	1.8
		0925	4.1			1040	3.6
		1607	1.1			1658	1.6
		2221	3.9			2341	3.6
2	F	0439	1.3	17	SA	0541	1.7
		1054	4.0			1200	3.6
		1722	1.1			1803	1.5
		2349	4.1				
3	SA	0557	1.2	18	SU	0040	3.8
		1218	4.2			0645	1.5
		1836	0.9			1259	3.8
						1903	1.3
4	SU	0058	4.4	19	M	0127	4.1
		0709	0.9			0735	1.2
		1323	4.5			1343	4.1
		1939	0.6			1951	1.1
5	M	0154	4.8	20	TU	0204	4.4
		0808	0.5			0820	0.9
		1418	4.8			1423	4.3
		2033	0.3			2033	0.9
6	TU	0242	5.1	21	W	0237	4.6
		0900	0.1			0900	0.6
		1508	5.1			1457	4.5
		2121	0.0			2111	0.6
7	W o	0329	5.3	22	TH	0311	4.8
		0943	-0.2			0937	0.3
		1553	5.2			1537	4.7 •
		2203	-0.1			2148	0.4
8	TH	0412	5.4	23	F	0346	5.0
		1025	-0.3			1011	0.1
		1638	5.2			1611	4.9
		2243	-0.1			2223	0.3
9	F	0454	5.4	24	SA	0423	5.1
		1105	-0.3			1047	0.0
		1721	5.2			1651	4.9
		2322	0.0			2300	0.3
10	SA	0535	5.3	25	SU	0500	5.1
		1144	-0.1			1125	0.0
		1803	4.9			1732	4.9
						2338	0.4
11	SU	0002	0.3	26	M	0542	5.0
		0617	5.0			1205	0.1
		1225	0.2			1814	4.8
		1844	4.7				
12	M	0044	0.6	27	TU	0020	0.5
		0655	4.7			0625	4.9
		1307	0.5			1249	0.3
		1925	4.4			1900	4.7
13	TU	0126	1.0	28	W	0110	0.8
		0737	4.4			0713	4.7
		1353	0.9			1342	0.6
		2011	4.1			1953	4.5
14	W	0217	1.4	29	TH	0206	1.0
		0822	4.0			0811	4.4
		1448	1.2			1443	0.9
		2103	3.8			2056	4.3
15	TH	0320	1.6	30	F	0315	1.2
		0921	3.7			0923	4.2
		1549	1.5			1552	1.0
		2218	3.6			2213	4.2
				31	SA	0430	1.2
						1047	4.1
						1704	1.0
						2332	4.3

NOVEMBER

Day		TIME	M	Day		TIME	M
1	SU	0544	1.1	16	M	0552	1.5
		1205	4.2			1205	3.7
		1814	0.9			1808	1.4
2	M	0037	4.5	17	TU	0032	4.1
		0650	0.8			0648	1.3
		1309	4.5			1258	3.9
		1915	0.7			1901	1.2
3	TU	0133	4.8	18	W	0114	4.3
		0748	0.5			0738	0.4
		1402	4.7			1341	4.5
		2010	0.5			1949	0.7
4	W	0220	5.1	19	TH	0154	4.5
		0839	0.2			0823	0.7
		1449	4.9			1421	4.4
		2057	0.3			2036	0.8
5	TH o	0305	5.2	20	F	0231	4.8
		0924	0.0			0905	0.4
		1533	5.0			1503	4.6
		2140	0.2			2118	0.6
6	F	0346	5.2	21	SA	0313	5.0
		1004	-0.1			0945	0.1
		1617	5.0			1545	4.8 •
		2221	0.2			2158	0.4
7	SA	0428	5.2	22	SU	0353	5.1
		1043	-0.1			1026	-0.1
		1700	4.9			1628	4.9
		2300	0.3			2240	0.3
8	SU	0508	5.0	23	M	0438	5.1
		1122	0.1			1108	0.1
		1742	4.8			1717	4.8
		2338	0.5			2324	0.5
9	M	0547	4.8	24	TU	0526	5.1
		1159	0.3			1150	0.3
		1821	4.6			1804	4.9
10	TU	0018	0.8	25	W	0010	0.4
		0628	4.6			0617	5.0
		1239	0.6			1238	0.1
		1900	4.4			1855	4.8
11	W	0100	1.1	26	TH	0102	0.6
		0709	4.3			0710	4.8
		1324	0.9			1332	0.3
		1945	4.1			1950	4.7
12	TH	0150	1.3	27	F	0201	0.8
		0753	4.0			0809	4.5
		1411	1.2			1430	0.6
		2030	4.0			2048	4.6
13	F	0246	1.6	28	SA	0304	0.9
		0843	3.8			0912	4.3
		1507	1.4			1533	0.8
		2128	3.8			2155	4.5
14	SA	0349	1.7	29	SU	0413	0.9
		0947	3.6			1027	4.2
		1608	1.5			1640	0.9
		2305	3.8			2305	4.5
15	SU	0453	1.6	30	M	0519	0.9
		1102	3.6			1141	4.2
		1709	1.5			1745	0.9
		2340	3.9				

DECEMBER

Day		TIME	M	Day		TIME	M
1	TU	0011	4.5	16	W	0552	1.2
		0624	0.8			1201	3.7
		1245	4.3			1805	1.3
		1845	0.9				
2	W	0106	4.7	17	TH	0019	4.0
		0722	0.6			0650	1.0
		1343	4.4			1256	3.9
		1943	0.8			1903	1.2
3	TH	0156	4.8	18	F	0111	4.4
		0816	0.4			0743	0.7
		1431	4.5			1346	4.2
		2036	0.7			1959	0.9
4	F	0241	4.8	19	SA	0157	4.6
		0904	0.3			0836	0.4
		1516	4.6			1433	4.4
		2121	0.6			2049	0.7
5	SA o	0326	4.8	20	SU	0244	4.8
		0945	0.2			0921	0.1
		1559	4.6			1524	4.7 •
		2203	0.5			2140	0.4
6	SU	0406	4.8	21	M	0334	4.9
		1025	0.2			1007	-0.1
		1641	4.6			1614	4.8
		2241	0.6			2225	0.2
7	M	0448	4.7	22	TU	0425	5.1
		1102	0.2			1052	-0.3
		1722	4.6			1706	5.0
		2320	0.7			2312	0.1
8	TU	0527	4.6	23	W	0520	5.1
		1139	0.3			1138	-0.3
		1804	4.5			1757	5.1
		2359	0.8				
9	W	0608	4.5	24	TH	0001	0.1
		1217	0.5			0612	5.0
		1841	4.4			1225	-0.2
						1847	5.1
10	TH	0039	0.9	25	F	0052	0.2
		0647	4.3			0705	4.9
		1257	0.7			1316	0.0
		1921	4.3			1939	5.0
11	F	0124	1.1	26	SA	0147	0.3
		0726	4.1			0758	4.7
		1340	0.9			1411	0.2
		2001	4.1			2030	4.8
12	SA	0214	1.3	27	SU	0246	0.5
		0809	3.9			0854	4.5
		1427	1.1			1510	0.5
		2043	4.0			2128	4.6
13	SU	0307	1.4	28	M	0347	0.6
		0856	3.7			1000	4.2
		1519	1.2			1610	0.7
		2130	3.9			2229	4.5
14	M	0400	1.4	29	TU	0449	0.7
		0955	3.6			1109	4.0
		1610	1.3			1710	0.9
		2224	3.9			2336	4.4
15	TU	0457	1.3	30	W	0552	0.8
		1100	3.6			1218	4.0
		1706	1.4			1813	1.0
		2324	4.0				
				31	TH	0037	4.3
						0656	0.8
						1319	4.0
						1917	1.0

TIME ZONE +0400

YEAR 1987

JANUARY

Day	TIME	M	Day	TIME	M
1 TH	0212	7.1	16 F	0224	6.1
	0842	0.3		0848	1.3
	1448	6.1		1454	5.3
	2048	0.9		2048	1.7
2 F	0300	7.1	17 SA	0254	6.0
	0930	0.3		0918	1.3
	1536	6.1		1524	5.3
	2136	1.0		2118	1.8
3 SA	0348	6.9	18 SU	0324	5.9
	1018	0.5		0948	1.4
	1630	5.9		1554	5.3
	2224	1.2		2148	1.8
4 SU	0442	6.5	19 M	0354	5.8
	1106	0.8		1018	1.5
	1718	5.7		1624	5.2
	2318	1.5		2218	1.9
5 M	0530	6.1	20 TU	0430	5.6
	1200	1.2		1048	1.7
	1812	5.5		1654	5.2
				2300	2.1
6 TU	0018	1.8	21 W	0506	5.4
	0624	5.6		1124	1.9
	1254	1.6		1736	5.1
	1906	5.2		2348	2.2
7 W	0118	2.1	22 TH	0554	5.1
	0724	5.2		1212	2.1
	1348	1.9		1830	5.1
	2012	5.1			
8 TH	0230	2.3	23 F	0048	2.4
	0836	4.8		0654	4.8
	1454	2.2		1312	2.3
	2118	5.1		1942	5.1
9 F	0342	2.3	24 SA	0218	2.4
	0948	4.6		0818	4.6
	1554	2.3		1436	2.3
	2218	5.2		2106	5.2
10 SA	0448	2.2	25 SU	0348	2.1
	1054	4.6		0954	4.7
	1654	2.2		1600	2.2
	2312	5.3		2224	5.6
11 SU	0542	2.0	26 M	0506	1.8
	1148	4.7		1112	5.0
	1742	2.1		1712	1.8
	2359	5.6		2330	6.1
12 M	0630	1.8	27 TU	0606	1.1
	1236	4.9		1212	5.5
	1824	2.0		1812	1.4
13 TU	0042	5.8	28 W	0030	6.6
	0712	1.6		0700	0.6
	1312	5.0		1306	5.9
	1906	1.9		1900	1.0
14 W	0118	5.9	29 TH	0118	7.0
	0748	1.4		0742	0.2
	1348	5.2		● 1354	6.2
	1942	1.8		1948	0.7
15 TH	0154	6.0	30 F	0206	7.3
	0818	1.3		0830	0.1
	○ 1418	5.3		1436	6.4
	2012	1.7		2036	0.5
			31 SA	0248	7.3
				0912	0.1
				1518	6.5
				2118	0.6

FEBRUARY

Day	TIME	M	Day	TIME	M
1 SU	0330	7.1	16 M	0300	6.2
	0954	0.3		0918	1.1
	1600	6.3		1518	5.8
	2206	0.8		2124	1.3
2 M	0412	6.7	17 TU	0330	6.1
	1030	0.7		0942	1.2
	1642	6.1		1548	5.8
	2248	1.2		2154	1.5
3 TU	0454	6.1	18 W	0400	5.8
	1112	1.2		1012	1.4
	1724	5.7		1618	5.7
	2330	1.6		2230	1.7
4 W	0536	5.5	19 TH	0436	5.5
	1154	1.7		1042	1.7
	1806	5.3		1700	5.5
				2312	1.9
5 TH	0024	2.1	20 F	0518	5.1
	0624	4.9		1124	2.1
	1242	2.2		1748	5.3
	1900	5.0			
6 F	0130	2.5	21 SA	0018	2.3
	0730	4.4		0618	4.6
	1342	2.6		1230	2.4
	2018	4.7		1900	5.0
7 SA	0300	2.7	22 SU	0154	2.4
	0906	4.1		0800	4.3
	1506	2.8		1412	2.6
	2148	4.7		2048	5.1
8 SU	0436	2.6	23 M	0348	2.2
	1042	4.2		1000	4.5
	1630	2.7		1600	2.4
	2300	5.0		2224	5.5
9 M	0536	2.3	24 TU	0506	1.7
	1148	4.5		1118	5.0
	1730	2.5		1712	1.9
				2330	6.1
10 TU	0024	5.3	25 W	0600	1.1
	0624	1.9		1212	5.6
	1230	4.8		1806	1.3
	1818	2.2			
11 W	0030	5.6	26 TH	0024	6.6
	0654	1.6		0648	0.5
	1300	5.1		1254	6.1
	1854	1.8		1854	0.9
12 TH	0106	5.9	27 F	0106	7.1
	0730	1.3		0730	0.2
	1330	5.4		1336	6.5
	1924	1.6		1936	0.4
13 F	0136	6.1	28 SA	0148	7.3
	0754	1.1		0806	0.0
	○ 1400	5.6		● 1412	6.7
	1954	1.4		2018	0.3
14 SA	0206	6.3			
	0824	1.0			
	1430	5.8			
	2024	1.3			
15 SU	0230	6.3			
	0854	1.0			
	1454	5.8			
	2054	1.3			

MARCH

Day	TIME	M	Day	TIME	M
1 SU	0230	7.2	16 M	0206	6.4
	0842	0.1		0818	0.8
	1448	6.7		1424	6.3
	2054	0.4		2030	0.9
2 M	0306	7.0	17 TU	0236	6.4
	0918	0.4		0842	0.9
	1524	6.6		1448	6.3
	2136	0.6		2100	0.9
3 TU	0342	6.5	18 W	0306	6.2
	0954	0.8		0912	1.1
	1600	6.3		1518	6.2
	2212	1.1		2130	1.1
4 W	0418	6.0	19 TH	0336	5.8
	1024	1.3		0942	1.4
	1636	5.8		1554	6.0
	2248	1.6		2212	1.4
5 TH	0454	5.3	20 F	0406	5.3
	1100	1.9		1018	1.7
	1712	5.4		1630	5.7
	2330	2.1		2300	1.8
6 F	0530	4.7	21 SA	0506	4.9
	1136	2.4		1106	2.2
	1754	4.9		1724	5.5
7 SA	0030	2.7	22 SU	0006	2.2
	0618	4.2		0612	4.4
	1224	2.9		1212	2.6
	1900	4.5		1848	5.0
8 SU	0218	2.9	23 M	0200	2.4
	0818	3.8		0818	4.2
	1418	3.2		1418	2.8
	2112	4.4		2048	5.0
9 M	0424	2.8	24 TU	0348	2.1
	1048	4.0		1006	4.6
	1618	3.0		1600	2.4
	2242	4.7		2218	5.5
10 TU	0524	2.4	25 W	0454	1.5
	1136	4.4		1112	5.2
	1718	2.6		1706	1.8
	2330	5.1		2318	6.1
11 W	0600	1.9	26 TH	0542	1.0
	1206	4.9		1154	5.8
	1800	2.2		1754	1.2
12 TH	0006	5.5	27 F	0006	6.6
	0630	1.5		0624	0.5
	1236	5.3		1236	6.3
	1830	1.8		1836	0.7
13 F	0036	5.9	28 SA	0048	6.9
	0654	1.2		0706	0.3
	1300	5.7		1312	6.7
	1900	1.4		1918	0.4
14 SA	0106	6.2	29 SU	0124	7.0
	0724	1.0		0742	0.2
	1330	6.0		● 1348	6.8
	1930	1.1		1954	0.3
15 SU	0136	6.4	30 M	0206	6.9
	0748	0.8		0812	0.3
	○ 1354	6.2		1418	5.8
	2000	1.0		2030	0.4
			31 TU	0236	6.6
				0848	0.6
				1454	6.6
				2106	0.7

APRIL

Day	TIME	M	Day	TIME	M
1 W	0312	6.2	16 TH	0242	6.1
	0918	1.0		0848	1.1
	1524	6.3		1500	6.5
	2142	1.1		2118	0.9
2 TH	0342	5.6	17 F	0324	5.7
	0942	1.5		0924	1.4
	1554	5.9		1536	6.2
	2206	1.9		2206	1.9
3 F	0412	5.1	18 SA	0406	5.3
	1012	2.0		1006	1.9
	1624	5.4		1624	5.8
	2254	2.2		2300	1.7
4 SA	0448	4.5	19 SU	0506	4.8
	1042	2.5		1106	2.2
	1706	4.9		1724	5.4
	2342	2.6			
5 SU	0536	4.1	20 M	0018	2.0
	1124	3.0		0636	4.4
	1800	4.5		1230	2.7
				1900	5.1
6 M	0124	2.9	21 TU	0200	2.1
	0736	3.8		0830	4.5
	1324	3.3		1424	2.6
	2006	4.3		2042	5.2
7 TU	0336	2.8	22 W	0330	1.8
	1018	4.0		0954	4.9
	1548	3.1		1548	2.2
	2200	4.6		2200	5.6
8 W	0436	2.4	23 TH	0430	1.4
	1100	4.4		1048	5.5
	1648	2.7		1648	1.7
	2254	5.0		2300	6.0
9 TH	0518	2.0	24 F	0518	1.0
	1130	4.9		1130	6.0
	1724	2.2		1736	1.2
	2330	6.0		2342	6.3
10 F	0548	1.6	25 SA	0600	0.7
	1154	5.4		1212	6.4
	1754	1.7		1818	0.8
	2359	5.8			
11 SA	0618	1.2	26 SU	0024	6.5
	1224	5.8		0636	0.6
	1830	1.3		1248	6.6
				1854	0.6
12 SU	0030	6.1	27 M	0106	6.5
	0642	0.8		0712	0.6
	1248	6.2		1318	6.7
	1900	1.0		1936	0.5
13 M	0106	6.5	28 TU	0136	6.4
	0712	0.8		0742	0.8
	1318	6.4		● 1354	6.7
	1930	0.8		2006	0.7
14 TU	0136	6.4	29 W	0212	6.1
	0742	0.8		0812	1.0
	○ 1348	6.6		1424	6.5
	2006	0.7		2042	0.9
15 W	0206	6.3	30 TH	0242	5.7
	0812	0.9		0842	1.4
	1424	6.6		1454	6.2
	2042	0.7		2118	1.3

No.	PLACE		Lat. S.		Long. E.		TIME DIFFERENCES MHW	MLW	HEIGHT DIFFERENCES (IN METRES) MHWS	MHWN	MLWN	MLWS	M.L. Z_0 m.	
							(Zone −0800)							
6244	**LEARMONTH**		See page 120						**2.5**	**1.8**	**1.2**	**0.4**		
	Australia, North Coast													
6243	Point Murat	A	21	49	114	11	−0008	−0002	−0.5	−0.4	−0.2	−0.1	1.19	
6244	LEARMONTH		22	11	114	05	Standard port		See table v				1.49	
6246	Long Island		21	36	114	41	−0041	−0049	−0.7	−0.5	+0.4	−0.1	1.04	
6247	Onslow, Beadon Point	A	21	38	115	06	−0019	−0030	−0.3	−0.2	+0.2	0.0	1.27	
6248	Large Islet		21	18	115	30	+0016	+0021	+0.8	+0.4	+0.2	−0.1	1.80	
6249	Barrow Island	A	20	43	115	28	−0018	−0017	+0.8	+0.4	+0.3	+0.3	1.87	
6249a	Tanker Mooring	A	20	49	115	33	−0015	−0014	+0.6	+0.4	+0.3	+0.2	1.87	
6250	Trimouille Island		20	23	115	33	−0011	−0005	+0.5	+0.3	+0.2	+0.1	1.74	
6250a	North West Island		20	22	115	31	−0025	−0025	+0.6	+0.3	+0.6	+0.4	1.94	
6259	**PORT HEDLAND**		See page 123						**6.8**	**4.7**	**3.3**	**1.0**		
6252	Fortescue Road		21	00	116	06	+0019	+0021	−3.0	−2.1	−1.8	−0.8	2.0	x
6252a	Steamboat Island		20	49	116	04	+0005	+0005	−3.0	−2.0	−1.7	−0.5	2.17	
6254	Dampier (Hampton Harbour)	A	20	39	116	43	−0016	−0004	−2.3	−1.5	−1.1	−0.1	2.66	
6254a	Cape Legendre		20	21	116	50	−0015	−0015	−2.5	−1.5	−1.5	−0.3	2.47	
6256	Point Samson	A	20	38	117	12	−0010	−0002	−1.6	−1.3	−1.0	−0.4	2.87	
6257	Depuch Island		20	37	117	45	+0007	+0009	−1.3	−1.1	−1.1	−0.8	2.9	
6259	PORT HEDLAND		20	18	118	35	Standard port		See table v				4.12	
6261	Bedout Island		19	36	119	06	+0102	+0105	−0.9	−1.3	−0.8	−1.0	2.94	
6263	Lagrange Bay		18	42	121	44	−0027	−0023	+1.3	+0.5	−0.2	−0.8	4.1	
6265	Broome	A	18	00	122	13	−0022	−0012	+1.7	+0.7	+0.3	−0.7	4.47	
6268	Red Bluff			04	122	19	+0001	+0010	−0.1	−0.6	−0.5	−0.9	3.47	
6271	Pender Bay		16	42	122	43	+0019	+0021	+0.4	−0.4	−0.9	−1.2	3.4	x
6273	Scott Reef		14	03	121	48	P	P	−3.1	−2.5	−1.8	−0.1	2.1	ax
6274	West Island		12	13	123	01	−0116	−0114	−2.1	−1.7	−1.4	−0.8	2.4	x
	King Sound													
6275	Karrakatta Bay		16	22	123	02	+0101	+0054	+0.3	0.0	−0.8	−1.0	3.63	
6275a	Sunday Island		16	23	123	11	+0150	+0123	−0.3	−0.1	−1.1	−0.8	3.39	
6278	Derby	A	17	17	123	39	*	*	+3.2	+2.3	−0.6	−0.8	4.40	*
6279	Escape point		17	25	123	33	+0505	⊙	⊙	⊙	⊙	⊙	⊙	
	Buccaneer Archipelago													
6284	Bedford Island		16	09	123	19	+0040	+0045	+1.6	+0.7	0.0	−0.8	4.32	
6286	Yampi Sound	A	16	05	123	35	+0028	+0033	+3.0	+1.9	+1.3	+0.5	5.63	
	(Cockatoo Island)													
6288	Macleay Island		15	57	123	41	+0024	+0025	+2.5	+1.1	+0.1	−0.9	4.63	
6289	Adele Island		15	31	123	09	+0013	+0011	+0.3	−0.1	−0.4	−0.4	3.78	
6290	Browse Island		14	06	123	33	−0134	−0132	−2.2	−1.7	−1.6	−0.8	2.41	
	Collier Bay													
6291	Shale Island		16	23	124	20	+0041	+0052	+5.0	+3.3	+1.8	+0.3	6.52	
6292	Hall Point		15	40	124	24	+0018	+0023	+2.0	+0.7	0.0	−1.1	4.36	
6293	Degerando Island		15	21	124	11	0000	+0003	+1.6	+0.5	0.0	−0.8	4.24	
6300	Baudin Island		14	08	125	36	−0026	−0019	−0.5	−0.4	−0.5	−0.3	3.5	x
	Admiralty Gulf													
6301	Port Warrender	A	14	32	125	49	−0012	−0012	+0.4	−0.3	+0.2	−0.2	3.96	
6301a	Troughton Island		13	45	126	08	−0023	−0023	−2.4	−2.1	−0.8	−0.3	2.54	x
6302	Vansittart Bay		14	04	126	16	−0030	−0030	−2.2	−2.0	−1.2	−0.8	2.4	ax
6323	**DARWIN**		See page 126											
	Napier Broome Bay									**6.9**	**5.1**	**3.2**	**1.4**		
6303	Geranium Harbour		13	56	126	35	p	P	−4.4	−3.5	−2.1	−1.2	1.37	
6303a	West Bay		14	05	126	29	p	p	−3.5	−2.9	−1.4	−0.8	2.0	ax
6304	Lesueur Island		13	50	127	17	P	p	−4.8	−3.6	−1.9	−0.7	1.40	
6305	Cape Domett	A	14	49	128	23	−0021	−0025	−0.2	−0.1	−0.1	0.0	4.04	
	Cambridge Gulf													
6306	Lacrosse Island		14	45	128	20	−0022	−0028	−0.4	−0.1	−0.4	−0.1	3.89	
6307	Webster Bluff		15	05	128	08	+0025	⊙	⊙	⊙	⊙	⊙	⊙	
6309	Wyndham	A	15	27	128	06	+0104	+0104	+0.5	+0.5	−0.5	−0.6	4.36	

⊙ No data.

Tide is usually diurnal.

* See notes on page 386.
A Tides predicted in Australian National Tide Tables.
a Data approximate.
p For predictions use harmonic constants (Part III) and N.P. 159.
x M.L. inferred.

AUSTRALIA

No.	PLACE	Lat. S.		Long. E.		TIME DIFFERNCES		HEIGHT DIFFERENCES (IN METRES)				M.L.	
						MHW	MLW	MHWS	MHWN	MLWN	MLWS	Z_0 m.	
						(Zone −0930)							
6323	**DARWIN**	See page 126						6.9	5.1	3.2	1.4		
	Victoria River												
6311	Turtle Point	14	50	129	14	+0140	+0140	−0.8	−0.6	−0.7	−0.5	3.5	
6312	Holdfast Reach	15	14	129	49	+0320	⊙	⊙	⊙	⊙	⊙	⊙	x
6313	Mosquito Flat	15	25	130	08	+0005	⊙	⊙	⊙	⊙	⊙	⊙	
6314	Sandy Island	15	28	130	22	+0700	⊙	⊙	⊙	⊙	⊙	⊙	
6316	Pearce Point	14	26	129	22	+0120	+0120	−0.4	−0.3	−0.7	−0.6	3.7	x
6317	Port Keats	14	03	129	34	+0020	+0020	−0.4	−0.3	−0.7	−0.6	3.7	x
6319	Daly River	13	22	130	19	−0040	−0040	−0.4	−0.5	−0.5	−0.5	3.7	x
	Bynae Harbour												
6320	Tapa Bay	12	27	130	36	−0004	−0004	−0.7	−0.7	−0.3	−0.3	3.66	
6321	East Point	12	35	130	34	+0004	−0.4	−0.5	−0.4	−0.2	−0.2	3.83	
	Port Darwin												
6322	Night Cliff	12	23	130	50	+0006	+0006	−0.4	−0.4	−0.2	−0.2	3.85	
6323	DARWIN	12	28	130	51	STANDARD PORT		See table v				4.10	
6325	Cape Hotham	12	03	131	17	+0114	+0114	−2.5	−1.7	−1.2	−0.4	2.69	
6328	Two Hills Bay	11	31	132	04	+0145	+0145	−3.0	−2.1	−1.5	−0.6	2.33	
	Melville Island												
6330	Cape Keith	11	36	131	28	+0133	+0133	−3.2	−2.2	−1.6	−0.6	2.26	
6330a	Camp Point	11	37	131	26	+0110	+0110	−3.1	−2.1	−1.6	−0.7	2.29	
6333	St. Asaph Bay	11	18	130	26	+0005	+0005	−2.9	−2.0	−1.7	−0.8	2.3	x
6334	Snake Bay	11	23	130	41	+0345	+0345	−5.1	−3.7	−2.4	−1.0	⊙	a
5435	**BANDA**	See page 57				HHW	LLW	MHHW	MLHW	MHLW	MLLW		
								2.0	1.8	1.0	0.4		
6336	Cape Don Boat Harbour	11	19	131	46	+0427	+0415	0.0	+0.2	+0.1	+0.1	1.41	t
6337	Port Essington	11	22	132	11	+0310	+0310	+0.5	+0.5	+0.2	+0.4	1.7	tx
6339	Cape Croker	11	00	132	34	+0307	+0305	+0.1	+0.1	+0.1	+0.2	1.43	
6323	**DARWIN**	See page 126				MHW	MLW	MHWS	MHWN	MLWN	MLWS		
								6.9	5.1	3.2	1.4		
6341	North Goulburn Island	11	33	133	26	+0045	+0043	−4.5	−3.2	−2.3	−1.0	1.4	tx
	Liverpool River												
6343	Entrance Island	11	58	134	15	+0126	+0126	−3.7	−2.5	−2.1	−1.0	1.8	tx
	Millingimbi Inlet												
6344	Yabooma Island	11	58	134	54	+0232	+0232	−3.0	−2.1	−2.1	−1.1	2.07	at
6345	Guluwuru Island	11	30	136	20	+0255	+0255	−3.8	−2.7	−2.0	−0.9	1.86	x
	Marchinbar Island												
6346	Jensen Bay	11	11	136	45	+0210	+0205	−3.5	−2.5	−1.8	−0.8	2.00	t
6347	Arnhem Bay	11	11	136	06	+0415	+0410	−2.2	−1.1	−1.9	−0.9	2.63	t
5435	**BANDA**	See page 57				HHW	LLW	MHHW	MLHW	MHLW	MLLW		
								2.0	1.8	1.0	0.4		
	GULF OF CARPENTARIA												
6348	Gove A	12	13	136	48	p	p	+1.2	+0.4	+0.6	+0.2	1.90	
6349	Caledon Bay	12	57	136	40	p	p	+0.1	−0.1	−0.1	0.0	1.2	x
6351	Port Langdon	13	52	136	50	p	p	−1.0	−0.6	−0.6	−0.2	0.58	
6351a	Milner Bay	13	50	136	30	p	p	−1.0	Δ	Δ	−0.2	0.61	
	Sir Edward Pellew Group												
6352	Purl McArthur	15	45	136	49	p	p	+1.5	+0.8	+1.3	+1.6	2.53	
6353	Turtle Island	15	35	137	06	p	p	+0.6	+0.5	+1.0	+1.0	2.07	
6996	**CUA CAM**	See page 165				(Zone −1000)		2.9	Δ	Δ	0.9		
6355	Bayley Island	16	54	139	04	p	p	+0.1	Δ	Δ	−0.2	2.0	x
6356	Sweers Isalnd	17	07	139	38	p	p	+0.1	Δ	Δ	+0.1	2.0	x
6357	Kangaroo Point	17	35	139	45	−0940	−0940	−0.3	Δ	Δ	+0.6	2.0	dx
6358	Karumba (Kimberley) A	17	30	140	50	−0945	−0945	+0.1	Δ	Δ	+0.2	1.81	t
6359	Van Diemen Inlet	16	58	140	58	−1045	−1045	−0.6	Δ	Δ	+0.3	1.7	dx

SEASONAL CHANGES IN MEAN LEVEL

No.	Jan 1	Feb 1	Mar 1	Apr 1	May 1	June 1	July 1	Aug 1	Sep 1	Oct 1	Nov 1	Dec 1	Jan 1
5435	0.0	+0.1	+0.1	0.0	0.0	0.0	−0.1	−0.1	−0.1	0.0	0.0	0.0	0.0
6243–6302	0.0	0.0	+0.1	+0.1	+0.1	+0.1	0.0	−0.1	−0.2	−0.1	−0.1	0.0	0.0
6303–6339	0.0	+0.1	+0.2	+0.2	0.0	−0.1	−0.2	−0.2	−0.1	0.0	0.0	0.0	0.0
6341–6353	+0.1	+0.2	+0.2	+0.1	0.0	−0.1	−0.2	−0.2	−0.1	0.0	0.0	+0.1	+0.1
6355–6359	+0.2	+0.3	+0.3	+0.2	0.0	−0.2	−0.3	−0.3	−0.3	−0.1	0.0	+0.1	+0.2
6996	0.0	−0.1	−0.1	−0.1	0.0	0.0	0.0	0.0	0.0	+0.1	+0.1	+0.1	0.0

No.	PLACE		Lat. S.		Long. E.		TIME DIFFERNCES		HEIGHT DIFFERENCES (IN METRES)				M.L.	
							MHW	MLW	MHWS	MHWN	MLWN	MLWS	Z_0 m.	
							(Zone −1200)							
6400	**AUCKLAND**		see page 132						3.1	2.8	0.8	0.4		
6408	Port Jackson		36	29	175	20	−0030	−0012	−0.5	−0.5	−0.1	0.0	1.49	a
6411	Mercury Bay(Whitianga)		36	50	175	43	−0012	−0002	−1.0	−0.9	−0.1	+0.1	1.27	
6412	Slipper Island		37	04	175	57	−0048	−0014	−1.1	−1.0	−0.3	0.0	1.17	a
6415	Tauranga		37	39	176	11	−0017	0000	−1.3	−1.2	−0.4	−0.1	1.13	
6417	Whale Island		37	54	176	58	−0045	−0045	−0.8	−0.8	−0.1	+0.1	1.37	
6418	Ohiwa		37	59	177	07	+0013	−0004	−1.2	−1.0	−0.3	0.0	1.16	
6419	Motunui Island		37	47	177	39	−0040	−0040	−1.0	−0.9	−0.1	+0.1	1.29	
6421	Hicks Bay		37	35	178	19	−0055	−0035	−0.9	−0.8	0.0	+0.1	1.33	
6422	East Cape		37	41	178	33	−0055	−0045	−1.0	−0.9	−0.1	+0.1	1.3	x
6423	Waipiro Bay		38	02	178	20	−0102	−0102	−1.1	−1.0	−0.2	+0.1	1.23	
6424	Tolaga Bay		38	22	178	19	−0114	−0114	−1.2	−1.0	−0.1	+0.1	1.2	
6490	**LYTTELTON**		see page 138						2.2	2.1	0.4	0.3		
6425	Gisborne		38	41	178	02	+0125	+0128	−0.4	−0.4	+0.1	+0.1	1.15	
6428	Portland Island		39	17	177	52	+0115	+0115	⊙	⊙	⊙	⊙	⊙	
6429	Waikokopu *Wairoa River*		39	04	177	50	+0125	+0125	−0.6	−0.6	0.0	−0.1	0.9	x
6430	Clyde		39	03	177	26	+0120	+0120	−0.6	−0.6	−0.1	−0.1	09	x
6439	**WELLINGTON** *Napier*		see page 135						1.4	1.3	0.5	0.4		
6432	No 3 Wharf	Z	39	29	176	55	+0056	+0056	+0.2	+0.2	−0.2	−0.2	0.90	
6436	Castle Point		40	55	176	13	+0015	+0015	−0.2	−0.2	−0.4	−0.4	0.61	
6438	Cape Palliser		41	37	175	15	+0010	+0010	+0.1	+0.1	0.0	−0.1	0.9	x
6439	**WELLINGTON**		41	17	174	47	STANDARD PORT		See Table V				0.89	
6526	WESTPORT		see page 144						3.2	2.6	0.9	0.3		
6441	Oteranga Bay		41	18	174	37	p	p	−2.3	−1.9	−0.4	0.0	0.62	
6442	Makara Bay		41	13	174	42	p	p	−1.9	−1.7	−0.2	0.0	0.79	
6443	Porirua Harbour		41	04	174	51	p	p	−1.8	−1.6	−0.3	−0.1	0.78	
6445	Manawatu River Entrance		40	28	175	13	−0125	−0125	−0.8	−0.8	0.0	−0.1	1.3	x
6447	Wanganui River Entrance		39	57	174	59	−0030	−0030	−0.5	−0.6	+0.1	0.0	1.5	x
6449	Patea		39	47	174	29	−0115	−0115	−0.7	−0.7	−0.1	−0.1	1.3	x
6451	Opunake Bay		39	28	173	51	−0110	−0110	+0.1	+0.1	+0.1	0.0	1.8	x
	New Zealand, South Island *Golden Bay*													
6453	Collingwood		40	40	172	40	−0120	−0130	+1.2	+0.9	+0.2	−0.1	2.3	x
6454	Motupipi River *Tasman Bay*		40	50	172	51	−0120	−0130	+1.5	+1.2	+0.3	0.0	2.5	x
6455	Astrolabe Road		40	58	173	03	−0115	−0125	+1.6	+1.2	+0.3	−0.1	2.5	x
6458	Nelson	Z	41	16	174	16	−0053	−0107	+0.5	+0.3	+0.2	0.0	1.99	
6460	Croisilles Harbour		41	05	173	42	−0110	−0120	+0.9	+0.7	+0.1	−0.1	2.2	x
6462	Greville Harbour		40	52	173	49	−0115	−0125	0.0	−0.2	0.0	−0.3	1.63	t
6464	Stephen's Island		40	40	174	01	−0125	−0135	−0.7	−0.6	−0.1	0.0	1.4	x
6466	Elmslie Bay		40		173	51	−0145	−0205	−0.8	−0.8	−0.2	−0.2	1.26	d
6467	Pelorus Sound Entrance		40		173	59	−0155	−0225	−0.7	−0.7	−0.1	−0.1	1.3	x
6471	Havelock *Queen Charlotte Sound*		41		173	46	−0050	−0100	−0.6	−0.6	0.0	−0.1	1.42	
6474	Long Island		41	07	174	17	−0200	−0248	−1.7	−1.6	−0.5	−0.2	0.73	t
6476	East Bay		41	14	174	08	−0210	−0305	−1.8	−1.6	−0.5	−0.3	0.70	t
6477	Picton	Z	41	17	174	00	−0157	−0245	−1.7	−1.6	−0.4	−0.3	0.73	t
6478	Okiwa Bay		41	17	173	55	−0200	−0250	−1.7	−1.6	−0.4	−0.3	0.73	t
6478a	Whekenui *Tory Channel*		41	12	174	18	p	p	−2.0	−1.7	−0.4	−0.1	0.7	x
6479	Te Iro Bay		41	14	174	11	−0215	−0255	−1.9	−1.7	−0.5	−0.3	0.66	t

SEASONAL CHANGES IN MEAN LEVEL

No	Jan 1	Feb 1	Mar 1	Apr 1	May 1	June 1	July 1	Aug 1	Sep 1	Oct 1	Nov 1	Dec 1	Jan 1
6362, 6363	+0.2	+0.3	+0.3	+0.2	0.0	−0.2	−0.3	−0.3	−0.3	−0.1	0.0	+0.1	+0.2
6366–6373	Negligible												
6374, 6375	0.0	0.0	0.0	+0.1	+0.1	+0.1	0.0	0.0	−0.1	−0.1	−0.1	0.0	0.0
6377–6526	Negligible												

No.	PLACE	Lat. S.		Long. E.		TIME DIFFERNCES		HEIGHT DIFFERENCES (IN METRES)				M.L.
						MHW	MLW	MHWS	MHWN	MLWN	MLWS	Z_0 m.
						(Zone −1200)						
6490	**LYTTELTON**	See page 138						2.2	2.1	0.4	0.3	
6481	Lucky Bay	41	16	174	17	p	p	−0.9	−0.9	+0.1	+0.1	0.89
6484	Lake Grassmere Entrance	41	42	174	11	p	p	−0.5	−0.6	0.0	0.0	1.0	x
6485	Cape Campbell	41	44	174	15	p	p	−0.5	−0.5	+0.1	+0.1	1.05
6487	Kaikoura Peninsula	42	25	173	42	p	p	−0.5	−0.5	+0.1	+0.1	1.07
6490	LYTTELTON	43	36	172	43	STANDARD PORT		See Table V				1.23
6491	Akaroa	43	48	172	55	−0045	−0040	+0.2	+0.1	+0.3	+0.2	1.49
6492	Timaru	44	24	171	15	−0138	−0141	0.0	−0.1	+0.3	+0.2	1.37
6504	**BLUFF**	See page 141						2.6	2.3	0.9	0.6	
6494	Oararu	45	07	170	59	+0105	+0127	−0.5	−0.5	−0.2	−0.1	1.28
	Otago Harbour											
6496	Entrance	45	47	170	44	+0054	+0134	−0.8	−0.7	−0.6	−0.5	0.98
6497	Port Chalmers	45	49	170	39	+0152	+0201	−0.8	−0.7	−0.6	−0.5	0.98
6498	Dunedin Z	45	53	170	30	+0227	+0313	−0.8	−0.7	−0.6	−0.4	0.98
6500	Nugget Point	46	26	169	48	+0100	+0100	−0.3	−0.2	0.0	0.0	1.49
6502	Waipapa Point	46	39	168	51	+0017	+0016	+0.1	+0.2	0.0	0.0	1.68
6504	BLUFF	46	36	168	21	STANDARD PORT		See Table V				1.61
6505	New River Entrance	46	32	168	15	−0022	−022	+0.2	+0.2	−0.2	−0.2	1.6	x
6506	Colac Bay	46	22	167	54	−0144	−0139	−0.2	−0.3	−0.2	−0.3	1.35
	Stewart Island											
6507	Paterson Inlet	46	54	168	07	−0008	−0010	0.0	0.0	0.0	0.0	1.58
6526	**WESTPORT**	see page 144						3.2	2.6	0.9	0.3	
6511	Preservation Inlet	46	04	166	41	+0130	+0130	−0.9	−0.8	−0.4	−0.1	1.2	x
6512	Dusky Sound	45	47	166	32	+0120	+0120	−1.1	−0.8	−0.4	0.0	1.2	x
6513	Deep Cove	45	27	167	10	+0112	+0112	−1.2	−0.8	−0.3	+0.1	1.19
6515	Bligh Sound	44	53	167	32	+0100	+0100	−1.1	−0.8	−0.3	0.0	1.2	x
6516	Milford Sound	44	40	167	55	+0030	+0055	−1.1	−0.8	−0.3	0.0	1.2	x
6518	Jackson Bay	43	59	168	37	+0040	+0040	−1.1	−0.8	−0.3	0.0	1.2	x
6519	Haast River Entrance	43	50	169	03	+0030	+0030	−1.0	−0.8	−0.3	−0.1	1.2	x
6520	Bruce Bay	43	35	169	35	+0020	+0020	−1.0	−0.8	−0.3	−0.1	1.2	x
6521	Okarito	43	13	170	11	+0015	+0015	−0.9	−0.8	−0.3	−0.1	1.2	x
6523	Hokitika River Bar	42	43	170	58	+0010	+0010	−0.9	−0.8	−0.3	−0.1	1.2	x
6524	Greymouth	42	26	171	13	+0005	+0005	0.0	+0.1	+0.4	+0.5	2.03
6526	WESTPORT	41	45	171	36	STANDARD PORT		See Table V				1.77
6527	Karamea River Entrance	41	15	172	06	−0035	−0035	⊙	⊙	⊙	⊙	⊙
6528	Whanganui Inlet	40	35	172	32	−0105	−0105	−0.3	−0.3	−0.1	−0.3	1.4	x
						(Zone −1130)						
6530	Macquarie Island	54	31	158	58	P	p	−2.0	−1.6	−0.5	−0.1	0.7
6504	**BLUFF**	see page 141						2.6	2.3	0.9	0.6	
						(Zone −1200)						
6532	Campbell Island	52	33	169	13	+0013	+0020	−1.4	−1.3	−0.5	−0.4	0.7	x
	Auckland Islands											
6534	Carnley Harbour	50	52	166	05	−0036	−0043	−1.5	−1.5	−0.6	−0.5	0.6	x
6535	Antipodes Islands	49	40	178	50	+0210	⊙	−0.8	⊙	⊙	⊙	1.1	dx
6400	**AUCKLAND**	see page 132						3.1	2.8	0.8	0.4	
		S.		W.		(Zone −1245)						
6536	Chatham Islands	43	57	176	34	−0150	−0140	−2.2	−2.0	−0.4	−0.1	0.58	tx
						(Zone −1200)						
6537	Raoul Island	29	15	177	55	−0033	−0025	−1.7	−1.6	−0.4	−0.2	0.58	x

⊙ No data
C Tides predicted in Chilean Tide Tables
F Tides predicted in French Tide Tables
Z Tides predicted in New Zealand Tide Tables
d Differences approximate
p For predictions use harmonic constants (Part III) and N.P. [159]
t Time differences approximate
x M.L. inferred

No.	PLACE		Lat. N.		Long. E.		TIME DIFFERNCES		HEIGHT DIFFERENCES (IN METRES)				M.L.
							MHW	MLW	MHWS	MHWN	MLWN	MLWS	Z_0 m.
							(Zone −1200)						
5062	**DAVAO**		see page 24						1.6	1.0	0.5	−0.2	
6771	Ailinglapalap Atoll		7	17	168	45	−0117	−0117	+0.1	+0.1	+0.2	+0.4	0.92
6772	Maloelap Atoll		8	43	171	14	−0129	−0129	+0.1	+0.2	+0.2	+0.3	0.90
6775	Wotje Atoll		9	28	170	14	−0135	−0135	0.0	+0.1	+0.1	+0.4	0.85
6776	Kwajalein Atoll	U	8	44	167	44	−0135	−0130	+0.1	+0.1	+0.2	+0.3	0.91
6776a	Nimuru To		9	27	167	29	−0131	−0132	+0.1	+0.2	+0.2	+0.4	0.92
6777	Likiep Atoll		9	49	169	17	−0125	−0125	0.0	+0.2	+0.2	+0.4	0.92
6782	Rongerik Atoll		11	29	167	31	−0138	−0138	0.0	+0.2	+0.2	+0.4	0.90
6783	Rongelap Atoll		11	09	166	54	−0131	−0131	0.0	+0.1	−0.2	+0.4	0.86
6786	Bikini Atoll		11	36	165	33	−0142	−0142	0.0	+0.1	+0.3	+0.4	0.92
6787	Eniwetok Atoll		11	26	162	23	−0128	−0129	−0.3	−0.1	+0.1	+0.4	0.79
6787a	Runit Island		11	33	162	21	−0133	−0133	−0.2	0.0	+0.1	+0.4	0.79
6788	Ujelang Atoll		9	46	160	58	−0121	−0121	−0.2	−0.1	+0.1	+0.4	0.80
6790	Wake Island		19	17	166	37	−0200	−0200	−0.7	−0.4	−0.2	+0.2	0.43
	Caroline												
	Islands												
6792	Kusaie Island		5	20	163	01	−0125	−0125	0.0	+0.1	+0.3	+0.4	0.92
							(Zone −1100)						
6795	Ponape Island	U	6	59	158	13	−0243	−0243	−0.4	−0.2	+0.1	+0.4	0.70	t
6795a	Metaranimo Ko		6	52	158	23	−0239	−0240	−0.2	−0.1	+0.2	+0.5	0.82	
9644	**VALPARAISO**		see page 231				HHW	LLW	MHHW	MLHW	MHLW	MLLW	
									1.5	1.2	0.5	0.4	
6797	Oroluk Island		7	40	155	10	−0737	−0707	−0.6	−0.6	−0.1	0.0	0.59	t
6798	Nomoi Islands		5	20	153	44	p	p	−0.5	−0.6	−0.1	0.0	0.62	
6798a	Moro Tu		5	29	153	33	p	p	−0.5	0.6	−0.1	0.0	0.61	
6800	Hall Islands		8	36	152	15	p	p	−0.7	−0.7	−0.1	0.0	0.51	
							(Zone −1000)						
6800a	Nomuuin To		8	27	151	47	p	p	−0.8	−0.8	−0.1	−0.1	0.47	
5599	**DREGER**		see page 63						1.5	Δ	Δ	0.9	
	HARBOUR												
6801	Truk Islands	JU	7	22	151	33	+0035	−0052	−0.8	Δ	Δ	−0.6	0.46	
6802	Namonuito		8	35	149	39	−0122	+0139	−0.9	Δ	Δ	−0.7	0.37	
	Islands												
6802a	Onari To		8	45	150	20	−0057	+0123	−0.9	Δ	Δ	−0.7	0.39	
6803	Pulap Island		7	39	149	25	−0033	+0110	−0.9	Δ	Δ	−0.6	0.41	
6804	Puluwat Island		7	22	149	13	p	P	−0.9	Δ	Δ	−0.7	0.4	x
7940	**YOKOHAMA**		see page 198				MHW	LLW	1.7	1.6	0.9	0.3	
6807	Lamotrek		7	28	146	23	p	p	−1.0	−0.9	−0.5	−0.2	0.52	
6811	Woleai Island		7	22	143	54	p	p	−1.0	−0.9	−0.4	−0.2	0.51	
6814	Ulithi Islands		9	55	139	40	+0313	+0307	−0.5	−0.4	−0.3	−0.1	0.80	
6814a	Yasoro Island		10	02	139	46	+0255	+0256	−0.5	−0.5	−0.2	0.0	0.82	
6815	Yap Island		9	30	138	08	+0315	+0310	−0.3	−0.6	−0.1	+0.1	1.0	x
6816	Ngulu Islet		8	18	137	29	+0324	+0312	−0.4	−0.3	−0.3	0.0	0.86	
7716	**NAHA KO**		see page 192				MHW	MLW	MHWS	MHWN	MLWN	MLWS	
									2.0	1.5	0.8	0.3	
	Palau Islands						(Zone −0900)						
6818	Garukoru		7	45	134	38	−0022	−0022	−0.2	−0.2	−0.1	0.0	1.05	
	(Ngaregur)												
6819	Toagel Mlungui		7	30	134	31	−0015	−0015	−0.2	−0.2	−0.1	+0.1	1.08	
6820	Malakal Harbour	JU	7	20	134	28	+0012	+0012	−0.1	−0.1	0.0	0.0	1.11	
6821	Ngesebus		7	03	134	16	+0012	+0011	−0.4	−0.2	−0.1	+0.1	1.00	
6825	Helen Reef		2	59	131	49	−0003	−0003	−0.2	−0.2	0.0	+0.1	1.07	

SEASONAL CHANGES IN MEAN LEVEL

No	Jan 1	Feb 1	Mar 1	Apr 1	May 1	June 1	July 1	Aug 1	Sep 1	Oct 1	Nov 1	Dec 1	Jan 1
5062	−0.1	−0.1	−0.1	0.0	0.0	0.0	0.0	0.0	+0.1	+0.1	0.0	0.0	−0.1
6705−6764	Negligible												
6766−6788	0.0	0.0	0.0	0.0	−0.1	−0.1	−0.1	0.0	0.0	+0.1	+0.1	+0.1	0.0
6790−6798a	Negligible												
6800−6825	0.0	−0.1	−0.1	0.0	0.0	0.0	0.0	0.0	0.0	+0.1	+0.1	0.0	0.0
7716	−0.1	−0.1	−0.1	−0.1	0.0	0.0	+0.1	+0.1	+0.1	+0.1	0.0	−0.1	−0.1
7940	−0.1	−0.1	−0.1	−0.1	0.0	0.0	+0.1	+0.1	+0.1	+0.1	0.0	0.0	−0.1
9644, 9700	Negligible												

No.	PLACE	Lat. S.		Long. E.		TIME DIFFERNCES MHW	MLW	HEIGHT DIFFERENCES (IN METRES) MHWS	MHWN	MLWN	MLWS	M.L. Z_0 m.
						(Zone −1200)						
6705	**SUVA**	see page 153						1.6	1.4	0.5	0.3	
6724b	Navatu Island	16	55	179	02	−0010	−0010	−0.3	−0.2	−0.1	0.0	0.82
6725	Muanithula	16	53	178	55	+0011	+0022	0.0	0.0	−0.1	0.0	0.91
		S.		W.								
6725a	Thikombia	15	44	179	55	−0055	−0055	+0.2	−0.1	+0.1	−0.2	0.95
	Taveuni											
6726	Veitalathangi Point	16	47	179	50	−0021	−0029	−0.2	−0.2	−0.1	−0.1	0.79
6726a	Waiyevo	16	47	179	59	−0037	−0037	−0.1	0.0	0.0	0.0	0.94
6728	Wailagilala Island	16	45	179	06	−0020	−0005	−0.1	−0.1	−0.1	−0.1	0.83
6729	Vanubalav Island	17	17	178	59	−0037	−0036	−0.1	0.0	0.0	+0.1	0.96
	Lakemba Island											
6730	Tumbou	18	14	178	46	−0002	−0001	0.0	0.0	0.0	0.0	0.94
6730a	Wainiyabia	18	12	178	50	−0004	−0003	−0.1	0.0	0.0	0.0	0.94
	Totoya											
6732	Herald Sound	18	59	179	52	+0050	+0050	−0.3	−0.2	−0.3	−0.3	0.7
		S.		E.								
	Moala											
6733	Naroi	18	33	179	57	−0002	−0002	0.0	0.0	+0.1	+0.1	1.00
6734	Matuku Harbour	19	10	179	45	+0030	+0015	0.0	−0.1	0.0	0.0	0.9
	Kandavu											
6735	Ngaloa Harbour	19	05	178	11	0000	−0015	0.0	0.0	−0.3	−0.3	0.8
6735a	Namalata Bay	19	03	178	09	−0004	−0004	+0.1	+0.2	+0.2	+0.3	1.16
9700	**PUERTO MONTT**	see page 234						6.5	4.7	2.4	0.7	
		S.		W.								
6739	Iles Wallis (Ile Urea) **F**	13	22	176	11	−0815	−0815	−4.8	−3.3	−1.8	−0.4	0.98
		S.		E.								
6740	Rotuma Island	12	29	177	07	−0825	−0825	−4.8	−3.3	−1.6	−0.3	1.07
	Tuvalu											
6744	Funafuti	8	31	179	12	−0853	−0852	−4.7	−3.4	−1.7	−0.5	1.0
	Kiribati											
6750	Arorae	2	39	176	50	−0905	−0905	−4.7	−3.4	−1.7	−0.6	1.0
6752	North Beru	1	17	176	00	−0935	−0935	−5.2	−3.8	−2.0	−0.7	0.62
6754	Tabiteuea	1	28	175	13	−0935	−0935	−4.5	−3.3	−1.7	−0.6	1.05
6755	Nonouti	0	40	177	27	−0935	−0935	−4.5	−3.2	−1.8	−0.5	1.06
		N.		E.								
6756	Abemama	0	29	173	52	−0900	−0900	−4.5	−3.3	−1.6	−0.5	1.07
6759	Tarawa	1	22	172	56	−0935	−0938	−4.6	−3.4	−1.7	−0.6	1.00
6760	Abaiang Atoll	1	49	173	02	−0915	−0925	−4.4	−3.1	−1.4	−0.3	1.25
6761	Butaritari (Makin)	3	02	172	48	−0935	−0938	−4.6	−3.4	−1.7	−0.6	1.00
		S.		E.								
6763	Ocean Island	0	52	169	35	−0917	−0916	−4.7	−3.4	−1.6	−0.4	1.04
6764	Nauru	0	32	166	54	−0938	−0941	−4.7	−3.5	−1.6	−0.5	1.0
5062	**DAVAO**	see page 24						1.6	1.0	0.5	−0.2	
	Marshall Islands	N.		E.								
	Mili Atoll											
6766	Port Rhin	6	14	171	48	−0121	−0121	+0.3	+0.3	+0.2	+0.3	1.01
	Arno Atoll											
6767	Dodo Passage	7	08	171	42	−0129	−0129	+0.2	+0.3	+0.1	+0.4	0.97
	Majuro Atoll											
6768	Djarrit	7	08	171	21	−0129	−0129	+0.2	+0.2	+0.2	+0.4	0.98
	Faluit Atoll											
6769	South−east Pass	5	55	169	39	−0133	−0133	0.0	+0.1	+0.2	+0.4	0.88

⊙ No data.

Δ Tide is usually diurnal

* See notes on page 386

F Tides predicted in French Tide Tables

J Tides predicted in Japanese Tide Tables

U Tides predicted in U.S. Tide Tables

a Data approximate

p For predictions use harmonic constants (Part III) and N.P. 159

t Time differences approximate

x M.L.inferred.

No.	PLACE	Lat. N.		Long. W.		TIME DIFFERNCES MHW	MLW	HEIGHT DIFFERENCES (IN METRES) MHWS	MHWN	MLWN	MLWS	M.L. Z_0 m.	
						(Zone +1000)							
6684	APIA	see page 150						1.0	0.8	0.2	0.0		
6616	Jarvis Island	0	23	160	02	+1054	+1037	−0.3	−0.3	0.0	0.0	0.4	*x*
6617	Christmas Island	1	59	157	29	+1037	+1037	−0.3	−0.2	0.0	+0.1	0.4	*x*
6618	Fanning Island	3	51	159	22	+1023	+1010	−0.4	−0.3	−0.1	−0.1	0.3	*dx*
						(Zone +1100)							
6620	Palmyra Island	5	52	162	06	+1054	+1041	−0.1	−0.1	+0.1	+0.1	0.5	*tx*
						(Zone +1000)							
6621	Johnston Island	16	45	169	30	+1105	+1105	−0.3	−0.3	−0.1	−0.1	0.30	*t*
6636	**HONOLULU**	see page 147				HHW	LLW	MHHW 0.6	MLHW 0.3	MHLW 0.1	MLLW 0.0		
	Hawaiian Islands												
	Hawau												
6622	Hilo Bay	19	44	155	04	−0050	−0050	+0.1	+0.1	+0.1	0.0	0.34	
6623	Honuapo	19	05	155	33	−0030	−0020	+0.1	+0.1	+0.1	0.0	0.4	*x*
6624	Napoopoo (Napupu)	19	28	155	55	−0020	−0015	0.0	+0.1	0.0	0.0	0.3	*x*
6625	Mahukona	20	11	155	54	−0030	−0020	0.0	+0.1	0.0	0.0	0.3	*x*
6625a	Kawaihae	20	02	155	50	−0015	−0015	0.0	0.0	0.0	0.0	0.28	
	Mana												
6627	Kihei	20	47	156	28	0000	−0025	+0.1	+0.1	+0.1	0.0	0.3	*x*
6628	Lahaina	20	52	156	41	−0035	−0041	+0.1	+0.1	+0.1	0.0	0.3	*x*
6629	Kabului	20	54	156	28	−0121	−0139	+0.1	+0.1	+0.2	0.0	0.34	
	Lanai												
6631	Kaumalapau	20	47	157	00	0000	0000	0.0	+0.1	+0.1	0.0	0.3	*x*
	Molakai												
6632	Pukoo Harbour	21	04	156	47	−0105	−0050	0.0	+0.1	+0.1	0.0	0.3	*x*
6633	Kamola Harbour	21	03	156	53	−0040	−0020	0.0	+0.1	0.0	0.0	0.3	*x*
6634	Kolo	21	06	157	12	0000	0000	0.0	0.0	0.0	0.0	0.3	*x*
	Oahu												
6636	HONOLULU	21	18	157	52	STANDARD PORT		See Table V				0.24	
6637	Waianae	21	27	158	11	+0015	+0015	−0.1	−0.1	0.0	0.0	0.2	*x*
6638	Waialua Bay	21	36	158	07	−0105	−0210	−0.1	Δ	Δ	0.0	0.2	*x*
6639	Laie Bay	21	39	157	56	−0150	−0150	0.0	0.0	0.0	0.0	0.3	*x*
6640	Moku o Loe	21	26	157	48	−0104	−0124	0.0	0.0	+0.1	0.0	0.3	*x*
6641	Waimanalo Bay	21	20	157	42	−0115	−0115	−0.1	−0.1	0.0	0.0	0.2	*x*
	Kauai												
6642	Nawiliwili	21	58	159	21	−0014	−0026	−0.1	−0.1	0.0	0.0	0.2	*x*
6643	Waimea Bay	21	57	159	40	−0020	−0010	−0.1	0.0	+0.1	0.0	0.2	*x*
6644	Hanalei Bay	21	13	159	30	−0130	−0150	0.0	+0.1	+0.1	0.0	0.2	*x*
	Niihau												
6645	Nonopapa	21	52	160	14	−0015	−0015	−0.1	0.0	+0.1	0.0	0.2	*x*
6649	French Frigate Shoals	23	52	166	17	+0017	+0005	−0.1	0.0	+0.1	+0.1	0.3	*x*
6651	Laysan Island	25	46	171	45	+0105	+0105	−0.3	−0.1	0.0	0.0	0.2	*tx*
						(Zone +1100)							
	Midway Islands												
6654	Welles Harbour	28	12	177	22	*p*	*p*	−0.2	0.0	+0.1	0.0	0.21	
6705	**SUVA**	see page 153				MHW	MLW	MHWS 1.6	MHWN 1.4	MLWN 0.5	MLWS 0.3		
	Tonga Islands	S.		W.									
6657	Niue Island	19	02	169	55	+0200	+0200	−0.5	−0.4	0.0	+0.1	0.74	*t*
						(Zone −1300)							
6660	Nukualofa	21	08	175	13	+0050	+0044	−0.3	−0.2	−0.2	−0.1	0.75	*t*
6664	Nomuka	20	16	174	48	+0037	+0038	−0.1	0.0	0.0	+0.1	0.94	*tx*
6666	Lifuka Island, Pangai	19	48	174	21	+0032	+0034	−0.4	−0.3	−0.1	0.0	0.72	
6671	Neiafu	18	39	173	59	+0019	+0019	−0.3	−0.2	−0.1	0.0	0.81	

⊙   No data
Δ   Tide is usually diurnal
a   Data approximate
d   Differences approximate
p   For predictions use harmonic constants (Part III) and N.P. 159
t   Time differences approximate
x   M.I. inferred

No.	PLACE	Lat. S.		Long. W.		TIME DIFFERENCES		HEIGHT DIFFERENCES (IN METRES)				M.L.	
						MHW	MLW	MHWS	MHWN	MLWN	MLWS	Z_0 m.	
						(Zone +1100)							
6684	APIA	see page 150						1.0	0.8	0.2	0.0		
	Samoa Islands												
	Tau Island												
6675	Faleasau Bay	14	14	169	32	−0025	−0025	+0.4	+0.3	+0.1	0.0	0.7	x
	Tutuila Island												
6677	Pago Pago	14	17	170	41	+0021	+0004	−0.1	0.0	0.0	+0.1	0.5	x
6678	Leone Bay	14	20	170	47	+0050	⊙	0.0	⊙	⊙	⊙	⊙	
	Upolu Island												
6683	Mulifanua	13	49	172	01	0000	0000	0.0	+0.1	0.0	0.0	0.55	d
6684	APIA	13	49	171	46	STANDARD		See Table V				0.49	
						PORT							
6685	Saluafata Harbour	13	52	171	37	+0006	+0006	+0.3	+0.3	+0.3	0.8	0.8	x
6686	Fangaloa Bay	13	53	171	30	+0013	⊙	−0.1	⊙	⊙	⊙	⊙	
	Tokelau Islands												
	Savaii Island												
6689	Asau Harbour	13	30	172	38	−0008	−0008	+0.1	+0.1	0.0	+0.1	0.57	
6690	Swains Island	11	03	171	05	⊙	⊙	−0.1	−0.1	−0.1	−0.1	0.4	dx
6691	Pakaofo	9	23	171	15	−0020	⊙	−0.1	⊙	⊙	⊙	⊙	d
	Phoenix Islands												
6697	Hull Island	4	30	172	10	p	p	+0.3	+0.2	+0.2	+0.1	0.7	x
6701	Canton Island	2	50	171	43	p	p	+0.2	+0.1	+0.2	+0.1	0.64	
6705	**SUVA**	see page 153						1.6	1.4	0.5	0.3		
	Fiji Islands	S.		E.		(Zone −1200)							
	Viti Levu												
6705	SUVA	18	08	178	26	STANDARD		See Table V				0.97	
	HARBOUR					PORT							
6706	Rukua	18	23	178	06	+0014	+0015	+0.1	+0.2	+0.2	+0.3	1.14	a
6707	Lautoka	17	36	177	26	−0010	−0010	+0.3	+0.2	+0.2	+0.1	1.15	
6707a	Vunda Point	17	41	177	23	−0007	−0007	+0.1	+0.1	−0.1	−0.1	0.97	a
	Yasawa Island												
6708	Manunggila Bay	16	42	177	36	−0031	−0031	0.0	−0.1	+0.1	−0.1	0.93	
6708a	Nabukeru	16	51	177	28	−0031	−0031	+0.1	−0.1	0.0	−0.1	0.94	
	Viti Levu												
6708b	Vatia Wharf	17	24	177	46	−0020	−0022	+0.4	+0.3	+0.2	+0.1	1.24	
6709	Manava Cay	17	21	177	49	−0027	−0026	+0.3	+0.2	+0.2	+0.1	1.15	
6710	Ellington Wharf	17	20	178	13	−0017	−0008	+0.1	0.0	0.0	0.0	0.97	
6710a	Nanukuloa	17	27	178	14	−0010	−0015	+0.2	+0.2	+0.3	+0.2	1.2	x
6710b	Naingani	17	35	178	40	−0021	−0020	+0.1	0.0	+0.1	0.0	1.02	a
6711	Tai Levu	17	39	178	35	−0002	−0002	0.0	−0.1	0.0	0.0	0.92	
6711a	Leleuvia Island	17	48	178	43	+0030	+0030	−0.1	−0.2	−0.2	−0.2	0.80	
	Ovalau												
6713	Levuka	17	41	178	50	−0005	+0001	−0.1	0.0	0.0	+0.1	0.96	
	Nairai												
6714	Suthunilevu	17	48	179	23	0000	0000	−0.4	−0.2	−0.3	−0.2	0.7	tx
	Ngau												
6715	Nawaikama	18	01	179	16	+0006	+0007	0.0	0.0	+0.2	+0.2	1.04	at
6716	Koro Island	17	24	179	23	−0013	−0012	0.0	0.0	0.0	0.0.	0.96	a
	Vanua Levu												
6717	Nambouwalu	17	00	178	43	−0006	−0006	0.0	0.0	−0.1	−0.1	0.91	a
6717a	Ndama	16	53	178	38	−0021	−0020	+0.2	+0.2	+0.2	+0.1	1.12	A
6718	Yadua Island	16	48	178	20	−0017	−0025	0.0	−0.1	0.0	−0.1	0.92	
6718a	Koroinasolo Inlet	16	43	178	32	−0044	−0043	+0.1	−0.1	+0.1	−0.1	0.95	
6718b	Ndreketi	16	33	178	51	−0048	−0047	+0.1	0.0	0.0	0.0	0.98	
6719	Navindamu	16	31	179	55	−0048	−0047	0.0	−0.1	0.0	−0.1	0.90	a
6719a	Naduri	16	26	179	09	−0050	−0049	+0.1	−0.1	0.0	−0.2	0.91	a
6720	Malau	16	22	179	22	−0050	−0044	+0.1	0.0	+0.1	−0.1	0.96	t
6721	Visonggo	16	13	179	41	−0044	−0035	+0.2	0.0	+0.1	−0.1	1.00	
6722	Natewa	16	35	179	44	−0039	−0039	0.0	−0.1	0.0	0.0	0.91	t
6723	Vunikura	16	40	179	52	−0035	−0035	+0.1	+0.1	+0.3	+0.3	1.16	a
6724	Na Kama Creek	16	47	179	20	+0012	+0013	0.0	0.0	0.0	0.0	0.98	a
6724a	Nayavu	16	48	179	05	−0010	−0010	0.0	+0.1	0.0	0.0	0.96	a

SEASONAL CHANGES IN MEAN LEVEL

No	Jan 1	Feb 1	Mar 1	Apr 1	May 1	June 1	July 1	Aug 1	Sep 1	Oct 1	Nov 1	Dec 1	Jan 1
6616–6620	0.0	0.0	0.0	−0.1	−0.1	0.0	0.0	0.0	0.0	+0.1	+0.1	0.0	0.0
6621–6724a	Negligible												

No.	PLACE		Lat. N.		Long. E.		TIME DIFFERENCES		HEIGHT DIFFERENCES (IN METRES)				M.L.
							MHW	MLW	MHWS	MHWN	MLWN	MLWS	Z_0 m.
							(Zone −1200)						
5062	**DAVAO**		see page 24						1.6	1.0	0.5	−0.2	
6771	Ailinglapalap Atoll		7	17	168	45	−0117	−0117	+0.1	+0.1	+0.2	+0.4	0.92
6772	Maloelap Atoll		8	43	171	14	−0129	−0129	+0.1	+0.2	+0.2	+0.3	0.90
6775	Wotje Atoll		9	28	170	14	−0135	−0135	0.0	+0.1	+0.1	+0.4	0.85
6776	Kwajalein Atoll	U	8	44	167	44	−0135	−0130	+0.1	+0.1	+0.2	+0.3	0.91
6776a	Nimuru To		9	27	167	29	−0131	−0132	+0.1	+0.2	+0.2	+0.4	0.92
6777	Likiep Atoll		9	49	169	17	−0125	−0125	0.0	+0.2	+0.2	+0.4	0.92
6782	Rongerik Atoll		11	23	167	31	−0138	−0138	0.0	+0.2	+0.2	+0.4	0.90
6783	Rongelap Atoll		11	09	166	54	−0131	−0131	0.0	+0.1	−0.2	+0.4	0.86
6786	Bikini Atoll		11	36	165	33	−0142	−0142	0.0	+0.1	+0.3	+0.4	0.92
6787	Eniwetok Atoll		11	26	162	23	−0128	−0129	−0.3	−0.1	+0.1	+0.4	0.79
6787a	Runit Island		11	33	162	21	−0133	−0133	−0.2	0.0	+0.1	+0.4	0.79
6788	Ujelang Atoll		9	46	160	58	−0121	−0121	−0.2	−0.1	+0.1	+0.4	0.80
6790	Wake Island		9	17	166	37	−0200	−0200	−0.7	−0.4	−0.2	+0.2	0.43
	Caroline Islands												
6792	Kusaie Island		5	20	163	01	−0125	−0125	0.0	+0.1	+0.3	+0.4	0.92
							(Zone −1100)						
6795	Ponape Island	U	6	59	158	13	−0243	−0243	−0.4	−0.2	+0.1	+0.4	0.70 t
6795a	Metaranimo Ko		6	52	158	23	−0239	−0240	−0.2	−0.1	+0.2	+0.5	0.82
9644	**VALPARAISO**		see page 231				HHW	LLW	MHHW	MLHW	MHLW	MLLW	
									1.5	1.2	0.5	0.4	
6797	Oroluk Island		7	40	155	10	−0737	−0707	−0.6	−0.6	−0.1	0.0	0.59
6798	Nomoi Islands		5	20	153	44	*p*	*p*	−0.5	−0.6	−0.1	0.0	0.62
6798a	Moro Tu		5	29	153	33	*p*	*p*	−0.5	−0.6	−0.1	0.0	0.61
6800	Hall Islands		8	36	152	15	*p*	*p*	−0.7	−0.7	−0.1	0.0	0.51
							(Zone −1000)						
6800a	Nomuuin To		8	27	151	47	*p*	*p*	−0.8	−0.8	−0.1	−0.1	0.47
5599	**DREGER HARBOUR**		see page 63						1.5	Δ	Δ	0.9	
6801	Truk Islands	JU	7	22	151	33	+0035	−0052	−0.8	Δ	Δ	−0.6	0.46
6802	Namonuito Islands		8	35	149	39	−0122	+0139	−0.9	Δ	Δ	−0.7	0.37
6802a	Onari To		8	45	150	20	−0057	+0123	−0.9	Δ	Δ	−0.7	0.39
6803	Pulap Island		7	39	149	25	−0033	+0110	−0.9	Δ	Δ	−0.6	0.41
6804	Puluwat Island		7	22	149	13	*p*	*P*	−0.9	Δ	Δ	−0.7	0.4 x
7940	**YOKOHAMA**		see page 198				MHW	LLW	1.7	1.6	0.9	0.3	0.52
6807	Lamotrek		7	28	146	23	*p*	*p*	−1.0	−0.9	−0.5	−0.2	0.51
6811	Woleai Island		7	22	143	54	*p*	*p*	−1.0	−0.9	−0.4	−0.2	0.80
6814	Ulithi Islands		9	55	139	40	+0313	+0307	−0.5	−0.4	−0.3	−0.1	0.82
6814a	Yasoro To		10	02	139	46	+0255	+0256	−0.5	−0.5	−0.2	0.0	1.0 x
6815	Yap Island		9	30	138	08	+0315	+0310	−0.3	−0.6	−0.1	+0.1	0.86
6816	Ngulu Islet		8	18	137	29	+0324	+0312	−0.4	−0.3	−0.3	0.0	
7716	**NAHA KO**		see page 192				MHW	MLW	MHWS	MHWN	MLWN	MLWS	
									2.0	1.5	0.8	0.3	
	Palau Islands						(Zone −0900)						
6818	Garukoru (Ngaregur)		7	45	134	38	−0022	−0022	−0.2	−0.2	−0.1	0.0	1.05
6819	Toagel Mlungui		7	30	134	31	−0015	−0015	−0.2	−0.2	−0.1	+0.1	1.08
6820	Malakal Harbour	JU	7	20	134	28	+0012	+0012	−0.1	−0.1	0.0	0.0	1.11
6821	Ngesebus		7	03	134	16	+0012	+0011	−0.4	−0.2	−0.1	+0.1	1.00
6825	Helen Reef		2	59	131	49	−0003	−0003	−0.2	−0.2	0.0	+0.1	1.07

SEASONAL CHANGES IN MEAN LEVEL

No	Jan 1	Feb 1	Mar 1	Apr 1	May 1	June 1	July 1	Aug 1	Sep 1	Oct 1	Nov 1	Dec 1	Jan 1
5062	−0.1	−0.1	−0.1	0.0	0.0	0.0	0.0	0.0	+0.1	+0.1	0.0	0.0	−0.1
6705–6764	Negligible												
6766–6788	0.0	0.0	0.0	0.0	−0.1	−0.1	−0.1	0.0	0.0	+0.1	+0.1	+0.1	0.0
6790–6798a	Negligible												
6800–6825	0.0	−0.1	−0.1	0.0	0.0	0.0	0.0	0.0	0.0	+0.1	+0.1	+0.1	0.0
7716	−0.1	−0.1	−0.1	−0.1	0.0	0.0	+0.1	+0.1	+0.1	+0.1	0.0	−0.1	−0.1
7940	−0.1	−0.1	−0.1	−0.1	0.0	0.0	+0.1	+0.1	+0.1	+0.1	0.0	0.0	−0.1
9644, 9700	Negligible												

No.	PLACE	Lat.N.		Long.E		MHW	MLW	MHWS	MHWN	MLWN	MLWS	Z_0m.	
								(Zone−0800)					
7280	**LUHUASHAN**	See page 180						4.3	3.3	2.0	0.9		
7261	Xihou Men	30	06	121	54	+0020	+0032	−0.7	−0.5	−0.6	−0.3	2.1	x
7262	Yuxingnao	30	21	121	51	+0137	+0213	−0.5	−0.3	−0.4	−0.2	2.26	
7264	Changtu Gang	30	15	122	16	−0013	+0006	−0.6	−0.6	−0.5	−0.4	2.1	x
	Shengsi Liedao												
7269	Baijieshan	30	37	122	25	−0005	+0007	−0.1	−0.1	−0.2	−0.1	2.5	1x
	Hangzhou Wan												
7270	Dajishan(Gutzlaff Island)	30	49	122	10	0000	+0010	0.0	0.0	−0.3	−0.2	2.5	x
7273	Tanxushan	30	36	121	38	+0155	+0207	+0.8	+0.6	+0.1	0.0	3.0	x
7275	Baitashan	30	26	120	58	+0303	+0341	+4.9	+3.6	+2.4	+0.2	5.2	x
	Chien Tang Chiang												
7277	Haining	30	18	120	33	+0500	⊙	+2.5	⊙	⊙	⊙	⊙	*
7280	CHANGJIANG	30	49	122	38	STANDARD		See Table V				2.63	
	APPROACHES(LUHUASHAN)					PORT							
7284	**WUSONG KOU**	See page 183						3.5	2.5	1.4	1.0		
	Changjiang(Yangtze Kiang)												
7281	Sheshan	31	25	122	14	−0033	−0243	+0.6	+0.5	+0.2	−0.5	2.29	
7282	Tongsha Hangdao	31	06	122	01	−0144	−0225	+0.6	+0.4	+0.4	−0.4	2.3	x
7283	Heqing	31	16	121	43	−0025	−0120	0.0	+0.2	0.0	−0.4	2.0	x
7284	**WUSONG KOU**	31	24	121	31	STANDARD		See Table V				2.02	
						PORT							
7285	Shanghai	31	15	121	29	+0040	+0045	−0.5	−0.2	−0.1	−0.4	1.80	
7287	Leo Point	31	36	121	15	+0035	+0045	−0.3	−0.1	⊙	⊙	1.80	
7288	Plover Point	31	46	120	58	+0245	+0250	−0.7	−0.5	−0.2	−0.6	1.59	
7289	Langshan Crossing	31	53	120	46	+0240	+0320	−1.1	⊙	⊙	⊙	⊙	
7291	Jiang Yin	31	55	120	14	+0515	+0610	−1.6	−1.2	−0.6	−0.7	1.07	
7294	Zhenjiang	32	13	119	26	+0800	+0940	⊙	⊙	⊙	⊙	⊙	*
7296	Nanjing	32	06	118	45	+1040	⊙	⊙	⊙	⊙	⊙	⊙	*
7298	Wuhu	31	20	118	21	+1500	⊙	⊙	⊙	⊙	⊙	⊙	*
5062	**DAVAO**	See page 24						1.6	1.0	0.5	−0.2		
	HUANG HAI (YELLOW SEA)												
7301	San Chia Chun	32	01	121	42	−0638	−0638	+1.9	+1.7	+0.7	+0.6	2.0	1x
7804	**SHIMONOSEKI**	See page 195						2.4	1.8	1.0	0.2		
7303	San Se Tang	32	38	120	54	−1050	−1050	+1.0	+0.9	+0.2	+0.3	2.0	1x
7306	Sin Yang Kau	33	38	120	35	−1220	−1235	+0.8	+0.8	+0.2	+0.4	1.9	x
7310	Kaishan Dao	34	35	119	50	+1106	+1055	+1.1	+0.9	+0.5	+0.5	2.1	x
7312	Lianyun Gang c	34	44	119	27	+0919	+0946	+2.5	+2.2	+0.8	+0.6		
7315	Huangjiatang Wan	35	32	119	45	+0827	+0815	+1.5	+1.3	+0.5	+0.5	2.3	x
7318	Qingdao cj	36	05	120	19	+0808	+0735	+1.6	+1.4	+0.5	+0.5	2.39	
7319	Laoshan Gang	36	06	120	32	+0741	+0735	+1.4	+1.5	+0.3	+0.5	2.3	1x
7322	Nu Dao	36	23	120	50	+0720	+0715	+1.3	+1.3	+0.3	+0.5	2.2	1x
7324	Fengcheng	36	41	121	14	+0645	+0640	+0.9	+1.1	+0.1	+0.5	2.0	1x
7326	Gulang Zui	36	44	121	38	+0612	+0605	+0.7	+0.9	+0.1	+0.5	1.9	1x
7329	Jinghai Jaio	36	51	122	11	+0551	+0545	+0.5	+0.7	+0.1	+0.5	1.8	1x
7330	Wangjia Dao c	36	52	122	24	+0518	+0510	+0.3	+0.4	+0.2	+0.4	1.66	
7332	Sanggou Wan	37	03	122	29	+0439	+0433	−0.1	+0.3	−0.1	+0.5	1.5	1x

SEASONAL CHANGES IN MEAN LEVEL

No.	Jan.1	Feb.1	Mar.1	Apr.1	May 1	June 1	July 1	Aug.1	Sep.1	Oct.1	Nov.1	Dec.1	Jan.1
4738	+0.1	0.0	0.0	0.0	0.0	0.0	−0.1	−0.1	−0.1	0.0	+0.1	+0.1	+0.1
48962	+0.2	+0.2	+0.1	0.0	−0.1	−0.1	−0.1	−0.1	−0.1	0.0	+0.1	+0.2	+0.2
5062	−0.1	−0.1	−0.1	0.0	0.0	0.0	0.0	0.0	+0.1	+0.1	0.0	0.0	−0.1
5147	+0.1	0.0	−0.1	−0.1	−0.1	0.0	0.0	0.0	0.0	0.0	+0.1	+0.1	+0.1
5339							Negligible						
5435	0.0	+0.1	+0.1	0.0	0.0	0.0	−0.1	−0.1	−0.1	0.0	0.0	0.0	0.0
7056	0.0	−0.1	−0.1	−0.1	0.0	−0.1	−0.1	−0.1	0.0	+0.1	+0.2	+0.2	0.0
7261−7283	−0.1	−0.2	−0.2	−0.1	0.0	0.0	0.0	+0.1	+0.2	+0.2	+0.1	0.0	−0.1
7284	−0.2	−0.3	−0.2	−0.1	0.0	+0.1	+0.2	+0.2	+0.2	+0.1	0.0	−0.1	−0.2
7285−7289	−0.4	−0.5	−0.4	−0.1	+0.1	+0.2	+0.3	+0.3	+0.3	+0.2	+0.1	−0.1	−0.4
7291	−1.2	−1.2	−0.6	−0.2	+0.3	+0.6	+0.8	+1.0	+0.7	+0.4	0.0	−0.6	−1.2
7294−7298							See notes on page 386						
7301−7349	−0.2	−0.2	−0.2	−0.1	0.0	0.0	+0.1	+0.2	+0.2	+0.1	0.0	−0.1	−0.2
7353−7385	−0.2	−0.2	−0.2	−0.1	0.0	+0.1	+0.2	+0.3	+0.2	+0.1	−0.1	−0.2	−0.2
7486	−0.2	−0.2	−0.1	−0.1	0.0	0.0	+0.1	+0.2	+0.2	+0.1	0.0	−0.1	−0.2
7716	−0.1	−0.1	−0.1	−0.1	0.0	+0.1	+0.1	+0.1	+0.1	+0.1	0.0	−0.1	−0.1
7804	−0.1	−0.2	−0.2	−0.1	0.0	0.0	+0.1	+0.2	+0.2	+0.1	0.0	−0.1	−0.1
8287							Negligible						

No.	PLACE		Lat. N		Long. E		HHW	LLW	MHHW	MLHW	MHLW	MLLW	Z_0m.	
							TIME DIFFERENCES		HEIGHT DIFFERENCES (IN METRES)				M.L.	
							(Zone−0800)							
5998	**BRISBANE BAR**		See page 90						2.1		0.5	0.4		
7388	Huludao Gang		40	43	120	59	−0407	−0408	+1.0	+0.7	+0.4	+0.5	1.84	
	Liao He													
7392	Bar		40	32	122	04	−0449	−0448	+1.1	+0.7	+0.1	+0.1	1.7	tx
7393	Bar Signal Station		40	38	122	10	−0427	−0426	+1.2	+0.8	+0.1	+0.1	1.74	t
7394	Yingkou		40	41	122	16	−0458	−0504	+1.4	+0.8	−0.1	−0.1	1.7	tx
7397	Daitze Shan		40	18	122	06	−0509	−0511	+1.9	+1.3	+0.6	+0.6	2.30	t
7401	Changxing Dao		39	39	121	28	−0629	−0654	+0.1	−0.2	+0.3	+0.3	1.31	t
7403	Xishong Dao		39	24	121	17	−0824	−0850	−0.2	−0.3	+0.4	+0.1	1.18	t
	Pulandian Wan													
7405	Boji Dao		39	23	121	45	−0802	−0810	+0.2	+0.2	+0.4	+0.1	1.42	t
7407	Hulu Dao		39	16	121	36	−0847	−0906	+0.2	0.0	+0.5	+0.1	1.40	t
									MHWS	MHWN	MLWN	MLWS		
7486	**INCH'ON**		See page 189				MHW	MLW	8.6	6.5	2.8	0.4		
	BOHAI HAIXIA													
7411	Yingzhengzi Wan		38	58	121	18	p	p	−6.7	−4.9	−2.0	0.0	1.18	
7414	Chang Zui		38	47	121	08	+0613	+0613	−6.7	−5.0	−2.1	−0.1	1.07	t
7416	Lushun Gang		38	48	121	15	+0531	+0531	−6.0	−4.4	−1.9	0.0	1.49	t
7417	Xiaobing Dao		38	49	121	31	+0523	+0523	−5.9	−4.3	−1.9	0.0	1.57	t
7418	Yu Yan		38	34	121	39	+0511	+0515	−5.8	−4.3	−1.8	0.0	1.6	tx
	YALU GULF													
7421	Dalian Gang(Dairen Ko)	CJ	38	56	121	39	+0511	+0511	−5.7	−4.2	−1.9	−0.1	1.63	
7424	Changjiang Ao		39	08	122	06	+0442	+0442	−5.1	−3.7	−1.6	+0.1	2.01	t
7427	Dachangshan Dao		39	16	122	35	+0421	+0421	−4.5	−3.3	−1.4	+0.2	2.32	t
7429	Haiyang Dao		39	04	123	09	+0356	+0356	−4.6	−3.3	−1.3	+0.2	2.32	t
7432	Dwangjia Dao		39	27	123	03	+0402	+0406	−3.9	−2.7	−1.3	+0.2	2.64	t
7435	Dalu Dao		39	45	123	45	+0338	+0338	−3.1	−2.1	−1.1	+0.2	3.05	t
7436	Takushian		39	46	123	33	+0405	+0405	−3.3	−2.0	−1.3	+0.2	2.96	t
	Yalu Jiang(Amnok Kang)													
7439	Sin Do		39	48	124	16	+0342	+0342	−2.2	−1.6	−0.7	+0.2	3.50	t
7440	Chao−shih−kou		39	53	124	12	+0340	+0442	−3.0	−2.1	−1.1	+0.1	3.05	*d
7441	Tuyup'o		39	56	124	20	+0353	+0513	−3.7	−2.5	−1.2	+0.2	2.78	*d
7442	San−tao−lang−t'ou		40	03	124	20	+0430	+0620	⊙	⊙	⊙	⊙	⊙	*d
7443	An−tung		40	07	124	24	+0605	+0640	−5.8	−4.2	−1.8	+0.1	1.65	*d
7445	Suun Do		39	42	124	25	+0325	+0331	−2.5	−1.7	−0.8	+0.3	3.37	
	Korea, West Coast						*(Zone − 0900)*							
7448	Tan−do(Tan Do)		39	31	124	40	+0423	+0423	−2.4	−1.7	−0.8	+0.3	3.45	t
7450	Nap To		39	16	124	43	+0407	+0407	−2.7	−2.0	−0.7	+0.3	3.30	t
7452	Unmu Do		39	25	125	07	+0423	+0423	−1.9	−1.3	−0.5	+0.4	3.76	t
	Taedong Gang													
7457	Sok To		38	38	125	00	+0331	+0331	−3.8	−2.8	−1.1	+0.2	2.70	t
7458	P'I Do	K	38	40	125	10	+0346	+0346	−2.9	−1.8	−1.1	+0.3	3.20	t
7459	Chinnanp'o	K	38	41	125	24	+0413	+0413	−3.2	−2.2	−1.2	+0.2	3.04	t
7460	Oeampo(Gaiganho)		38	39	125	34	+0421	+0421	−2.4	−1.6	−0.7	+0.3	3.47	t
7462	Yohori		38	49	125	33	+0458	+0458	−2.1	−1.3	−0.8	+0.3	3.59	t
7463	P'yongyang		38	00	125	41	+0557	+0707	−5.0	−3.5	−1.6	+0.2	2.11	d
7468	Monggum P'o		38	11	124	47	+0232	+0233	−5.0	−3.7	−1.5	+0.1	2.14	t
	HUANG HAI(YELLOW SEA)													
7469	Wollae Do(Getsnai Tau)		38	03	124	49	+0108	+0110	−5.1	−3.8	−1.5	+0.2	2.05	t
7470	Taeryonggi Bong	K	35	57	124	44	+0122	+0124	−4.8	−3.6	−1.1	+0.4	2.30	
7471	Taechong Do	K	37	50	124	43	+0100	+0100	−5.2	−3.9	−1.4	+0.1	1.99	t
7474	Sunwi Do	K	37	45	125	20	+0033	+0033	−3.7	−2.8	−1.0	+0.2	2.88	t
7475	Mu Do		37	44	125	35	+0022	+0022	−2.5	−1.9	−0.7	+0.3	3.59	
7477	Taeyonp'yong Do	K	37	40	125	43	+0026	+0021	−2.0	−1.6	−0.3	+0.4	3.70	

⊙   See notes on page 386
C   Tides predicted in Chinese Tide Tables
J   Tides predicted in Japanese Tide Tables
K   Tides predicted in Korean Tide Tables
d   Differences approximate.
p   For predictions use harmonic constants(Part III) and N.P.159.
t   Time differences approximate
x   M.L. inferred

No.	PLACE		TIME DIFFERENCES						HEIGHT DIFFERENCES (IN METRES)				M.L.
			Lat.N		Long.E.		MHW	MLW	MHWS	MHWN	MLWN	MLWS	Z_0m.
7486	**INCH'ON**		See page 189				(Zone −0900)		**8.6**	**6.5**	**2.8**	**0.4**	
	Han Gan												
7479	Chumun Do		37	39	126	14	+0017	+0018	−0.5	−0.4	−0.1	+0.3	4.39
7480	Oepori	K	37	42	126	23	+0043	+0041	−0.5	−0.5	+0.2	+0.5	4.50
7482	Seoul		37	33	126	57	+0430	⊙	⊙	⊙	⊙	⊙	⊙
	Yom Ha												
7484	Sun Tol Mok		37	40	126	32	+0055	⊙	⊙	⊙	⊙	⊙	⊙
7486	INCH'ON(CHEMULPHO)		37	28	126	37	STANDARD			See Table V			4.68
							PORT						
7487	Taemuni Do		37	23	126	27	−0006	−0006	−0.2	−0.2	0.0	+0.3	4.53
7489	Tokchok To	K	37	15	126	07	−0018	−0014	−0.9	−0.7	0.0	+0.5	4.30
7492	Asan Myoji(Gazan Byochi)		36	58	126	47	−0008	−0008	+0.3	+0.3	+0.1	+0.3	4.82
7494	Umu Do		37	02	126	27	−0027	−0027	−0.8	−0.6	−0.3	+0.3	4.24
7497	Manli Po		36	47	126	08	−0055	−0055	−2.0	−1.5	−0.6	+0.2	3.63
7500	Sachang−po(Chung Do)		36	23	126	26	−0108	−0108	−1.7	−1.3	−0.5	+0.2	3.76
7501	Daecheon Hang	K	36	20	126	30	−0133	−0056	−2.9	−2.2	−0.7	+0.3	3.20
7503	Eocheong Do	K	36	07	125	59	−0141	−0141	−2.9	−2.3	−0.8	+0.1	3.10
7504	Kunsan(Outer Port)	K	35	58	126	38	−0203	−0130	−1.9	−1.4	−0.5	+0.5	3.61
7505	Kunsan	JK	35	59	126	43	−0109	−0108	−1.5	−1.8	−0.7	+0.3	3.41
7507	Gogunsan Gundo		35	59	126	24	−0153	−0153	−2.5	−1.8	−0.8	+0.2	3.36
7280	LUHUASHAN		See page 180						4.3	3.3	2.0	0.9	
7510	Amma Do	K	35	21	126	01	+0457	+0457	+0.7	+0.6	−0.3	−0.3	2.80
7512	Hampyeong Man		35	09	126	21	+0449	+0449	+1.3	+1.1	−0.1	−0.2	3.15
7515	Buggang Sudo		34	53	126	06	+0400	+0400	+0.1	+0.1	−0.5	−0.4	2.47
7517	Mokp'o Hang	K	34	47	126	23	+0452	+0552	−0.5	−0.3	−0.8	−0.4	2.15
7519	Siha Do		34	42	126	15	+0400	+0400	−0.5	−0.2	−0.7	−0.3	2.27
7521	Hatae Do		34	32	126	03	+0300	+0301	−0.8	−0.6	−0.7	−0.4	2.02
7523	Bigeum Sudo		34	43	125	56	+0334	+0334	−0.4	−0.3	−0.6	−0.4	2.18
	Daeheugsan Gundo												
7525	Daeheugsan Do	K	34	41	125	26	+0351	+0348	−1.0	−0.7	−0.8	−0.5	1.87
7529	Usuyong	K	34	35	126	19	+0336	+0333	−0.7	−0.4	−0.7	−0.3	2.10
7530	Hajo Do	K	34	18	126	03	+0326	+0309	−1.3	−1.0	−0.7	−0.3	1.80
7531	Pyokpajin		34	32	126	20	+0132	+0217	−0.8	−0.7	−0.6	−0.4	2.00
7532	Sangma Do(Joma To)		34	27	126	25	+0127	+0127	−0.6	−0.5	−0.5	−0.3	2.15
7533	Oran Jin		34	21	126	29	+0107	+0120	−0.7	−0.7	−0.6	−0.4	2.02
	KOREA STRAIT												
	Jeju Do												
7534	Chuk−to(Shaki To)		33	18	126	09	+0053	+0055	−1.7	−1.3	−0.9	−0.4	1.54
7535	Hoasgunpo		33	14	126	20	+0009	+0022	−1.6	−1.3	−0.8	−0.4	1.60
7536	Sogwi−po(Seikiho Ko)	K	33	14	126	33	−0015	−0014	−1.7	−1.3	−0.9	−0.5	1.53
7538	U Do		33	30	126	54	−0011	−0010	−2.0	−1.6	−1.0	−0.5	1.36
7539	Jeju Hang	K	33	31	126	32	+0057	+0056	−1.9	−1.5	−1.0	−0.5	1.42
	Chuja Gundo												
7541	Sangchuja Do		33	58	126	18	+0111	+0124	−1.4	−1.1	−0.8	−0.4	1.70
7542	Soan Do		34	09	126	38	+0032	+0044	−0.9	−0.8	−0.6	−0.4	1.94
7543	Wan Do	K	34	19	126	45	+0056	+0123	−0.8	−0.7	−0.5	−0.3	2.05
7543a	Cheongsan Do		34	11	126	51	+0015	+0028	−1.1	−0.9	−0.7	−0.4	1.82
7544	Geumdang Do		34	25	127	05	−0007	+0005	−0.7	−0.6	−0.5	−0.3	2.10
7545	Geomun Do	K	34	01	127	19	−0028	−0028	−1.2	−0.7	−0.7	−0.5	1.76 ∗
7548	Nogdong	K	34	31	127	08	−0010	−0010	−0.6	−0.6	−0.5	−0.4	2.11

SEASONAL CHANGES IN MEAN LEVEL

No.	Jan.1	Feb.1	Mar.1	Apr.1	May 1	June 1	July 1	Aug.1	Sep.1	Oct.1	Nov.1	Dec.1	Jan.1
5998	0.0	0.0	0.0	+0.1	+0.1	0.0	0.0	0.0	−0.1	−0.1	−0.1	0.0	0.0
7280	−0.1	−0.2	−0.2	−0.1	0.0	0.0	0.0	+0.1	+0.2	+0.2	+0.1	0.0	−0.1
7388−7468	−0.2	−0.2	−0.2	−0.1	0.0	+0.2	+0.2	+0.3	+0.2	+0.1	−0.1	−0.2	−0.2
7469−7533	−0.2	−0.2	−0.1	−0.1	0.0	+0.1	+0.1	+0.2	+0.2	+0.1	0.0	−0.1	−0.2
7534−7548	−0.1	−0.1	−0.1	−0.1	0.0	+0.1	+0.1	+0.1	+0.1	+0.1	0.0	−0.1	−0.1

JAPAN

No.	PLACE		Lat.N		Long.E.		MHW	MLW	MHWS	MHWN	MLWN	MLWS	Z_0m.	
							(Zone–0900)							
7804	**SHIMONOSEKI**		See page 195						2.4	1.8	1.0	0.2		
Imari Wan														
7649	Hase		33	19	129	48	+0022	+0014	0.0	0.0	0.0	+0.2	1.40	
Hirado Seto														
7652	Kuroko Shima		33	22	129	34	−0012	−0012	+0.3	+0.2	+0.1	+0.2	1.55	
7653	Shishiki Wan		33	12	129	23	−0014	−0014	+0.3	+0.2	+0.1	+0.2	1.57	
7656	Sasebo	J	33	09	129	44	−0033	−0041	+0.5	+0.3	+0.2	+0.2	1.65	⋆
Omura Wan														
7658	Inoura		33	03	129	45	−0017	−0017	−0.5	−0.4	−0.2	+0.1	1.14	
7659	Ogashi		33	04	129	49	+0217	+0214	−1.6	−1.2	−0.7	0.0	0.47	t
7660	Omura		32	54	129	57	+0221	+0218	−1.5	−1.1	−0.6	0.0	0.53	t
7663	Terashima Suido		33	02	129	37	−0046	−0046	+0.5	+0.3	+0.2	+0.2	1.68	
7666	Nagasaki		32	44	129	52	−0057	−0104	+0.4	+0.3	+0.2	+0.2	1.64	⋆
7667	Fukabori		32	41	129	49	−0050	−0050	+0.5	+0.3	+0.2	+0.2	1.67	
7669	Kaba Shima Suido		32	33	129	47	−0102	−0100	+0.5	+0.4	+0.2	+0.3	1.66	
Shimabara Kaiwan														
7672	Kuchinotsu Ko		32	36	130	12	+0010	−0007	+1.0	+0.7	+0.4	+0.3	1.94	
7673	Shimabara		32	47	130	22	−0001	−0008	+1.9	+1.4	+0.6	+0.3	2.37	
7674	Oura		32	59	130	13	+0022	+0014	+2.6	+1.8	+0.9	+0.3	2.75	
7675	Suminoe		33	12	130	13	+0030	+0048	+2.7	+1.9	+0.9	+0.3	2.79	
7676	Miike Ko	J	33	00	130	23	+0017	+0010	+2.6	+1.9	+0.9	+0.4	2.80	
7678	Misumi Ko		32	37	130	27	−0007	−0015	+1.5	+1.1	+0.5	+0.3	2.21	
7679	Yanagino Seto		32	32	130	25	−0012	−0019	+1.5	+1.1	+0.5	+0.3	2.19	
7682	Sakitsu Wan		32	19	130	01	−0112	−0112	+0.5	+0.3	+0.2	+0.3	1.66	
Yatsushiro Kaiwan														
7684	Ushibuka Ko		32	12	130	01	−0107	−0107	+0.5	+0.4	+0.2	+0.3	1.67	
7686	Zozono Seto		32	34	130	28	−0020	−0020	+1.4	+1.0	+0.5	+0.3	2.17	
7690	Akune		32	01	130	12	−0123	−0123	+0.4	+0.3	+0.1	+0.3	1.61	
7716	**NAHA KO**		See page 192						2.0	1.5	0.8	0.3		
7694	Tomari Ura		31	17	130	13	+0005	+0007	+0.5	+0.4	+0.3	+0.2	1.50	
Kagoshima Kaiwan														
7696	Yamagawa Ko		31	12	130	38	+0014	+0015	+0.6	+0.5	+0.3	+0.2	1.54	
7696a	Kiire		31	23	130	34	+0002	+0002	+0.6	+0.5	+0.3	+0.1	1.5	x
7697	Kagoshima Ko	J	31	36	130	34	+0002	+0002	+0.7	+0.5	+0.3	+0.1	1.55	
Koshiki Retto														
7700	Nakagawara Ura(Tatsu Maru)		31	51	129	51	+0036	+0034	+0.7	+0.5	+0.4	+0.2	1.60	
Nansei Shoto														
7704	Ti A Usu		25	56	123	41	+0042	+0042	−0.3	−0.2	0.0	+0.1	1.06	
Sakishima Gunto														
7706	Yonakuni Shima		24	25	123	00	−0002	−0003	−0.4	−0.3	−0.1	0.0	0.96	
7707	Funauke Ko		24	20	123	44	+0018	+0017	−0.3	−0.3	0.0	0.0	1.02	
7709	Ishigaki		24	20	124	10	+0023	+0023	−0.3	−0.2	0.0	+0.1	1.07	
7712	Miyako		24	48	125	17	+0025	+0025	−0.3	−0.1	−0.1	0.0	1.03	
7712a	Oagari Shima		25	49	131	14	−0023	−0023	−0.4	−0.3	−0.1	0.0	0.98	
Okinawa Gunto														
7713	Gima Ko		26	20	126	44	−0006	−0006	0.0	0.0	+0.1	+0.1	1.17	
7714	Kerama Kaikyo		26	13	127	18	−0007	−0006	0.0	0.0	0.0	0.0	1.16	
7716	NAHA KO		26	12	127	40	STANDARD PORT			See Table V			1.18		
7717	Ishikawa		26	25	127	51	−0034	−0034	0.0	0.0	+0.1	+0.1	1.18	
7718	Sesoko		26	38	127	52	+0013	+0015	0.0	0.0	0.0	0.0	1.14	

SEASONAL CHANGES IN MEAN LEVEL

No.	Jan.1	Feb.1	Mar.1	Apr.1	May 1	June 1	July 1	Aug.1	Sep.1	Oct.1	Nov.1	Dec.1	Jan.1
7280	−0.1	−0.2	−0.2	−0.1	0.0	0.0	0.0	+0.1	+0.2	+0.2	+0.1	0.0	−0.1
7550–7566	−0.1	−0.1	−0.1	0.0	0.0	0.0	+0.1	+0.1	+0.1	+0.2	0.0	−0.1	−0.1
7568–7603	0.0	−0.1	−0.1	−0.1	−0.1	0.0	+0.1	+0.1	+0.1	+0.1	0.0	0.0	0.0
7606–7690	−0.1	−0.2	−0.2	−0.1	0.0	0.0	+0.1	+0.2	+0.2	+0.1	0.0	−0.1	−0.1
7694–7718	−0.1	−0.1	−0.1	−0.1	0.0	0.0	+0.1	+0.1	+0.1	+0.1	0.0	−0.1	−0.1
7804	−0.1	−0.2	−0.2	−0.1	0.0	0.0	+0.1	+0.2	+0.2	+0.1	0.0	−0.1	−0.1
7994	0.0	−0.1	−0.1	−0.1	−0.1	0.0	0.0	+0.1	+0.1	+0.1	0.0	0.0	0.0

ALASKA

No.	PLACE	Lat.N		Long.W.		MHW	MLW	MHWS	MHWN	MLWN	MLWS	Z_0m.	
						(Zone +0900)						M.L.	
8850	**PRINCE RUPERT**	See page 207						6.5	5.2	2.5	1.2		
Shelikof Strait													
8538	Kanatak Lagoon	57	31	156	04	+0100	+0100	−2.8	−2.4	−1.6	−1.1	1.9	dx
8540	Katmai Bay	58	00	154	59	+0055	+0055	−2.4	−2.1	−1.5	−1.2	2.0	dx
Kodiak Island													
8543	Uganik Passage	57	48	153	18	+0100	+0100	−2.0	−1.8	−1.4	−1.2	2.3	dx
8544	Uyak Bay	57	38	154	00	+0055	+0055	−2.2	−1.9	−1.4	−1.2	2.2	dx
								MHHW	**MLHW**	**MHLW**	**MLLW**		
9050	**TOFINO**	See page 210				HHW	LLW	3.4	3.0	1.4	0.7		
8547	Lazy Bay	56	54	154	15	+0130	+0130	0.0	−0.1	−0.5	−0.5	1.9	dx
8548	Olga Bay	57	10	154	14	+0520	⊙	−3.0	−2.6	⊙	⊙	⊙	d
8551	Three Saints Bay	57	07	153	31	+0105	+0105	−0.9	−1.0	−0.7	−0.7	1.3	dx
8553	Ugak Bay	57	29	152	44	+0100	+0100	−0.8	−0.9	−0.8	−0.6	1.3	dx
8554	Kodiak **U**	57	47	152	24	+0120	+0120	−0.8	−0.9	−0.7	−0.6	1.31	
Afognak Island													
8557	Seal Bay	58	22	152	15	+0155	+0155	0.0	0.0	−0.6	−0.7	1.8	dx
								MHWS	**MHWN**	**MLWN**	**MLWS**		
8850	**PRINCE RUPERT**	See page 207				MHW	MLW	6.5	5.2	2.5	1.2		
8559	Shuyak Strait	58	27	152	36	+0115	+0115	−2.3	−2.0	−1.5	−1.2	2.1	dx
8561	Malina Bay	58	11	152	57	+0055	+0055	−1.7	−1.8	−1.2	−1.4	2.3	dx
Cook Inlet													
8562	Shaw Island	59	00	153	23	+0104	+0104	−1.9	−1.9	−1.2	−1.2	2.28	
8563	Iliamna Bay	59	36	153	37	+0112	+0112	−1.7	−1.8	−1.2	−1.2	2.37	
8564	Snug Harbour	60	06	152	34	+0206	+0206	−1.3	−1.5	−1.0	−1.2	2.58	
8565	Drift River	60	34	152	08	+0236	+0247	−0.4	−0.7	−0.8	−1.3	3.14	
8567	North Foreland	61	02	151	11	+0045	+0445	+0.2	−0.4	−0.6	−1.2	3.3	x
8569	Fire Island	61	09	150	15	+0535	+0559	+2.3	+1.5	−0.5	−1.7	4.41	
8570	Anchorage, **U** Knik Arm	61	14	149	54	+0547	+0613	+2.4	+2.3	−0.3	−1.2	5.00	
8573	Nikiski	60	41	151	24	+0320	+0331	+0.2	−0.3	−0.6	−1.2	3.41	
8574	Chinulna Point	60	31	151	18	+0250	+0311	+0.1	−0.4	−0.8	−1.6	3.28	
8575	Cape Ninilchik	60	01	151	44	+0150	+0150	0.0	−0.6	−0.7	−1.2	3.2	x
8576	Homer	59	36	151	25	+0106	+0106	−0.3	−0.8	−0.9	−1.1	3.04	
8577	Seldovia **U**	59	26	151	43	+0100	+0100	−0.6	−1.0	−1.1	−1.4	2.85	
8578	Port Chatham	59	13	151	44	+0037	+0037	−1.9	−1.8	−1.3	−1.2	2.34	
8579	Ushagat Island	58	57	152	16	+0055	+0055	−2.1	−1.9	−1.4	−1.2	2.2	dx
								MHHW	**MLHW**	**MHLW**	**MLLW**		
9050	**TOFINO**	See page 210				HHW	LLW	3.4	3.0	1.4	0.7		
Kenai Peninsula													
8581	Picnic Harbour	59	15	151	26	+0110	+0110	+0.5	−0.4	−0.5	−0.5	2.0	dx
8583	Nuka Passage	59	24	150	40	+0100	+0100	0.0	0.0	−0.5	−0.6	1.9	dx
8585	Two Arm Bay	59	40	150	07	+0050	+0050	−0.2	−0.3	−0.5	−0.6	1.7	dx
8587	Seward(Resurrection Bay)	60	06	149	27	+0045	+0040	−0.3	−0.4	−0.5	−0.6	1.68	
PRINCE WILLIAM SOUND													
8589	Port Baimbridge	60	04	148	12	+0045	+0045	−0.3	−0.3	−0.5	−0.6	1.7	dx
8590	Macleod Harbour	59	53	147	46	+0045	+0045	0.0	−0.1	−0.4	−0.6	1.8	dx
8591	Port Chalmers	60	14	147	14	+0055	+0055	+0.1	0.0	−0.4	−0.6	1.9	dx
8594	Chenega Island	60	20	148	09	+0055	+0055	0.0	0.0	−0.4	−0.6	1.9	dx
8596	Smith Island	60	32	147	19	+0050	+0050	+0.1	0.0	−0.5	−0.6	1.9	dx
8599	Whittier	60	47	148	40	+0050	+0050	+0.3	+0.2	−0.4	−0.6	1.92	

SEASONAL CHANGES IN MEAN LEVEL

No.	Jan.1	Feb.1	Mar.1	Apr.1	May 1	June 1	July 1	Aug.1	Sep.1	Oct.1	Nov.1	Dec.1	Jan.1
8427–8524	Negligible												
8538–8579	0.0	0.0	0.0	−0.1	−0.1	−0.1	−0.1	0.0	0.0	+0.1	+0.1	+0.1	0.0
8581–8599	+0.1	0.0	0.0	0.0	−0.1	−0.1	−0.1	−0.1	0.0	0.0	+0.1	+0.1	+0.1
8850	+0.1	+0.1	+0.1	0.0	0.0	0.0	−0.1	−0.1	−0.1	0.0	0.0	+0.1	+0.1
9050	0.0	+0.1	+0.1	+0.1	0.0	0.0	0.0	−0.1	−0.1	0.0	0.0	0.0	0.0
9065	0.0	+0.1	+0.1	+0.1	0.0	0.0	0.0	0.0	−0.1	−0.1	−0.1	0.0	0.0

CANADA

NO.	PLACE	Lat N.		Long W.		TIME DIFFERENCES HHW	LLW (Zone + 0800)	HEIGHT DIFFERENCES (IN METERS) MHHW	MLHW	MHLW	MLLW	M.L. Z_0m.	
9050	**TOFINO**	see page 210						**3.4**	**3.0**	**1.4**	**0.7**		
]8960	Egg Island	51	15	127	50	+0019	+0018	+0.6	+0.5	+0.4	+0.3	2.56	
8960a	Leroy Bay	51	16	127	41	+0017	+0016	+1.0	+1.0	+0.7	+0.6	2.95	
8961	Treadwell Bay	51	06	127	32	+0044	+0046	−0.1	−0.1	+0.2	+0.3	2.23	
8962	Johnson Point	51	07	127	32	+0243	+0242	−1.7	−1.7	−0.7	−0.2	1.05	
8964	Raynor Group	50	53	127	14	+0030	+0029	+1.1	+1.0	+0.7	+0.5	2.93	
8967	Jessie Point	50	57	126	48	+0047	+0053	+1.3	+1.2	+0.8	+0.5	3.05	
8971	Sunday Harbour Knight Inlet	50	43	126	42	+0030	+0036	+1.1	+1.0	0.6	+0.3	2.87	
8972	Lagoon Cove	50	36	126	19	+0045	+0050	+1.7	+1.6	+0.9	+0.7	3.35	
8973	Glendale Cove Chatham Channel	50	40	125	44	+0035	+0040	+1.5	+1.6	+1.0	+0.6	3.33	d
8974	Root Point Swanson Island	50	35	126	12	+0054	+0104	+1.2	+1.1	+0.6	+0.3	2.9	dx
8978	Farewell Harbour Johnstone Strait	50	36	126	42	+0048	+0105	+0.9	+0.9	+0.6	+0.3	2.8	dx
8984	Port Harvey	50	34	126	16	+0057	+0055	+0.7	+0.7	+0.6	+0.6	2.80	D
8986	Port Neville	50	30	126	05	+0101	+0113	+1.1	+0.9	+0.9	+0.5	2.97	
8988	Yorke Island	50	27	125	59	+0100	+0105	+0.9	+0.9	+0.8	+0.5	2.91	D
8989	Salmon River	50	24	125	58	+0110	+0121	+0.8	+0.8	+0.9	+0.7	2.86	d
8990	Helmcken Island	50	24	125	52	+0116	+0134	+0.3	+0.2	+0.5	+0.2	2.40	d
8993	Knox Bay	50	24	125	36	+0134	+0135	+0.3	+0.3	+0.8	+0.4	2.55	
8998	Chatham Point Wellbore Channel	50	20	125	26	+0151	+0251	+0.4	+0.4	+0.9	+0.4	2.62	
8999	Carterer Point Loughborough Inlet	50	27	125	45	+0123	+0130	+0.9	+0.8	+0.6	+0.2	2.7	dx
9000	Sydney Bay Cordero Channel	50	31	125	35	+0119	+0137	+0.2	+0.1	+0.4	0.0	2.32	
9003	Cordero Islands	50	26	125	29	+0155	+0220	+0.4	+0.4	+1.1	+0.6	2.75	d
9004	Shoal Bay	50	28	125	22	+0155	+0225	+0.4	+0.5	+1.1	+0.4	2.70	d
9005	Mermaid Bay	50	24	125	11	P	P	+0.3	+0.4	+1.4	+0.7	2.79	
9065	**VICTORIA** Okisolio Channel	See page 213						**2.6**	**2.3**	**2.0**	**0.8**		
9009	Owen Bay C	50	19	125	13	+0103	+0003	+1.1	+1.1	+0.5	+0.4	2.69	
9012	Brown Bay	50	10	125	22	+0053	−0006	+1.4	+1.2	+0.5	+0.4	2.77	
9050	**TOFINO** Vancouver Island Cormorant Island	See page 210						**3.4**	**3.0**	**1.4**	**0.7**		
9016	Alert Bay C	50	35	126	56	+0039	+0040	+1.0	+0.9	+0.7	+0.4	2.88	
9018	Port McNeill	50	36	127	05	+0035	+0040	+1.1	+1.1	+0.7	+0.5	2.98	
8850	**PRINCE RUPERT**	see page 207				MHW	MLW	MHWS **6.5**	MHWN **5.2**	MLWN **2.5**	MLWS **1.2**		
9020	Port Hardy	50	43	127	29	−0033	−0022	−1.8	−1.4	−0.5	0.0	2.93	
9022	Shusharite Bay	50	51	127	51	−0030	−0035	−2.1	−1.6	−0.6	−0.1	2.75	

SEASONAL CHANGES IN MEAN LEVEL

No.	Jan 1	Feb 1	Mar 1	Apr 1	May 1	June 1	July 1	Aug 1	Sep 1	Oct 1	Nov 1	Dec 1	Jan 1
8850–8954	+0.1	+0.1	0.0	0.0	0.0	0.0	−0.1	−0.1	−0.1	0.0	0.0	+0.1	+0.1
8960–9022	+0.1	+0.1	+0.1	0.0	0.0	0.0	0.0	−0.1	−0.1	−0.1	0.0	0.0	+0.1
9050	0.0	+0.1	+0.1	+0.1	0.0	0.0	0.0	−0.1	−0.1	−0.1	0.0	0.0	0.0
9065	0.0	+0.1	+0.1	+0.1	0.0	0.0	0.0	0.0	−0.1	−0.1	−0.1	0.0	0.0

U.S.A

NO.	PLACE		Lat		Long		TIME DIFFERENCES		HEIGHT DIFFERENCES (IN METERS)				M.L.	
			N.		W.		HHW	LLW	MHHW	MLHW	MHLW	MLLW	Z_0m.	
							(Zone + 0800)							
9133	**VANCOUVER**		see page 216						4.4	3.9	2.9	1.2		
	PUGET SOUND													
9172	Everett		47	59	122	13	−0100	−0050	−1.2	−0.9	−1.1	−1.1	1.98	
9174	Seattle	U	47	36	122	20	−0100	−0040	−1.1	−0.9	−1.1	−1.1	2.01	
9176	Tacoma		47	17	122	25	−0050	−0100	−1.0	−0.8	−1.2	−1.1	2.07	
9178	Steilacoom		47	10	122	36	−0030	−0015	−0.5	−0.4	−1.1	−1.0	2.3	dx
9180	Olympia		47	03	122	54	−0025	−0005	−0.3	0.0	−1.0	−1.0	2.50	d
9183	Vaughn		47	20	122	46	−0015	−0025	−0.2	−0.1	−0.9	−1.0	2.5	dx
9186	Bremerton		47	33	122	38	−0050	−0033	−1.1	−0.7	−1.2	−1.0	2.07	
9187	Tracyton		47	37	122	40	−0025	+0005	−0.7	−0.6	−1.1	−1.0	2.2	X
9191	Point No Point		47	55	122	32	−0110	−0105	−1.4	−1.1	−1.2	−1.0	1.9	x
	Hood Canal													
9194	Seabeck		47	38	122	50	−0055	−0045	−1.0	−0.8	−1.1	−1.0	2.1	x
9196	Union		47	21	123	06	−0100	−0040	−1.0	−0.7	−1.1	−1.0	2.1	x
	Admiralty Inlet													
9197	Port Ludlow		47	55	122	41	−0120	−0110	−1.6	−1.2	−1.2	−1.1	1.8	x
9199	Port Townsend	U	48	08	122	46	−0117	−0129	−2.2	−1.7	−1.3	−1.1	1.46	
9201	Port Discovery		48	04	122	55	−0200	−0145	−2.4	−2.0	−1.6	−1.3	1.2	x
9203	New Dungeness Bay		48	11	123	07	−0205	−0205	−2.5	−2.1	−1.7	−1.3	1.2	dx
9065	**VICTORIA**		See page 213						2.6	2.3	2.0	0.8		
	Fuan de Fuca Strait													
9204	Port Angeles		48	07	123	26	0000	−0025	−0.6	−0.4	−0.5	−0.7	1.34	
9206	Crescent Bay		48	10	123	43	−0108	−0104	−0.5	−0.6	−0.6	−0.6	1.3	
9050	**TOFINO**		See page 210						3.4	3.0	1.4	0.7		
9208	Clallam Bay		48	16	124	18	+0035	+0045	−1.2	−1.4	−0.4	−0.7	1.2	x
9210	Neah Bay		48	22	124	37	0000	+0015	−1.0	−1.0	−0.5	−0.6	1.34	t
9211	Cape Flattery		48	23	124	44	−0005	+0005	−1.1	−1.2	−0.6	−0.6	1.2	x
9214	Union Seamount		49	32	132	41	+0019	+0024	−0.4	−0.4	0.0	−0.1	1.9	x
9215	Destruction Island		47	40	124	20	−0020	−0010	−0.9	−1.0	−0.6	−0.6	1.3	x
9218	Point Grenville		47	18	124	16	−0020	−0010	−1.0	−1.0	−0.6	−0.6	1.3	x
	Grays Harbour													
9220	Point Chehalis		46	55	124	07	−0013	+0002	−0.7	−0.7	−0.5	−0.6	1.3	x
9222	Aberdeen	U	46	58	123	51	+0045	+0045	−0.5	−0.5	−0.3	−0.5	1.68	
	Willapa Bay													
9223	Toke Point		46	42	123	58	+0027	+0026	−0.9	−0.9	−0.6	−0.7	1.37	
9224	Raymond		46	41	123	45	+0051	+0051	−0.5	−0.5	−0.5	−0.5	1.62	
9225	Cobb Searmount		46	46	130	49	−0006	+0006	−0.5	−0.5	+0.1	0.0	1.9	x
	Columbia River													
9226	Entrance (North Jetty)		46	16	124	02	−0002	−0025	−1.2	−1.1	−0.6	−0.7	1.2	x
9227	Youngs Bay		46	10	123	51	+0038	+0040	−0.9	−0.9	−0.6	−0.7	1.37	
9228	Astoria (Tongue Point)	U	46	13	123	46	+0055	+0100	−1.0	−1.0	−0.7	−0.7	1.31	★
9229	Harrington Point		46	16	123	39	+0115	+0142	−1.2	−1.1	−0.8	−0.7	1.2	★ x
9230	Eagle Cliff		46	10	123	14	+0240	+0400	−1.9	◉	◉	◉	◉	★
9231	Saint Helens		45	52	122	48	+0430	+0645	◉	◉	◉	◉	◉	★
9233	Portland		45	31	120	40	+0605	+0835	◉	◉	◉	◉	◉	★
	Tillamook Bay													
9236	Tillamook		45	28	123	51	−0024	−0024	−1.0	−1.0	−0.6	−0.7	1.34	
	Nestucca Bay Entrance		45	09	123	59	+0005	+0005	−1.1	−1.0	−0.7	−0.7	1.2	x
	Yaquina River													
9241	Newport		44	38	124	03	−0039	−0007	−1.0	−1.0	−0.6	−0.7	1.33	
9242	Yaquina		44	38	124	02	−0021	−0015	−1.1	−1.0	−0.5	−0.7	1.3	x
9243	Waldport		44	26	124	04	−0002	−0018	−1.2	−1.1	−0.6	−0.7	1.25	

◉ No data.

See notes on page 386.

U Tides predicted in U.S. Tide Tables

d Differences approximate.

t Time differences approximate.

x M.L. inferred.

NO.	PLACE		Lat N.		Long W.	TIME DIFFERENCES HHW (Zone + 0800)	LLW	HEIGHT DIFFERENCES (IN METERS) MHHW	MLHW	MHLW	MLLW	M.L. Z_0m.	
9305	**SAN FRANCISCO**		see page 219					1.7	1.3	0.7	−0.1		
	Monterey Bay												
9318	Santa Cruz		36	58	122	01	−0110	−0110	−0.1	−0.1	−0.1	+0.2	0.8	x
9320	Monterey		36	36	121	54	−0106	−0100	−0.1	−0.1	0.0	+0.1	0.86	
9321	Carmel Cove		36	31	121	56	−0115	−0115	−0.2	−0.2	−0.2	+0.2	0.8	x
9326	San Simeon		35	38	121	11	−0125	−0120	−0.2	−0.2	−0.2	+0.2	0.8	x
9329	Morro Bay		35	22	120	52	−0050	−0040	−0.4	−0.4	−0.1	+0.2	0.73	
9331	Avila, San Luis		35	10	120	44	−0135	−0130	−0.1	−0.1	0.0	+0.2	0.85	
	Obispo Bay												
9335	Point Arguello		34	35	120	39	−0145	−0135	−0.2	−0.2	−0.1	+0.1	0.8	x
9365	**SAN DIEGO**		See page 222					1.8	1.3	0.5	0.0		
9338	Santa Barbara		34	25	119	41	−0154	−0154	−0.2	−0.2	+0.1	0.0	0.85	
9342	Port Hueneme		34	09	119	12	−0005	+0002	−0.2	−0.2	+0.1	0.0	0.85	
	San Miguel Island												
9345	Cuyler Harbour		34	03	120	21	+0025	+0025	−0.3	−0.2	+0.1	0.0	0.8	x
	Santa Cruz Island												
9347	Prisoners Harbour		34	01	119	41	+0020	−0020	−0.4	−0.3	+0.1	0.0	0.8	x
9349	Santa Monica		34	00	118	30	−0010	−0005	−0.2	−0.1	+0.1	0.0	0.85	
9351	Los Angeles Harbour	U	33	43	118	16	+0013	+0014	−0.2	−0.1	+0.1	0.0	0.85	
9352	San Nicholas Island		33	16	119	30	+0010	+0010	−0.4	−0.3	+0.1	0.0	0.8	x
	Santa Catalina Island												
9354	Avalon		33	21	118	19	−0003	−0003	−0.3	−0.1	+0.1	0.0	0.82	
	San Clemente Island												
9355	Wilson Cove		33	00	118	33	−0003	−0009	−0.4	−0.2	0.0	0.0	0.81	
9357	Newport Bay		33	36	117	54	+0006	+0006	−0.2	−0.1	+0.1	+0.1	0.88	
9358	Dana Point		33	27	117	43	0000	0000	−0.3	−0.2	+0.1	0.0	0.8	x
9362	La Jolla		32	52	117	15	−0019	−0014	−0.2	−0.2	+0.1	+0.1	0.82	
	San Diego Bay												
9364	Point Loma		32	40	117	14	−0012	−0006	−0.2	−0.2	+0.1	+0.1	0.8	x
9365	SAN DIEGO		32	43	117	10	STANDARD PORT			See Table V			0.88	
MEXICO													
	Bahia de Todos Santos												
9371	Ensenada	M	31	51	116	38	−0030	−0030	−0.3	−0.2	+0.1	+0.1	0.82	
9374	Bahia Colnett		30	57	116	15	−0015	−0015	0.0	0.0	−0.1	0.0	0.9	dx
9377	Bahia San Quintin		30	25	115	54	−0020	−0020	−0.4	−0.3	−0.1	0.0	0.7	x
9379	Bahia Rosario		20	54	115	43	−0025	−0025	0.0	0.0	+0.1	0.0	0.9	dx
9380	Guadalupe Island	M	28	53	116	18	−0037	−0037	−0.4	−0.4	0.0	0.0	0.70	
							MHWS	MHWN	MLWN	MLWS				
9700	**PUERTO MONTT**		See page 234				**MHW**	**MLW**	6.5	4.7	2.4	0.7		
9381	Bahia Playa Maria		28	55	114	32	+0715	+0715	−4.2	−3.0	−1.8	−0.7	1.1	dx
9384	Isla Cerros		28	12	115	14	+0720	+0720	−4.1	−3.0	−1.8	−0.7	1.2	dx
(Zone +0700)													
9385	Puerto San Bartolome		27	40	114	51	+0815	+0815	−4.0	−3.0	−1.6	−0.7	1.2	x
9387	Punta Abreojos		26	43	113	34	+0820	+0816	−4.5	−3.4	−1.7	−0.7	1.01	
9389	Bahia San Juanico		26	15	112	28	+0704	+0744	−4.8	−3.6	−1.8	−0.7	0.85	
9391	San Carlos	M	24	47	112	07	+0745	+0745	−4.8	−3.6	−1.8	−0.7	0.88	
9392	Bahia Magdalena		24	38	112	09	P	P	−5.0	−3.6	−1.9	−0.6	0.82	
							MHHW	MLHW	MHLW	MLLW				
9365	**SAN DIEGO**		See page 222				**HHW**	**LLW**	1.8	1.3	0.5	0.0		
9393	Isla Santa Margarita		24	24	111	49	−0040	−0040	−0.4	−0.3	−0.2	0.0	0.6	dx
9395	Cabo San Lucas		22	53	109	54	−0031	−0034	−0.7	−0.5	−0.1	0.0	0.6	x

No data.

Δ Tide is usually diurnal.

See notes on page 386.

M Tides predicted in Mexican Tide Tables.

U Tides predicted in U.S. Tide Tables

d Differences approximate.

P For predictions use harmonic constants (part III) and N.P.159.

t Time differences approximate.

x **M.L.** inferred.

NO.	PLACE	Lat N.		Long W.		TIME DIFFERENCES		HEIGHT DIFFERENCES (IN METERS)				M.L.	
						HHW	LLW	MHHW	MLHW	MHLW	MLLW	$Z_0 m.$	
						(Zone + 0800)							
9133	**VANCOUVER**	see page 216						4.4	3.9	2.9	1.2		
	PUGET SOUND												
9172	Everett	47	59	122	13	−0100	−0050	−1.2	−0.9	−1.1	−1.1	1.98	
9174	Seattle U	47	36	122	20	−0100	−0040	−1.1	−0.9	−1.1	−1.1	2.01	
9176	Tacoma	47	17	122	25	−0050	−0100	−1.0	−0.8	−1.2	−1.1	2.07	
9178	Steilacoom	47	10	122	36	−0030	−0015	−0.5	−0.4	−1.1	−1.0	2.3	dx
9180	Olympia	47	03	122	54	−0025	−0005	−0.3	0.0	−1.0	−1.0	2.50	d
9183	Vaughn	47	20	122	46	−0015	−0025	−0.2	−0.1	−0.9	−1.0	2.5	dx
9186	Bremerton	47	33	122	38	−0050	−0033	−1.1	−0.7	−1.2	−1.0	2.07	
9187	Tracyton	47	37	122	40	−0025	+0005	−0.7	−0.6	−1.1	−1.0	2.2	x
9191	Point No Point	47	55	122	32	−0110	−0105	−1.4	−1.1	−1.2	−1.0	1.9	x
	Hood Canal												
9194	Seabeck	47	38	122	50	−0055	−0045	−1.0	−0.8	−1.1	−1.0	2.1	x
9196	Union	47	21	123	06	−0100	−0040	−1.0	−0.7	−1.1	−1.0	2.1	x
	Admiralty Inlet												
9197	Port Ludlow	47	55	122	41	−0120	−0110	−1.6	−1.2	−1.2	−1.1	1.8	x
9199	Port Townsend U	48	08	122	46	−0117	−0129	−2.2	−1.7	−1.3	−1.1	1.46	
9201	Port Discovery	48	04	122	55	−0200	−0145	−2.4	−2.0	−1.6	−1.3	1.2	x
9203	New Dungeness	48	11	123	07	−0205	−0205	−2.5	−2.1	−1.7	−1.3	1.2	dx
	Bay												
9065	**VICTORIA**	See page 213						2.6	2.3	2.0	0.8		
	Fuan de Fuca Strait												
9204	Port Angeles	48	07	123	26	0000	−0025	−0.6	−0.4	−0.5	−0.7	1.34	
9206	Crescent Bay	48	10	123	43	−0108	−0104	−0.5	−0.6	−0.6	−0.6	1.3	
9050	**TOFINO**	See page 210						3.4	3.0	1.4	0.7		
9208	Clallam Bay	48	16	124	18	+0035	+0045	−1.2	−1.4	−0.4	−0.7	1.2	x
9210	Neah Bay	48	22	124	37	0000	+0015	−1.0	−1.0	−0.5	−0.6	1.34	t
9211	Cape Flattery	48	23	124	44	−0005	+0005	−1.1	−1.2	−0.6	−0.6	1.2	x
9214	Union Seamount	49	32	132	41	+0019	+0024	−0.4	−0.4	0.0	−0.1	1.9	x
9215	Destruction	47	40	124	29	−0020	−0010	−0.9	−1.0	−0.6	−0.6	1.3	x
	Island												
9218	Point Grenville	47	18	124	16	−0020	−0010	−1.0	−1.0	−0.6	−0.6	1.3	x
	Grays Harbour												
9220	Point Chehalis	46	55	124	07	−0013	+0002	−0.7	−0.7	−0.5	−0.6	1.3	x
9222	Aberdeen U	46	58	123	51	+0045	+0045	−0.5	−0.5	−0.3	−0.5	1.68	
	Willapa Bay												
9223	Toke Point	46	42	123	58	+0027	+0026	−0.9	−0.9	−0.6	−0.7	1.37	
9224	Raymond	46	41	123	45	+0051	+0051	−0.5	−0.5	−0.5	−0.5	1.62	
9225	Cobb Searmount	46	46	130	49	−0006	+0006	−0.5	−0.5	+0.1	0.0	1.9	x
	Columbia River												
9226	Entrance (North	46	16	124	02	−0002	−0025	−1.2	−1.1	−0.6	−0.7	1.2	x
	Jetty)												
9227	Youngs Bay	46	10	123	51	+0038	+0040	−0.9	−0.9	−0.6	−0.7	1.37	
9228	Astoria (Tongue U	46	13	123	46	+0055	+0100	−1.0	−1.0	−0.7	−0.7	1.31	
	Point)												
9229	Harrington Point	46	16	123	39	+0115	+0142	−1.2	−1.1	−0.8	−0.7	1.2	x
9230	Eagle Cliff	46	10	123	14	+0240	+0400	−1.9	⊙	⊙	⊙	⊙	★
9231	Saint Helens	45	52	122	48	+0430	+0645	⊙	⊙	⊙	⊙	⊙	★
9233	Portland	45	31	120	40	+0605	+0835	⊙	⊙	⊙	⊙	⊙	★
	Tillamook Bay												
9236	Tillamook	45	28	123	51	−0024	−0024	−1.0	−1.0	−0.6	−0.7	1.34	
	Nestucca Bay	45	09	123	59	+0005	+0005	−1.1	−1.0	−0.7	−0.7	1.2	x
	Entrance												
	Yaquina River												
9241	Newport	44	38	124	03	−0039	−0007	−1.0	−1.0	−0.6	−0.7	1.33	
9242	Yaquina	44	38	124	02	−0021	−0015	−1.1	−1.0	−0.5	−0.7	1.3	x
9243	Waldport	44	26	124	04	−0002	−0018	−1.2	−1.1	−0.6	−0.7	1.25	

SEASONAL CHANGES IN MEAN LEVEL

No.	Jan 1	Feb 1	Mar 1	Apr 1	May 1	June 1	July 1	Aug 1	Sep 1	Oct 1	Nov 1	Dec 1	Jan 1
9050	0.0	+0.1	+0.1	+0.1	0.0	0.0	0.0	−0.1	−0.1	−0.1	0.0	0.0	0.0
9065	0.0	+0.1	+0.1	+0.1	0.0	0.0	0.0	0.0	−0.1	−0.1	−0.1	0.0	0.0
9133	+0.1	+0.1	0.0	0.0	0.0	0.0	0.0	0.0	−0.1	−0.1	0.0	0.0	+0.1
9172–9197	+0.1	+0.1	0.0	0.0	0.0	0.0	0.0	−0.1	−0.1	0.0	0.0	0.0	+0.1
9199–9206	0.0	+0.1	+0.1	0.0	0.0	0.0	0.0	0.0	−0.1	−0.1	−0.1	0.0	0.0
9208–9220	+0.1	+0.1	+0.1	0.0	0.0	0.0	−0.1	−0.1	−0.1	−0.1	0.0	+0.1	+0.1
9222	+0.2	+0.2	+0.1	0.0	−0.1	−0.1	−0.2	−0.2	−0.1	0.0	+0.1	+0.2	+0.2
9223–9256	+0.1	+0.1	0.0	0.0	0.0	0.0	−0.1	−0.1	−0.1	0.0	0.0	+0.1	+0.1

NO.	PLACE		Lat N.		Long W.		TIME DIFFERENCES HHW (Zone + 0700)	LLW	HEIGHT DIFFERENCES (IN METERS) MHHW	MLHW	MHLW	MLLW	M.L. Z_0m.	
9365	**SAN DIEGO** GULF OF CALIFORNIA		see page 222						1.8	1.3	0.5	0.0		
9396	Bahia San Jose del Cabo Bahia La Paz		23	03	109	04	−0030	−0030	−0.7	−0.3	−0.1	+0.0	0.6	x
9399	La Paz	M	24	10	110	19	p	p	−0.9	−0.7	0.0	+0.1	0.52	
9402	Babia Conception		26	43	111	54	⊙	⊙	−0.1	+0.3	⊙	⊙	0.8	dx
9404	Santa Rosalia		27	20	112	16	⊙	⊙	−0.2	+0.3	⊙	+0.2	0.6	dx
9407	Bahia de los Angeles		28	57	113	33	p	p	+0.2	+0.3	+0.3	0.0	1.1	x
									MHWS	MHWN	MLWN	MLWS		
6323	**DARWIN**		See page 126				MHW	MLW	6.9	5.1	3.2	1.4		
9410	Rio Colorado Entrance		31	46	114	44	⊙	⊙	+0.4	+1.3	⊙	⊙	⊙	★d
9411	Puerto Penasco	M	31	21	113	34	−0340	−340	−1.9	−2.0	−1.3	−1.4	2.5	x
									MHHW	MLHW	MHLW	MLLW		
9365	**SAN DIEGO**		See page 222				HHW	LLW	1.8	1.3	0.5	0.0		
9412	Tepoca Bay		30	16	112	52	⊙	⊙	+2.5	+2.2	⊙	⊙	⊙	d
9415	Guaymas	U	27	55	110	54	p	p	−1.1	−0.6	+0.1	0.0	0.46	
9417	Yavaros		26	42	109	30	p	p	−0.9	−0.5	+0.1	+0.1	0.6	x
9418	Topolobampo	M	25	36	109	03	+0024	+0042	−0.7	−0.5	0.0	+0.1	0.61	
9420	Mazatlan PACIFIC COAST	M	23	11	106	26	−0050	−0045	−0.7	−0.5	−0.1	+0.1	0.61	
		(Zone +0600)												
9422	Puerto San Blas		21	32	105	19	+0045	+0045	−0.9	−0.6	−0.1	0.0	0.5	dx
9424	Puerto Vallarta	M	20	37	105	15	p	p	−0.8	−0.6	−0.1	+0.1	0.55	
9425	Bahia de Charnela		19	35	105	09	⊙	⊙	−1.1	−0.8	−0.2	0.0	0.4	dx
9426	Manzanillo	M	19	03	104	20	p	p	−1.2	−0.8	−0.1	+0.1	0.40	
9427	Isla Socorro	M	18	43	110	57	+0015	+0021	−0.8	−0.6	−0.1	+0.1	0.55	
9429	Melchor Ocampo (Port Lazaro Cardenas)	M	17	55	102	10	p	p	−1.3	Δ	Δ	+0.1	0.30	
9430	Bahia Sihuatanejo		17	36	101	33	⊙	⊙	−1.1	−0.8	−0.4	0.0	0.3	dx
9432	Acapulco	M	16	50	99	55	p	p	−1.2	−0.9	−0.3	0.0	0.30	
9433	Maldonado		16	20	98	35	⊙	⊙	−0.0	−0.6	−0.4	0.0	0.4	dx
9434	Puerto Angel	M	15	40	96	30	p	p	−0.8	−0.5	−0.4	+0.1	0.49	
9435	Puerto Sacrificios		15	41	96	15	⊙	⊙	−0.7	−0.4	−0.3	0.0	0.5	x
									MHWS	MHWN	MLWN	MLWS		
9448	**LA UNOIN**		See page 225				MHW	MLW	3.0	2.5	0.6	−0.1		
9436	Puerto Salina Cruz	U	16	10	95	12	p	p	−1.9	−1.5	−0.4	+0.1	0.58	
9437	Puerto Arista (La Puerta)		15	56	93	50	−0015	−0010	−1.7	−1.3	−0.4	+0.1	0.7	x
9438	Soconusco Bar Guatemala		15	09	92	55	−0015	−0010	−1.5	⊙	⊙	⊙	⊙	d
9440	Champerico		14	18	91	56	+0005	+0005	−0.9	−0.5	−0.1	+0.1	1.2	x
9442	San Jose		13	55	90	50	−0015	−0015	−1.2	−1.0	−0.3	+0.2	0.91	
	Salvador													
9444	Acajutla		13	35	89	51	−0015	−0015	−1.0	−0.9	−0.2	+0.1	1.0	x
9445	La Libertad Golfo de Fonseca		13	29	89	19	−0015	−0015	−1.0	−0.9	−0.2	+0.1	0.9	x
9448	LA UNION		13	20	87	49	STANDARD PORT		See Table V				1.55	
	Honduras													
9449	Puerto Amapala		13	18	87	39	−0005	−0005	0.0	−0.2	0.0	+0.1	1.5	x

SEASONAL CHANGES IN MEAN LEVEL

No.	Jan.1	Feb.1	Mar.1	Apr.1	May.1	June.1	July.1	Aug.1	Sep.1	Oct.1	Nov.1	Dec.1	Jan.1
6323	0.0	+0.1	+0.2	+0.2	0.0	−0.1	−0.2	−0.2	−0.1	0.0	0.0	0.0	0.0
9305–9389	Negligible												
9391–9399	−0.1	−0.1	−0.1	−0.1	0.0	0.0	+0.1	+0.1	+0.1	+0.1	0.0	0.0	−0.1
9402–9415	−0.2	−0.2	−0.2	−0.1	0.0	0.0	+0.2	+0.2	+0.2	+0.1	0.0	−0.1	−0.2
9417–9435	−0.1	−0.1	−0.1	−0.1	0.0	0.0	+0.1	+0.1	+0.1	+0.1	0.0	0.0	−0.1
9436–9449	0.0	0.0	−0.1	−0.1	0.0	0.0	+0.1	+0.1	0.0	0.0	0.0	0.0	0.0
9700	Negligible												

NO.	PLACE		Lat N.		Long W.		TIME DIFFERENCES HHW (Zone + 0600)	LLW	HEIGHT DIFFERENCES (IN METERS) MHHW	MLHW	MHLW	MLLW	M.L. $Z_0m.$	
9448	**LA UNION**		see page 225						3.0	2.5	0.6	−0.1		
	Nicaragua													
9453	Puerto Corinto		12	29	87	10	−0010	−0010	−0.7	−0.4	−0.4	+0.1	1.2	x
9455	San Juan del Sur		11	15	85	53	−0021	−0006	−0.7	−0.6	−0.1	+0.2	1.16	
	Costa Rica													
9456	Bahia Elena		10	56	85	49	−0010	−0010	−0.2	−0.3	−0.1	+0.1	1.4	x
9457	Puerto Culebra		10	38	85	40	−0010	−0010	−0.2	−0.3	0.0	+0.1	1.4	x
	Golfo de Nicaya													
9460	Puntarenas	U	9	58	84	50	−0014	−0008	−0.2	−0.3	0.0	+0.1	1.40	
9462	Puerto Herradura		9	38	84	39	−0015	−0015	−0.2	−0.3	0.0	+0.1	.1.4	
9464	Bahia Uvita		9	09	83	45	−0030	−0030	−0.2	−0.3	0.0	+0.1	1.4	x
	Golfo Dulce													
9467	Bahia del Rineon		8	42	83	29	−0010	−0010	−0.2	−0.3	−0.0	+0.1	1.4	xx
	Isla del Coco													
9470	Chatham Bay		5	33	87	03	−0035	−0035	−0.5	−0.5	+0.2	+0.3	1.4	x
9487	BALBOA		See page 228						4.9	3.8	1.1	−0.1		
	Panama						(Zone +0500)							
9471	Puerto Armuelles		8	16	82	51	−0010	−0010	−1.9	−1.5	−0.3	+0.2	1.46	
9474	Isla Parida		8	08	82	19	−0005	−0005	−1.6	−1.3	−0.2	+0.2	1.7	x
9477	Bahia Honda		7	46	81	31	−0005	−0005	−1.6	−1.3	−0.2	+0.1	1.7	x
9478	Isla Coiba		7	24	81	39	−0005	−0005	−1.4	−1.1	−0.2	+0.2	1.8	x
9480	Isla Cebaco		7	31	81	13	−0005	−0005	−1.6	−1.3	−0.2	+0.1	1.7	x
9484	Cabo Maia		7	18	80	00	−0005	−0005	−1.6	−1.4	−0.2	+0.2	1.7	x
9487	BALBOA		8	57	79	34	STANDARD PORT		See Table v				2.56	
9488	Rio Chepo		8	59	79	07	0000	0000	−0.1	+0.1	+0.1	+0.3	2.5	x
9492	Punta Garachine		8	05	78	25	−0005	−0005	−0.7	−0.5	0.0	+0.3	2.2	x
	Isla det Ray													
9493	St. Elmo Bay		8	18	78	54	−0005	−0005	−0.7	−0.5	0.0	+0.3	2.2	x
9496	Bahia Pinas		7	34	78	11	−0005	−0005	−0.8	−0.6	0.0	+0.3	2.1	x
	Colombia													
9498	Bahia Octavia		6	52	77	40	+0005	−0005	−0.9	−0.7	−0.2	+0.1	2.0	x
9501	Puerto Utria		6	00	77	21	+0015	0000	−1.0	−0.8	−0.2	+0.1	2.0	x
9503	Puerto Cuevita (Cabita)		5	28	77	31	+0020	0000	−1.0	−0.8	−0.2	+0.1	2.0	x
9505	Punta Chirambira		4	17	77	30	+0020	0000	−1.1	−0.8	−0.2	+0.1	1.9	x
9507	Buenaventura	EU	3	54	77	05	+0025	+0004	−0.9	−0.7	−0.2	+0.1	1.98	
9510	Rio Sanguianga		2	40	78	19	+0020	+0006	−1.0	−0.8	−0.1	+0.2	2.0	x
9511	Puerto Tumaco		1	50	78	44	+0015	0000	−1.1	−0.8	−0.2	+0.1	1.9	x
9448	**LA UNION**		see page 225						3.0	2.5	0.6	−0.1		
	Ecuador		(Zone +0600)											
	Archipelago de colon (Islas Galapagos)													
9424	Bahia Darwin		0	19	89	57	−0020	−0008	−1.0	−1.0	−0.2	+0.1	0.9	dx
			S.		W.									
9519	Caleta Iguana		0	58	91	27	−0020	⊙	−1.2	⊙	⊙	⊙	⊙	d
9521	Caleta Aeolian		0	26	90	17	−0019	−0003	−1.2	−1.1	−0.2	+0.1	0.91	d
9522	Bahia Post Office		1	15	90	27	−0015	−0007	−1.5	−1.3	−0.3	+0.1	0.8	dx
9524	Bahia Agua Dulce	E	0	54	89	37	−0017	−0001	−1.0	−0.9	0.0	+0.3	1.13	

⊙ No data.
See notes on page 386.
C Tides predicted in Chilean Tide Tables.
E Tides predicted in Ecuador Tide Tables.
P Tides predicted in Peruvian Tide Tables.
U Tides predicted in U.S. Tide Tables
d Differences approximate.
t Time differences approximate.
x **M.L.** inferred.

NO.	PLACE		Lat. N.		Long. W.		TIME DIFFERENCES HHW (Zone + 0500)	LLW	HEIGHT DIFFERENCES (IN METERS) MHHW	MLHW	MHLW	MLLW	M.L. Z_0m.	
9487	**BALBOA**		see page 228						4.9	3.8	1.1	-0.1		
9525	San Lorenzo	E	1	18	78	51	+0029	+0023	-1.2	-0.8	0.0	+0.5	2.02	
9527	Esmeralds	E	1	00	79	39	-0008	-0010	-1.7	-1.2	-0.2	+0.4	1.76	
9528	Bahia Atacames	E	0	53	79	52	+0025	+0025	-1.0	-0.8	-0.1	+0.2	2.0	x
9530	Muisne	E	0	37	80	01	+0017	+0017	-1.9	-1.5	-0.2	+0.4	1.63	
			S.		**W.**									
9533	Cabo Pasado		0	22	80	30	+0005	+0005	-1.8	-1.3	-0.5	+0.1	1.6	x
9534	Bahia de Caraquez	E	0	35	80	26	+0011	+0009	-2.2	-1.6	-0.3	+0.4	1.51	
9535	Manta	E	0	57	80	44	-0002	-0003	-2.3	-1.7	-0.3	+0.4	1.46	
9538	Puerto Lopez	E	1	34	80	50	+0004	+0003	-2.0	-1.5	-0.2	+0.4	1.62	
9539	La Libertad *Golfo de Guayaquil*	E	2	12	80	55	-0008	-0007	-2.8	-2.2	-0.6	0.1	1.07	
9540	Posorja	E	2	43	80	14	+0130	+0127	-2.4	-1.7	-0.5	+0.3	1.34	
9540a	Puerto Maritimo de Guayaquil	E	2	17	79	55	+0301	+0252	-0.9	-0.5	+0.1	+0.5	2.30	
9541	Isla Santa Clara	E	3	10	80	26	0000	⊙	-3.0	-2.4	-0.8	-0.1	0.9	dx
9543	Puna	EU	2	44	79	55	+0156	+0155	-1.3	-0.8	-0.1	+0.5	2.0	x
9544	Guayaquil	E	2	12	79	52	+0423	+0511	-0.8	-0.4	-0.1	+0.4	2.20	
9545	Puerto Bolivar	E	3	18	80	00	+0154	+0139	-2.0	-1.5	-0.2	+0.4	1.61	
	Peru													
9547	Caleta Zorritos	P	3	39	80	40	+0104	+0054	-3.2	-2.4	-0.6	+0.3	0.97	
7280	**LUHUASHAN**		See page 180						4.3	3.3	2.0	0.9		
9548	Lobitos	P	4	27	81	17	+0616	+0628	-2.7	-2.1	-0.3	-0.8	0.86	*
9549	Bahia de Talara	EPU	4	35	81	17	+0616	+0628	-2.8	-2.1	-1.6	-0.9	0.79	*t
9551	Puerto de Paita	P	5	05	81	07	+0624	+0636	-2.9	-2.2	-1.6	-0.8	0.77	*
9553	Puerto Bayovar	P	5	48	81	02	+0634	+0647	-3.0	-2.3	-1.6	-0.8	0.73	*
9554	Lobos de Tierra		6	26	80	50	+0710	⊙	-2.9	⊙	⊙	⊙	0.7	dx*

NO.	PLACE		Lat. N.		Long. W.		HHW	LLW	MHHW	MLHW	MHLW	MLLW	M.L. Z_0m.	
9644	**VALPARAISO**		See page 231				HHW	LLW	1.5	1.2	0.5	0.4		
9557	Puerto Eten	P	6	57	79	52	-0603	-0620	-0.4	-0.4	-0.2	-0.2	0.62	*
9560	Puerto Chicama	P	7	42	79	27	-0548	-0605	-0.5	-0.4	-0.2	-0.2	0.56	*
9562	*Salaverry*	P	8	13	78	59	-0534	-0637	-0.5	-0.4	-0.2	-0.2	0.58	*
9563	Chimbote	P	9	00	78	35	-0521	-0540	-0.4	-0.4	0.0	-0.1	0.68	*
9564	Bahia Ferrol		9	08	78	36	-0506	-0526	-0.4	-0.3	-0.1	-0.2	0.67	*
9566	Bahia Huarmey	P	10	05	78	09	-0508	-0527	-0.7	-0.6	-0.2	-0.3	0.48	*
9568	Puerto Bermejo		10	33	77	53	-0500	⊙	-0.4	⊙	⊙	⊙	0.7	dx*
9570	Puerto Huacho	P	11	08	77	37	-0443	-0553	-0.7	-0.7	-0.3	-.03	0.42	*
9573	Callao	EPU	12	04	77	10	-0418	-0531	-0.6	-0.6	-0.2	-0.2	0.52	t*
9575	Caleta Pucusana (Chilca)		12	30	76	50	-0410	⊙	-0.7	⊙	⊙	⊙	0.5	dx*
9577	Puerto Cerro Azul		13	03	76	31	-0400	⊙	-0.7	⊙	⊙	⊙	0.5	dx*
9579	PIsco	P	13	43	76	14	-0342	-0400	-0.8	-0.7	-0.3	-0.3	0.40	*
9580	Bahia Independencia		14	18	76	08	-0445	⊙	-0.7	⊙	⊙	⊙	0.5	dx*
9583	San Juan	P	15	21	75	09	-0249	-0240	-0.7	-0.8	-0.4	-0.3	0.37	
9584	Punta Lomas		15	33	74	52	-0300	⊙	-0.7	⊙	⊙	⊙	0.5	dx
9586	Rada Atico		16	13	73	43	-0210	⊙	-0.7	⊙	⊙	⊙	0.5	dx
9588	Caleta Quilca		16	42	72	27	-0205	⊙	-0.6	⊙	⊙	⊙	0.5	dx
9589	Matarani	EPU	17	00	72	07	-0214	-0219	-0.6	-0.6	-0.4	-0.3	0.43	
9591	Puerto de Ho	P	17	38	71	21	-0215	-0216	-0.6	-0.6	-0.4	-0.3	0.44	
	Chile	(Zone +0400)												
9594	Arica	C	18	28	70	20	-0108	-0117	-0.3	-0.4	-0.2	-0.1	0.65	
9597	Caleta Junin		19	39	70	11	-0110	-0110	-0.3	-0.3	-0.3	-0.3	0.6	dx
9599	Iquique	C	20	13	70	09	-0100	-0102	-0.3	-0.3	-0.1	0.0	0.76	

SEASONAL CHANGES IN MEAN LEVEL

No.	Jan.1	Feb.1	Mar.1	Apr.1	May.1	June.1	July.1	Aug.1	Sep.1	Oct.1	Nov.1	Dec.1	Jan.1
7280	-0.1	-0.2	-0.2	-0.1	0.0	0.0	0.0	+0.1	+0.2	+0.2	+0.1	0.0	-0.1
9448–9480	0.0	0.0	-0.1	-0.1	0.0	0.0	+0.1	+0.1	0.0	0.0	0.0	0.0	0.0
9484–9496	0.0	-0.1	-0.2	-0.1	0.0	0.0	+0.1	0.0	0.0	+0.1	+0.1	+0.1	0.0
9498–9511	0.0	-0.1	-0.1	-0.1	0.0	0.0	0.0	0.0	0.0	+0.1	+0.1	0.0	0.0
9415–9644	Negligible												

BIBLIOGRAPHY

Anwar, N. (2006). *Navigation: Advanced Mates/Masters*. Seamanship International Ltd. Strathclyde.

Bowditch, N. (1995). *The American Practical Navigator*. Defence Mapping Agency Hydrographic/Topographic centre. Bethesda.

Frost, A. (1981). *Practical Navigation for Second Mates* (5th Edition). Brown, Son & Ferguson Ltd. Glasgow.

Moore, D.A. (1975). *Basic Principles of Marine Navigation* (4th Edition). Stanford Maritime. London.

Mortzer-Bruyns, W. (1994). *Cross-staff: History and Development of a Navigational Instrument*. Available [Online] http://www.bighistory.info/bhi_005_032.htm.

Pugh, C. (2004). *Effects of Tides, Weather and Climate*. Cambridge University Press. Cambridge.

Royal Navy (2008). *The Principles of Navigation: The Admiralty Manual Of Navigation* (10th Edition). The Nautical Institute. London.

UKHO (1987). Admiralty Tide Tables *Volume 1 & Volume 3* (1987). United Kingdom Hydrographic Office. Taunton.

UKHO (2017). Admiralty Tide Tables (NP201a) (2017). United Kingdom Hydrographic Office. Taunton.

UKHO (2017) Admiralty Tide Tables (NP201b) (2017). United Kingdom Hydrographic Office. Taunton.

UKHO (1995). Tidal Stream Atlas (HP 233). *Dover Strait*. (Edition 3 Reprinted 2003). United Kingdom Hydrographic Office. Taunton.

INDEX